Instructor's Solutions Ma

MW00678759

A First Course in Abstract Algebra

Fifth Edition

Instructor's Solutions Manual

A First Course in Abstract Algebra

Fifth Edition

John B. Fraleigh

Addison-Wesley Publishing Company

Reading, Massachusetts • Menlo Park, California • New York
Don Mills, Ontario • Wokingham, England • Amsterdam • Bonn
Sydney • Singapore • Tokyo • Madrid • San Juan • Milan • Paris

Reproduced by Addison-Wesley from camera-ready copy supplied by the author.

ISBN 0-201-53468-1

Copyright © 1994 by Addison-Wesley Publishing Company, Inc. All rights reserved. No part of this publication may be reproduced, stored in a retrieval system, or transmitted, in any form or by any means, electronic, mechanical, photocopying, recording, or otherwise, without the prior written permission of the publisher. Printed in the United States of America.

3 4 5 6 7 8 9 10 CRC 9695

PREFACE

This manual contains complete solutions to all exercises in the text, except those odd-numbered exercises for which fairly lengthy complete solutions are given in the answers at the back of the text. Then reference is simply given to the text answers to save typing.

I prepared these solutions myself. While I tried to be accurate, there are sure to be the inevitable mistakes and typos. An author proofreading tends to see what he or she wants to see. However, the instructor should find this manual adequate for the purpose for which it is intended.

Morgan, VT. J.B.F.

August, 19932

CONTENTS

CHAPTER 5 Introduction to Rings and Fields

CHAPTER 6 Factor Rings and Ideals

CHAPTER 7 Factorization

CHAPTER 8 Extension Fields

CHAPTER 9 Automorphisms and Galois Theory

Instructor's Solutions Manual

A First Course in Abstract Algebra

Fifth Edition

CHAPTER 0

A FEW PRELIMINARIES

SECTION 0.1 - Mathematics and Proofs

1. proving theorems

2. set

3. efficiency

4. definition

5. Let the even integers be $2n$ and $2m$. Now $2n + 2m = 2(n + m)$, so $2n + 2m$ is even.

6. Let the even integers be $2n$ and $2m$. Now $(2n)(2m) = 4mn$ which is a multiple of 4 .

7. An integer m is odd if there exists an integer n such that $m = 2n + 1$. For the proof requested, let $r = 2m$ and $s = 2n + 1$. Then $r + s = 2(m + n) + 1$ which is an odd integer.

8. counterexample

9. ABEFGM, CDJ, HKN, I, L, O

10. a) 2 pieces
    b) 4 pieces. Either 7 or 8 would be a natural conjecture from the present list 1, 2, 4 .
    c) 8 pieces. We conjecture the list would go 1, 2, 4, 8, 16, 32, 64, 128 with nth term 2^{n-1}.
    d) There are 16 pieces, consistent with the conjecture.
    e) There are 31 pieces, so the conjecture is not a theorem. We have no new conjecture based on just this data.

11. True. Note that $(2n + 1)^2 = 2(2n^2 + 2n) + 1$, so the square of an odd integer is not even. Thus if an even integer is a square, it must be the square of an even integer, and thus of the form $(2n)^2 = 4n^2$, which is a multiple of 4.

12. False. $n = 0$ gives a counterexample.

13. False. The equation is satisfied by two integers, 1 and -1.

14. True. There actually exist two such integers, 2 and -2, and if there exist two such things, there certainly exists one of them. The problem did not say "exactly one."

15. True. $n = 3$ is such an integer.

16. False. The equation is satisfied by both 3 and -3.

17. False. $n = 1$ gives a counterexample.

Section 0.2

18. True. If n is negative, then $n^2 > 0 > n$ so $n^2 > n$.

19. True. $(1/2)^2 = 1/4 < 1/2$.

20. True. $2^2 > 2$.

21. False. Both 0 and 1 satisfy this equation.

22. True. If $n = 2m + 1$, then $n^2 = 4(m^2 + m) + 1$.

23. True. If $n = 3m + 1$, then $n^2 = 3(3m^2 + 2m) + 1$.

24. True. $(-2)^3 = -8 < -2$.

25. False. Let $n = -2$ and $m = 1$. Then $(-2/1)^2 = 4 > 1$.

26. False. Let $n = -1$ and $m = 1$. Then $(-1/1)^2 = 1 > -1$.

27. False. Let $n = -2$ and $m = -1$. Then $(n/m)^3 = 8$ and $(n/m)^2 = 4$, so $(n/m)^3 > (n/m)^2$.

SECTION 0.2 Sets and Equivalence Relations

1. $\{-\sqrt{3}, \sqrt{3}\}$ 2. The set is empty.

3. $\{1,-1,2,-2,3,-3,4,-4,5,-5,6,-6,10,-10,12,-12,15,-15,20,-20, 30,-30,60,-60\}$

4. $\{-10,-9,-8,-7,-6,-5,-4,-3,-2,-1,0,1,2,3,4,5,6,7,8,9,10,11\}$

5. Not a set (not well defined)

6. The set is empty.

7. The set is empty.

8. Not a set (not well defined as stated, $2/3 = 200/300$)

9. $\mathbf{Q}$

10. The set containing all numbers that are (positive, negative, or zero) integral multiples of 1, of 1/2, or of 1/3.

11. Not an equivalence relation, 0 is not related to 0.

12. Not an equivalence relation, not symmetric

13. An equivalence relation; $\bar{0} = \{0\}$, $\bar{a} = \{a,-a\}$ for $a \in \mathbf{R}$, $a \neq 0$

14. Not an equivalence relation, not transitive

15. (See the text answer.)

16. An equivalence relation; $\bar{1} = \{1, 11, 21, 31, \cdots\}$, $\bar{2} = \{2, 12, 22, 32, \cdots\}$, $\cdots$, $\overline{10} = \{10, 20, 30, \cdots\}$

17. An equivalence relation; $\bar{1} = \{1, 3, 5, 7, 9, \cdots\}$, $\bar{2} = \{2, 4, 6, 8, 10, \cdots\}$

18. Let h, k, and m be integers. We check the three criteria.
Reflexive: $h - h = n0$ so $h \equiv h \pmod{n}$.
Symmetric: If $h - k = ns$, then $k - h = n(-s)$.
Transitive: If $h - k = ns$ and $k - m = nt$, then $h - m = (h - k) + (k - m) = ns + nt = n(s + t)$.

19. $\mathbb{Z}$

20. $\{2n \mid n \in \mathbb{Z}\} = \{\cdots, -8, -6, -4, -2, 0, 2, 4, 6, 8, \cdots\}$
$\{2n + 1 \mid n \in \mathbb{Z}\} = \{\cdots, -7, -5, -3, -1, 1, 3, 5, 7, \cdots\}$

21. (See the text answer.)

22. $\{4n \mid n \in \mathbb{Z}\}$, $\{4n + 1 \mid n \in \mathbb{Z}\}$, $\{4n + 2 \mid n \in \mathbb{Z}\}$, $\{4n + 3 \mid n \in \mathbb{Z}\}$

23. (See the text answer.)

24. It may be that $a \in S$ is not related to any $b \in S$. For example, we might take $S = \mathbb{Z}$ and define m to be related to n if $mn \neq 0$. Then 0 is not related to 0.

25. 1. There is only one partition of a one-element set.

26. 2. $\{a, b\}$ has partitions $\{\{a, b\}\}$ and $\{\{a\}, \{b\}\}$.

27. 5. $\{a, b, c\}$ has partitions $\{a, b, c\}$; $\{\{a\}, \{b, c\}\}$; $\{\{b\}, \{a, c\}\}$; $\{\{c\}, \{a, b\}\}$; and $\{\{a\}, \{b\}, \{c\}\}$.

28. 15. $\{a, b, c, d\}$ has 1 partition into one cell, 7 partitions into two cells, 6 partitions into three cells, and 1 partition into four cells for a total of 15 partitions.

29. 52. $\{a, b, c, d, e\}$ has 1 partition into one cell, 15 into two cells, 25 into three cells, 10 into four cells, and 1 into five cells for a total of 52. (Do a combinatorics count for each possible case, such as a 1,2,2 split where there are 15 possible partitions.)

SECTION 0.3 Mathematical Induction

1. Let $P(n)$ be the formula to be proved. Since $\frac{1(1 + 1)(2 + 1)}{6} = \frac{6}{6} = 1 = 1^2$, we see that $P(1)$ is true. Proceeding by induction, suppose $P(k)$ is true. Then $1^2 + 2^2 + \cdots + k^2 + (k + 1)^2 = \frac{k(k + 1)(2k + 1)}{6} + (k + 1)^2$

$$= \frac{k(k + 1)(2k + 1) + 6(k + 1)^2}{6} = \frac{(k + 1)(2k^2 + k + 6k + 6)}{6}$$

$$= \frac{(k + 1)(2k^2 + 7k + 6)}{6} = \frac{(k + 1)(k + 2)(2k + 3)}{6}$$

$$= \frac{(k + 1)[(k + 1) + 1][2(k + 1) + 1]}{6}$$ so $P(k + 1)$ is true.

2. Let $P(n)$ be the formula to be proved. Since $\frac{1^2(1 + 1)^2}{4} = \frac{1(4)}{4} = 1 = 1^3$, we see that $P(1)$ is true. Proceeding by induction, suppose $P(k)$ is true. Then

$$1^3 + 2^3 + \cdots + k^3 + (k + 1)^3 = \frac{k^2(k + 1)^2}{4} + (k + 1)^3$$

$$= \frac{k^2(k + 1)^2 + 4(k + 1)^3}{4} = \frac{(k + 1)^2(k^2 + 4k + 4)}{4}$$

$$= \frac{(k + 1)^2(k + 2)^2}{4} = \frac{(k + 1)^2[(k + 1) + 1]^2}{4}$$ so $P(k + 1)$ is true.

3. Let $P(n)$ be the formula to be proved. Since $1 = 1^2$, we see that $P(1)$ is true. Proceeding by induction, suppose that $P(k)$ is true. Then

$1 + 3 + 5 + \cdots + (2k - 1) + [2(k + 1) - 1]$
$= k^2 + [2(k + 1) - 1] = k^2 + 2k + 1 = (k + 1)^2$, so $P(k + 1)$ is true.

4. Let $P(n)$ be the formula to be proved. Since $\frac{1}{1\cdot 2} = \frac{1}{1 + 1}$, we see that $P(1)$ is true. Proceeding by induction, suppose that $P(k)$ is true. Then

$$\frac{1}{1\cdot 2} + \frac{1}{2\cdot 3} + \cdots + \frac{1}{k(k + 1)} + \frac{1}{(k + 1)[(k + 1) + 1]}$$

$$= \frac{k}{k + 1} + \frac{1}{(k + 1)(k + 2)} = \frac{k(k + 2) + 1}{(k + 1)(k + 2)} = \frac{k^2 + 2k + 1}{(k + 1)(k + 2)}$$

$$= \frac{(k + 1)^2}{(k + 1)(k + 2)} = \frac{k + 1}{k + 2} = \frac{k + 1}{[(k + 1) + 1]},$$ so $P(k + 1)$ is true.

5. Let $P(n)$ be the formula to be proved. Since $\frac{a(1 - r^{1+1})}{1 - r} = \frac{a(1 - r^2)}{1 - r} = \frac{a(1 - r)(1 + r)}{1 - r} = a(1 + r)$ $= a + ar$, we see that $P(1)$ is true. Proceeding by induction, suppose that $P(k)$ is true. Then

$$a + ar + ar^2 + \cdots + ar^k + ar^{k+1} = \frac{a(1 - r^{k+1})}{1 - r} + ar^{k+1}$$

$$= \frac{a(1 - r^{k+1}) + ar^{k+1}(1 - r)}{1 - r} = \frac{a - ar^{k+2}}{1 - r} = \frac{a(1 - r^{(k+1)+1})}{1 - r},$$

so $P(k + 1)$ is true.

6. It is in the final paragraph. While i and j are in $\mathbb{Z}^+$, it need not be true that $i - 1$ and $j - 1$ are in $\mathbb{Z}^+$. For example, in trying to prove $P(2)$, we might have $i = 1$ and $j = 2$. Then $i - 1 = 0$, which is outside $\mathbb{Z}^+$, and $P(1)$ does not cover this case.

7. (See the text answer.)

8. a) As I said, I don't know just where the flaw is. I have asked a few logicians, and as best I can understand them, the argument depends on knowing what you know, and apparently that is logically taboo.
   b) I don't know the answer to this either. Whenever I have asked logicians what the last possible day is, they say there is such a day, and that it is not Friday. Some say it is Thursday, but then when I go through the argument, they agree it can't be Thursday and back off to Wednesday, etc.

SECTION 0.4 Complex and Matrix Algebra

1. $6 + 2i$
2. $5 - 2i$
3. $2 + 9i$
4. $5 - 5i$
5. $-i$
6. 1
7. $i^{23} = (i^4)^5 i^3 = 1^5 i^3 = 1(-i) = -i$
8. $(-i)^{35} = (-i^4)^8(-i^3) = 1^8(-i)^3 = 1(-1)(i^3) = -(-i) = i$

Section 0.4

9. $(4 - i)(5 + 3i) = 20 - 3i^2 + 12i - 5i = 20 - 3(-1) + 7i = 23 + 7i$

10. $(8 + 2i)(3 - i) = 24 - 2i^2 + 6i - 8i = 24 - 2(-1) - 2i = 26 - 2i$

11. $(2 - 3i)(4 + i) + (6 - 5i) = 8 - 3i^2 - 12i + 2i + 6 - 5i$
$= 8 - 3(-1) + 6 - 15i = 17 - 15i$

12. $(1 + i)^3 = 1^3 + 3i + 3i^2 + i^3 = 1 + 3i - 3 - i = -2 + 2i$

13. $(1 - i)^5 = 1^5 - 5i + 10i^2 - 10i^3 + 5i^4 - i^5$
$= 1 - 5i - 10 + 10i + 5 - i = -4 + 4i$

14. $\frac{7 - 5i}{1 + 6i} = \frac{7 - 5i}{1 + 6i}\cdot\frac{1 - 6i}{1 - 6i} = \frac{7 + 30i^2 - 5i - 42i}{1 + 36} = \frac{-23 - 47i}{37} = -\frac{23}{37} - \frac{47}{37}i$

15. $\frac{i}{1 + i} = \frac{i}{1 + i}\cdot\frac{1 - i}{1 - i} = \frac{i - i^2}{1 + 1} = \frac{i - (-1)}{2} = \frac{1 + i}{2} = \frac{1}{2} + \frac{1}{2}i$

16. $\frac{1 - i}{i} = \frac{1 - i}{i}\cdot\frac{-i}{-i} = \frac{-i + i^2}{1} = \frac{-i - 1}{1} = -1 - i$

17. $\frac{i(3 + i)}{2 - 4i} = \frac{3i - 1}{2 - 4i}\cdot\frac{2 + 4i}{2 + 4i} = \frac{6i - 12 - 2 - 4i}{4 + 16} = \frac{-14 + 2i}{20} = -\frac{7}{10} + \frac{1}{10}i$

18. $\frac{3 + 7i}{(1 + i)(2 - 3i)} = \frac{3 + 7i}{5 - i} = \frac{3 + 7i}{5 - i}\cdot\frac{5 + i}{5 + i} =$
$\frac{15 + 3i + 35i + 7i^2}{25 + 1} = \frac{8 + 38i}{26} = \frac{4}{13} + \frac{19}{13}i$

19. $\frac{(1 - i)(2 + i)}{(1 - 2i)(1 + i)} = \frac{3 - i}{3 - i} = 1$

20. $|3 - 4i| = \sqrt{3^3 + 4^4} = \sqrt{25} = 5$

21. $|6 + 4i| = \sqrt{6^2 + 4^2} = \sqrt{52} = 2\sqrt{13}$

22. $5(\frac{3}{5} - \frac{4}{5}i)$

23. $\sqrt{2}(\frac{-1}{\sqrt{2}} + \frac{1}{\sqrt{2}}i)$

24. $13(\frac{12}{13} + \frac{5}{13}i)$

25. $\sqrt{34}(\frac{-3}{\sqrt{34}} + \frac{5}{\sqrt{34}}i)$

26. $|z_1/z_2| = |z_1|/|z_2|$. The polar angle of z_1/z_2 is (polar angle of z_1) - (polar angle of z_2). Of course if this difference should be negative, it can be increased by 2π to obtain a positive polar angle.

27. $1, i, -1, -i$

28. $|z|^4(\cos 4\theta + i \sin 4\theta) = -1 = 1(\cos \pi + i \sin \pi)$ so $|z| = 1$ and $\theta = \pi/4, 3\pi/4, 5\pi/4, 7\pi/4$. Thus z can be $(1/\sqrt{2}) \pm (1/\sqrt{2})i$ or $-(1/\sqrt{2}) \pm (1/\sqrt{2})i$.

29. $|z|^3(\cos 3\theta + i \sin 3\theta) = -8 = 8(\cos \pi + i \sin \pi)$ so $|z| = 2$ and $\theta = \pi/3, \pi, 5\pi/3$. Thus z can be $1 + \sqrt{3}i$, -2 or $1 - \sqrt{3}i$.

30. $|z|^3(\cos 3\theta + i \sin 3\theta) = 27[\cos(3\pi/2) + i \sin(3\pi/2)]$ so $|z| = 3$ and $\theta = \pi/2, 7\pi/6, 11\pi/6$. Thus z can be $3i$, $-(3\sqrt{3}/2) - (3/2)i$ or $(3\sqrt{3}/2) - (3/2)i$.

31. $|z|^6(\cos 6\theta + i \sin 6\theta) = 1 = 1(\cos 0 + i \sin 0)$ so $|z| = 1$ and $\theta = 0, \pi/3, 2\pi/3, \pi, 4\pi/3, 5\pi/3$. Thus z can be ± 1, $(1/2) \pm (\sqrt{3}/2)i$ or $-(1/2) \pm (\sqrt{3}/2)i$.

32. $|z|^6(\cos 6\theta + i \sin 6\theta) = -64 = 64(\cos \pi + i \sin \pi)$ so $|z| = 2$ and $\theta = \pi/6, \pi/2, 5\pi/6, 7\pi/6, 3\pi/2, 11\pi/6$. Thus z can be $\sqrt{3} \pm i$, $\pm 2i$ or $-\sqrt{3} \pm i$.

33. $\begin{bmatrix} 2 & 1 \\ 2 & 7 \end{bmatrix}$

34. $\begin{bmatrix} 4 + i & -3 + i & 1 \\ 7 - i & 1 + 2i & 2 - i \end{bmatrix}$

35. $\begin{bmatrix} -3 + 2i & -1 - 4i \\ 2 & -i \\ 0 & -i \end{bmatrix}$

36. $\begin{bmatrix} 3 & 1 \\ 5 & 15 \end{bmatrix}$

37. $\begin{bmatrix} 5 & 16 & -3 \\ 0 & -18 & 24 \end{bmatrix}$

38. Undefined

39. $\begin{bmatrix} 1 & -i \\ 4 - 6i & -2 - 2i \end{bmatrix}$

Section 0.4

40. $= \begin{bmatrix} 0 & -1 \\ 1 & -1 \end{bmatrix}^2 = \begin{bmatrix} -1 & 1 \\ -1 & 0 \end{bmatrix}$

41. $= \begin{bmatrix} 2 & -2i \\ 2i & 2 \end{bmatrix}^2 = \begin{bmatrix} 8 & -8i \\ 8i & 8 \end{bmatrix}$

42. $\begin{bmatrix} 1 & 0 \\ 1 & 1 \end{bmatrix}\begin{bmatrix} 0 & 1 \\ 2 & 1 \end{bmatrix} = \begin{bmatrix} 0 & 1 \\ 2 & 2 \end{bmatrix}$, $\begin{bmatrix} 0 & 1 \\ 2 & 1 \end{bmatrix}\begin{bmatrix} 1 & 0 \\ 1 & 1 \end{bmatrix} = \begin{bmatrix} 1 & 1 \\ 3 & 1 \end{bmatrix}$

43. $\begin{bmatrix} 0 & -1 \\ 1 & 0 \end{bmatrix}$

44. $\begin{bmatrix} 1/2 & 0 & 0 \\ 0 & 1/4 & 0 \\ 0 & 0 & -1 \end{bmatrix}$

45. Given that A^{-1} and B^{-1} exist, the associative law for matrix multiplication yields $(B^{-1}A^{-1})(AB) = B^{-1}(A^{-1}A)B = B^{-1}I_nB = B^{-1}B = I_n$ so AB is invertible. Similarly, $(A^{-1}B^{-1})(BA) = A^{-1}(B^{-1}B)A = A^{-1}I_nA = A^{-1}A = I_n$ so BA is invertible also.

46. Let $P(n)$ be the given formula. Now $P(1)$ is simply the given notation for z. Suppose $P(k)$ is true. Then

$P(k + 1) = z^{k+1} = z^k z$

$= r^k(\cos k\theta + i \sin k\theta)\cdot r(\cos \theta + i \sin \theta)$

$= r^{k+1}[(\cos k\theta \cos \theta - \sin k\theta \sin \theta)$
$+ i(\sin k\theta \cos \theta + \cos k\theta \sin \theta)]$

$= r^{k+1}[\cos(k\theta + \theta) + i \sin(k\theta + \theta)]$ by familiar trigonometric identities. But this is $r^{k+1}[\cos (k + 1)\theta + i \sin (k + 1)\theta]$, so $P(k + 1)$ is true and our induction proof is complete.

47. a) $\phi((a + bi) + (c + di)) = \phi((a + c) + (b + d)i) =$

$\begin{bmatrix} a + c & -(b + d) \\ b + d & a + c \end{bmatrix} = \begin{bmatrix} a & -b \\ b & a \end{bmatrix} + \begin{bmatrix} c & -d \\ d & c \end{bmatrix} =$

$\phi(a + bi) + \phi(c + di)$.

b) $\phi((a + bi)(c + di)) = \phi((ac - bd) + (ad + bc)i) =$

$\begin{bmatrix} ac - bd & -(ad + bc) \\ ad + bc & ac - bd \end{bmatrix} = \begin{bmatrix} a & -b \\ b & a \end{bmatrix}\begin{bmatrix} c & -d \\ d & c \end{bmatrix} =$

$\phi(a + bi)\phi(c + di)$.

CHAPTER 1

GROUPS AND SUBGROUPS

SECTION 1.1 - Binary Operations

1. e, b, a 2. a, a. No, you can't say.

3. a, c. Yes, you know the operation is not associative.

4. No, since $b * e = c$ but $e * b = b$.

5. First row: d. Second row: a. Fourth row: c, b.

6. $d * a = (c * b) * a = c * (b * a) = c * b = d$. In a similar fashion, substituting $c * b$ for d and using the associative property, we find that $d * b = c$, $d * c = c$, and $d * d = d$.

7. Not commutative since $1 - 2 \neq 2 - 1$. Not associative since $2 = 1 - (2 - 3) \neq (1 - 2) - 3 = -4$.

8. Commutative since $ab + 1 = ba + 1$ for all $a, b \in \mathbb{Q}$. Not associative since $(a * b) * c = (ab + 1) * c = abc + c + 1$ but $a * (b * c) = a * (bc + 1) = abc + a + 1$, and we need not have $a = c$.

9. Commutative since $ab/2 = ba/2$ for all $a, b \in \mathbb{Q}$. Associative since $a * (b * c) = a * (bc/2) = [a(bc/2)]/2 = abc/4$, and $(a * b) * c = (ab/2) * c = [(ab/2)c]/2 = abc/4$.

10. Commutative since $2^{ab} = 2^{ba}$. Not associative since

$(a * b) * c = 2^{ab} * c = 2^{(2^{ab})c}$ but $a * (b * c) = a * 2^{bc}$ $= 2^{a(2^{bc})}$.

11. Not commutative since $a^b \neq b^a$ for some $a, b \in \mathbb{Z}^+$. Not associative since $a * (b * c) = a * b^c = a^{(b^c)}$ but $(a * b) * c = a^b * c = (a^b)^c = a^{bc}$ and $bc \neq b^c$ for some $b, c \in \mathbb{Z}^+$.

12. 1 operation, for the only answer is the sole element of S. There are 16 operations if S has two elements, for there are four places to fill in a table, and each may be filled in two ways, and $2 \cdot 2 \cdot 2 \cdot 2 = 16$. There are 19,683 operations if S has three elements, for there are nine places to fill in a table, and $3^9 = 19{,}683$. With n elements, there are $n \cdot n$

places to fill, which can be done in $n^{(n^2)}$ ways.

13. A commutative binary operation on a set with n elements is completely determined by the elements on and above the main diagonal in its table. The number of such places is $n + \frac{n^2 - n}{2} = (n^2 + n)/2$. Thus there are $n^{(n^2+n)/2}$ possible commutative binary operations on an n-element set. For $n = 2$, we obtain $2^3 = 8$, and for $n = 3$ we obtain $3^6 = 729$.

14. No, Condition 2 is violated. 15. Yes 16. Yes

17. Yes 18. No, Condition 1 is violated.

19. No, (2) is violated; $1 * 1$ is not in $\mathbb{Z}^+$.

20. F T F F F T T T T F

21. (See the text answer.)

22. We have $(a * b) * (c * d) = (c * d) * (a * b) = (d * c) * (a * b) = [(d * c) * a] * b$, where we used commutativity for the first two steps and associativity for the last.

23. True. Commutativity and associativity assert the equality of certain computations. For a binary operation on a set with just one element, that element is the result of every computation involving the operation, so the operation must be commutative and associative.

24. No. Consider the operation on $\{a, b\}$ defined by the table

*	a	b
a	b	a
b	a	a

. Then $(a * a) * b = b * b = a$ but $a * (a * b) = a * a = b$.

25. Yes. $[(f + g) + h](x) = (f + g)(x) + h(x) = [f(x) + g(x)] + h(x) = f(x) + [g(x) + h(x)] = f(x) + [(g + h)(x)] = [f + (g + h)](x)$ since addition in $\mathbb{R}$ is associative.

26. No. Let $f(x) = 2x$ and $g(x) = 5x$. Then $(f - g)(x) = f(x) - g(x) = 2x - 5x = -3x$ while $(g - f)(x) = 5x - 2x = 3x$.

27. No. Let $f(x) = 2x$, $g(x) = 5x$, and $h(x) = 8x$. Then $[f - (g - h)](x) = f(x) - (g - h)(x) = f(x) - [g(x) - h(x)] = f(x) - g(x) + h(x) = 2x - 5x + 8x = 5x$, but $[(f - g) - h](x) = (f - g)(x) - h(x) = f(x) - g(x) - h(x) = 2x - 5x - 8x = -11x$.

28. Yes. $(f \cdot g)(x) = f(x) \cdot g(x) = g(x) \cdot f(x) = (g \cdot f)(x)$ since multiplication in $\mathbb{R}$ is commutative.

29. Yes. $[(f \cdot g) \cdot h](x) = (f \cdot g)(x) \cdot h(x) = [f(x) \cdot g(x)] \cdot h(x) = f(x) \cdot [g(x) \cdot h(x)] = [f \cdot (g \cdot h)](x)$ since multiplication in $\mathbb{R}$ is associative.

30. No. Let $f(x) = x^2$ and $g(x) = x + 1$. Then $(f \circ g)(3) = f(g(3)) = f(4) = 16$ but $(g \circ f)(3) = g(f(3)) = g(9) = 10$.

31. No. Let $*$ be $+$ and let $*'$ be $\cdot$ on $\mathbb{Z}$.

32. a) Two binary operations $*$ and $*'$ on the same set S give algebraic structures of the same type if each $x \in S$ has a counterpart $x' \in S$ such that the correspondence of x to x' is one to one and such that $(a * b)' = a' *' b'$ for all $a, b \in S$.

b) Let the set be $\{a, b\}$. We need to decide whether interchanging the names of the letters everywhere in the table and then writing the table again in the order a first and b second gives the same table or a different table. The same table is obtained if and only if in the body of the table, diagonally opposite entries are different. Four such tables exist, since there are four possible choices for the first row: namely, the table bodies

$$\begin{matrix} a & a \\ b & b \end{matrix} \qquad \begin{matrix} a & b \\ a & b \end{matrix} \qquad \begin{matrix} b & a \\ b & a \end{matrix} \qquad \text{and} \qquad \begin{matrix} b & b \\ a & a \end{matrix}.$$

The other 12 tables can be paired off into tables giving the same algebraic structure. The number of different algebraic structures is thus $12/2 + 4 = 10$.

33. (See the text answer.)

34. The input string $x_1x_2\cdots x_n$ applied to the machine means first input x_1, then input x_2, then x_3, $\cdots$, then x_n. For inputs a_i, b_j, and c_k, the concatenations $(a_1a_2\cdots a_r)[(b_1b_2\cdots b_s)(c_1c_2\cdots c_t)]$ and $[(a_1a_2\cdots a_r)(b_1b_2\cdots b_s)](c_1c_2\cdots c_t)$ both lead to the string $a_1a_2\cdots a_rb_1b_2\cdots b_sc_1c_2\cdots c_t$ having the meaning explained above.

35. (See the text answer.)

36. 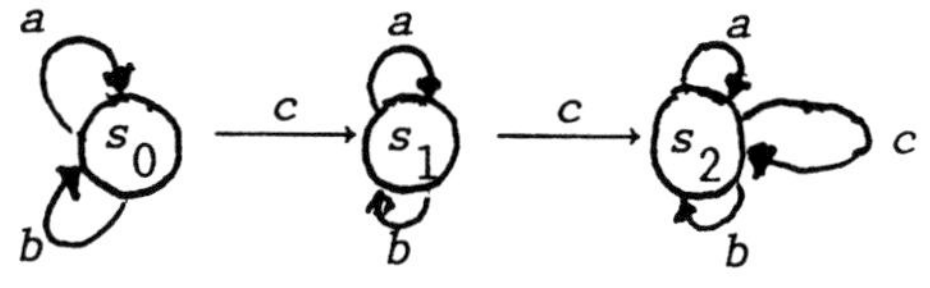

37. (See the text answer.)

Section 1.2

38.

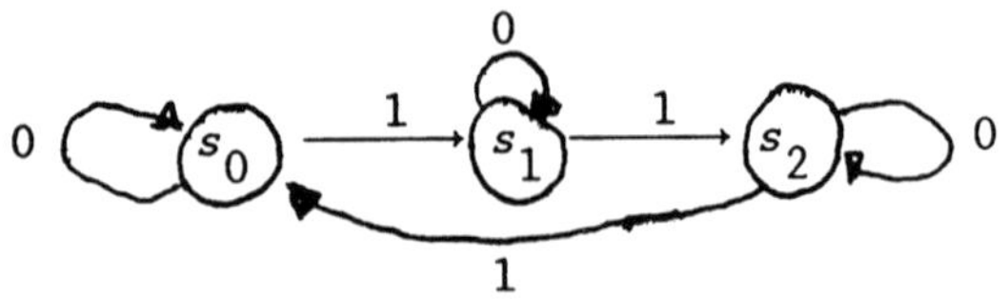

39. (See the text answer.)

40.

Present state	a	b	c (inputs)
s_0	s_0	s_0	s_1
s_1	s_1	s_1	s_2
s_2	s_2	s_2	s_2

SECTION 1.2 - Groups

1. No. G_3 fails.
2. Yes
3. No. G_1 fails.
4. Yes.
5. No. G_1 fails.
6. No. G_3 fails.
7. Section 0.4 showed, using the polar form of complex numbers, that $|z_1z_2| = |z_1||z_2|$, so the complex numbers of norm 1 are closed under multiplication. Since $|1| = 1$, we see that $1 = 1 + 0i$ is a multiplicative identity. If $|z| = 1$ then since $z(1/z) = 1$ and $|z||1/z| = |1|$, we have $1 \cdot |1/z| = 1$ so $1/z$ also has norm 1. Thus all group axioms are satisfied.
8. Yes. Addition of diagonal matrices amounts to adding entries in corresponding positions on the diagonal.
9. No. Multiplication of diagonal matrices amounts to multiplying entries in corresponding positions on the diagonal. The diagonal matrix with all diagonal entries 1 is a multiplicative identity, but a diagonal matrix having an entry 0 on the diagonal has no inverse.
10. Yes. See the answer to Exercise 9.
11. Yes. See the answer to Exercise 9.
12. No. The matrix with all entries 0 is upper triangular, but has no inverse.
13. Yes. The sum of upper-triangular matrices is again upper triangular, and addition amounts to just adding entries in $\mathbf{R}$ in corresponding positions.
14. Yes. Let A and B be upper triangular with determinant 1.

The entry in row i and column j in $C = AB$ is 0 if $i > j$, since for each product $a_{ik}b_{kj}$ appearing the computation of c_{ij}, either $k < i$ so that $a_{ik} = 0$ or $k \geq i > j$ so that $b_{kj} = 0$. Thus the product of two upper-triangular matrices is again upper triangular. Since I_n is upper triangular, we have a multiplicative identity. We are told that $\det(A) = 1$ implies that A is invertible. Since $\det(AB) = \det(A)\cdot\det(B)$, we see that $\det(A)\cdot\det(A^{-1}) = \det(I_n) = 1$, so $\det(A^{-1}) = 1$ also. Thus we have a group.

15. Yes. The relation $\det(AB) = \det(A)\cdot\det(B)$ shows that the set of matrices with determinant ± 1 is closed under multiplication. The set contains an identity since $\det(I_n) = 1$. As in the preceding solution, we see that $\det(A) = \pm 1$ implies that $\det(A^{-1}) = \pm 1$, so we have a group.

16. a) We must show that S is closed under $*$, that is, that $a + b + ab \neq -1$ for $a, b \in S$. Now $a + b + ab = -1$ if and only if $0 = ab + a + b + 1 = (a + 1)(b + 1)$. This is the case if and only if either $a = -1$ or $b = -1$, which is not the case for $a, b \in S$.

b) G_1: $a * (b * c) = a * (b + c + bc)$
$= a + (b + c + bc) + a(b + c + bc)$
$= a + b + c + ab + ac + bc + abc.$
$(a * b) * c = (a + b + ab) * c$
$= (a + b + ab) + c + (a + b + ab)c$
$= a + b + c + ab + ac + bc + abc.$

G_2: 0 acts as identity for $*$, for $0 * a = a * 0 = a$.

G_3: $-a/(a + 1)$ acts as inverse of a, for $a * (-a/(a + 1))$

$$= a + \frac{-a}{a+1} + a\frac{-a}{a+1} = \frac{a(a+1) - a - a^2}{a+1} = \frac{0}{a+1} = 0.$$

c) Since the operation is commutative, $2 * x * 3 = 2 * 3 * x = 11 * x$. Now the inverse of 11 is $-11/12$ by (b). From $11 * x = 7$, we obtain $x = (-11/12) * 7 =$

$$(-11/12) + 7 + (-11/12)7 = \frac{-11 + 84 - 77}{12} = \frac{-4}{12} = -\frac{1}{3}.$$

17. (See the text answer for the tables.) Table I is structurally different from the others since every element is its own inverse. Table II can be made to look just like Table III by interchanging the names a and b everywhere:

*	e	b	a	c
e	e	b	a	c
b	b	e	c	a
a	a	c	b	e
c	c	a	e	b

and rewriting this table in the order e, a, b, c.

a) All groups of order 4 are commutative.
b) Table III gives the group U_4, replacing e by 1, a by i, b by -1, and c by $-i$.
c) Take $n = 2$. There are four 2×2 diagonal matrices with diagonal entries ±1, namely

$$E = \begin{bmatrix} 1 & 0 \\ 0 & 1 \end{bmatrix}, \quad A = \begin{bmatrix} -1 & 0 \\ 0 & 1 \end{bmatrix}, \quad B = \begin{bmatrix} 1 & 0 \\ 0 & -1 \end{bmatrix} \text{ and } C = \begin{bmatrix} -1 & 0 \\ 0 & -1 \end{bmatrix}.$$

If we write the table for this group using the letters E, A, B, C in that order, we obtain Table I in the text answer with the letters capitalized.

18. 2, 3. (It gets harder for 4 elements, where the answer is *not* 4.)

19. (See the text answer.)

20. Ignoring spelling, punctuation and grammar, here are some of the mathematical errors.
a) The statement "x = identity" is wrong.
b) The identity element should be e, not (e). It would also be nice to give the properties satisfied by the identity and by inverse elements.
c) Associativity is missing. Logically, the identity should be mentioned before inverses. The statement "an inverse exists" is not quantified correctly: for each element of the set, an inverse exists. Again, it would be nice to give the properties satisfied by the identity and by inverse elements.
d) Replace "such that for all $a, b \in G$" by "if for all $a \in G$ ". Delete "under addition" in line 2. Replace "$= e$" by "$= a$" in line 3.

21. (See the text answer.)

22. F T T F F T T T F F

23. Let $S = \{x \in G \mid x' \neq x\}$. Then S has an even number of elements, since its elements can be grouped in pairs x, x'. Since G has an even number of elements, the number of elements in G but not in S (the set $G - S$) must be even. The set $G - S$ in nonempty since it contains e. Thus there is at

least one element of $G - S$ other than e, that is, at least one element other than e that is its own inverse.

24. a) $(a * b) * c = (|a|b) * c = \big||a|b\big|c = |ab|c$.

$a * (b * c) = a * (|b|c) = |a||b|c = |ab|c$.

b) $1 * a = |1|a = a$ for all $a \in R^*$ so 1 is a left identity. For $a \in R^*$, $1/|a|$ is a right inverse.

c) No, since both 1/2 and -1/2 are left inverses of 2.

d) A one-sided definition of a group, mentioned just before the exercises, must be all left sided or all right sided. We must not mix them.

25. Let G be a group and let $x \in G$ such that $x * x = x$. Then $x * x = x * e$, and by left cancellation, $x = e$, so e is the only idempotent element in a group.

26. We have $e = (a * b) * (a * b)$ and $(a * a) * (b * b) = e * e = e$. Thus $a * b * a * b = a * a * b * b$. Using left and right cancellation, we have $b * a = a * b$.

27. Since multiplication $\mathbb{C}$ is commutative, if $z_1{}^n = 1$ and $z_2{}^n = 1$, then $(z_1z_2)^n = z_1{}^nz_2{}^n = 1\cdot 1 = 1$, so the nth roots of unity are closed under multiplication. Multiplication in $\mathbb{C}$ is associative. The number 1 is an nth root of unity andis the multiplicative identity. If z_1 is an nth root of unity, then $(1/z_1)^n = 1/(z_1)^n = 1/1 = 1$, so $1/z_1$ is also an nth root of unity and is the inverse of z_1.

28. Let $P(n)$ be the given formula. Since $(a * b)^1 = a * b = a^1 * b^1$ we see that $P(1)$ is true. Suppose that $P(k)$ is true. Then $(a * b)^{k+1} = (a * b)^k * (a * b) = (a^k * b^k) *(a * b) = [a^k * (b^k * a)] * b = [a^k* (a * b^k)] * b = [(a^k* a) * b^k] * b = (a^{k+1} * b^k) * b = a^{k+1} * (b^k * b) = a^{k+1} * b^{k+1}$. This This completes the induction argument.

29. The elements $e, a, a^2, a^3, \cdots, a^m$ cannot all be different since G has only m elements. If one of $a, a^2, \cdots, a^m$ is e, then we are done. If not, then we must have $a^i = a^j$ where $i < j$. Repeated left cancellation of a yields $e = a^{j-1}$.

30. We have $(a * b) * (a * b) = (a * a) * (b * b)$, so

$a * [b * (a * b)] = a * [a * (b * b)]$ and left cancellation yields $b * (a * b) = a * (b * b)$. Then $(b * a) * b = (a * b) * b$ and right cancellation yields $b * a = a * b$.

31. Let $a * b = b * a$. Then $(a * b)' = (b * a)' = a' * b'$ by the corollary of Theorem 1.3. Conversely, if $(a * b)' = a' * b'$ then $b' * a' = a' * b'$. Then $(b' * a')' = (a' * b')'$ so $(a')' * (b')' = (b')' * (a')'$ and $a * b = b * a$.

32. $a * b * c = a * (b * c) = e$ implies that $b * c$ is the inverse of a. Therefore $(b * c) * a = b * c * a = e$ also.

33. We need to show that a left identity is a right identity and a left inverse is a right inverse. Note that $e * e = e$. Then $(x' * x) * e = x' * x$ so $(x')' * (x' * x) * e = (x')' * (x' * x)$. Using associativity, $[(x')' * x'] * x * e = [(x')' * x'] * x$, $(e * x) * e = e * x$ so $x * e = x$ and e is right identity also. If $a' * a = e$, then $(a' * a) * a' = e * a' = a'$. Multiplication of $a' * a * a' = a'$ on the left by $(a')'$ and associativity yield $a * a' = e$, so a' is also a right inverse of a.

34. Using the hint, we show there is a left identity and each element has a left inverse. Let $a \in G$; we are given that G is nonempty. Let e be a solution of $y * a = a$. We show that $e * b = b$ for any $b \in G$. Let c be a solution of the equation $a * x = b$. Then $e * b = e * (a * c) = (e * a) * c = a * c = b$. Thus e is a left identity. For each $a \in G$, let a' be a solution of $y * a = e$. Then a' is a left inverse of a. By Exercise 33, G is a group.

35. a) a semigroup b) a monoid

SECTION 1.3 - Subgroups

1. Yes
2. No. There is no identity.
3. Yes
4. Yes
5. Yes
6. No. The set is not closed under addition.
7. No. If $\det(A) = \det(B) = 2$, then $\det(AB) = 4$. The set is not closed under multiplication.
8. Yes
9. Yes. See Exercise 14 of the Section 1.2.
10. No. If $\det(A) = \det(B) = -1$, then $\det(AB) = 1$. The set is not closed under muultiplication.

11. Yes. See Exercises 14 and 15 of Section 1.2.

12. Yes. Suppose that $(A^T)A = I_n$ and $(B^T)B = I_n$. Then $(AB)^T AB = B^T A^T AB = B^T(A^T A)B = B^T I_n B = B^T B = I_n$, so the set of these matrices is closed under multiplication. Since $I_n^T = I_n$ and $I_n I_n = I_n$, the set contains the identity. For each A in the set, the equation $(A^T)A = I_n$ shows that A has an inverse A^T. The equation $(A^T)^T A^T = AA^T = I_n$ shows that A^T is in the given set. Thus we have a group.

13. a) No. Not closed under addition. b) Yes

14. a) Yes b) No. It is not even a subset of $\tilde{F}$.

15. a) No. Not closed under addition b) Yes

16. a) No. Not closed under addition. b) Yes

17. a) No. Not closed under addition.
b) No. Not closed under multiplication.

18. a) Yes b) No. The zero constant function has no inverse.

19. (See the text answer.)

20. a) -50, -25, 0, 25, 50 b) 4, 2, 1, 1/2, 1/3
c) 1, π, π^2, $1/\pi$, $1/\pi^2$

21. $\begin{bmatrix} 0 & -1 \\ -1 & 0 \end{bmatrix}$, $\begin{bmatrix} 1 & 0 \\ 0 & 1 \end{bmatrix}$

22. All matrices $\begin{bmatrix} 1 & n \\ 0 & 1 \end{bmatrix}$ for $n \in \mathbb{Z}$

23. All matrices $\begin{bmatrix} 3^n & 0 \\ 0 & 2^n \end{bmatrix}$ for $n \in \mathbb{Z}$

24. All matrices of the form $\begin{bmatrix} 4^n & 0 \\ 0 & 4^n \end{bmatrix}$ or $\begin{bmatrix} 0 & -2^{2n+1} \\ -2^{2n+1} & 0 \end{bmatrix}$ for $n \in \mathbb{Z}$

25. (See the text answer.)

26. 4   27. 2   28. 3   29. 5   30. 4   31. 8

32. 2   33. 4   34. 3

Section 1.3

35. (See the text answer.)

36. T F T F F F F F T F

37. In the Klein 4-group, the equation $x^2 = e$ has all four elements of the group as solutions.

38. Let $S = \{hk \mid h \in H, k \in K\}$ and let $x, y \in S$. Then $x = hk$ and $y = h'k'$ for some $h, h' \in H$ and $k, k' \in K$. Since G is abelian, we have $xy = hkh'k' = (hh')(kk')$. Since H and K are subgroups, we have $hh' \in H$ and $kk' \in K$, so $xy \in S$ and S is closed under the induced operation. Since H and K are subgroups, $e \in H$ and $e \in K$ so $e = ee \in S$. For $x = hk$, we have $h^{-1} \in H$ and $k^{-1} \in K$ since H and K are subgroups. Then $h^{-1}k^{-1} \in S$ and since G is abelian, $h^{-1}k^{-1} = k^{-1}h^{-1} = (hk)^{-1} = x^{-1}$, so the inverse of x is in S. Hence S is a subgroup.

39. If H is empty, then there is no $a \in H$.

40. Let H be a subgroup of G. Then for $a, b \in H$, we have $b^{-1} \in H$ and $ab^{-1} \in H$ since H must be closed under the induced operation.

Conversely, suppose $ab^{-1} \in H$ for all $a, b \in H$ and H is nonempty. Let $a \in H$. Then taking $b = a$, we see that $aa^{-1} = e$ is in H. Taking $a = e$ and $b = a$, we see that $ea^{-1} = a^{-1} \in H$. Thus H contains the identity and the inverse of each element. Then for $a, b \in H$, we also have $a, b^{-1} \in H$ and thus $a(b^{-1})^{-1} = a \in H$ so H is closed under the induced operation.

41. Let $G = \{e, a, a^2, a^3, \cdots, a^{n-1}\}$ be a cyclic group of $n \geq 2$ elements. Then $a^{-1} = a^{n-1}$ also generates G, since $(a^{-1})^i = (a^i)^{-1} = a^{n-i}$ for $i = 1, 2, \cdots, n - 1$. Thus if G has only one generator, we must have $a = a^{n-1}$ so $n - 1 = 1$ and $n = 2$. Of course, $G = \{e\}$ is also cyclic with one generator.

42. Let $a, b \in H$. Since G is abelian, $(ab)^2 = a^2b^2 = ee = e$ so $ab \in H$ and H is closed under the induced operation. Since $ee = e$, we see $e \in H$. Since $aa = e$, we see each element of H is its own inverse. Thus H is a subgroup.

43. Let $a, b \in H$. Since G is abelian, $(ab)^n = a^nb^n = ee = e$ so $ab \in H$ and H is closed under the induced operation. Since

$e^n = e$, we see $e \in H$. Since $a^n = e$, we see the inverse of a is a^{n-1} which is in H since H is closed under the induced operation. Thus H is a subgroup.

44. Let G have m elements. Then the elements
$$a, a^2, a^3, \cdots, a^{m+1}$$
cannot all be different, so $a^i = a^j$ for some $i < j$. Then multiplication by a^{-i} shows that $e = a^{j-i}$, and we can take $j - i$ as the desired n.

45. Let $a \in H$ and let H have n elements. Then the elements $a, a^2, a^3, \cdots, a^{n+1}$ are in H (since H is closed) but cannot all be different. Thus $a^i = a^j$ for some $i < j$. Multiplication by a^{-i} shows that $e = a^{j-i}$ so $e \in H$. Also, $a^{-1} \in H$ since $a^{-1} = a^{j-i-1}$ and H is closed. This shows that H is a subgroup of G.

46. Let $x, y \in H_a$. Then $xa = ax$ and $ya = ay$. We then have $(xy)a = x(ya) = x(ay) = (xa)y = (ax)y = a(xy)$, so $xy \in H_a$ and H_a is closed under the operation. Since $ea = ae = a$, we see that $e \in H_a$. From $xa = ax$, we obtain $xax^{-1} = a$ and then $ax^{-1} = x^{-1}a$, showing that $x^{-1} \in H_a$, which is thus a subgroup.

47. a) This is exactly like the proof of Exercise 46 with a replaced by s and with everything quantified by "for all $s \in S$".
    b) Let $a \in H_G$. Then $ag = ga$ for all $g \in G$; in particular, $ab = ba$ for all $b \in H_G$ since H_G is part of G. This shows that H_G is abelian.

48. *Reflexive:* Let $a \in G$. Then $aa^{-1} = e$ and $e \in H$ since H is a subgroup. Thus $a \sim a$.

Symmetric: Let $a \sim b$, so that $ab^{-1} \in H$. Since H is a subgroup, we have $(ab^{-1})^{-1} \in H$, and $(ab^{-1})^{-1} = ba^{-1}$ so $b \sim a$.

Transitive: Let $a \sim b$ and $b \sim c$. Then $ab^{-1} \in H$ and $bc^{-1} \in H$ so $(ab^{-1})(bc^{-1}) = ac^{-1} \in H$ and $a \sim c$.

Section 1.4

49. Let $a, b \in H \cap K$. Then $a, b \in H$ and $a, b \in K$. Since H and K are subgroups, we have $ab \in H$ and $ab \in K$ so $ab \in H \cap K$. Since H and K are subgroups, we have $e \in H$ and $e \in K$ so $e \in H \cap K$. Also, $a^{-1} \in H$ and $a^{-1} \in K$ so $a^{-1} \in H \cap K$. Thus $H \cap K$ is a subgroup.

50. Let G be cyclic and let a be a generator for G. For $x, y \in G$, there exist r and s such that $x = a^r$ and $y = a^s$. Then $xy = a^r a^s = a^{r+s} = a^{s+r} = a^s a^r = yx$, so G is abelian.

51. We can show it if G_n is abelian. Let $a, b \in G$, so that $a^n, b^n \in G_n$. Then $a^n b^n = (ab)^n \in G_n$ since G is abelian. Also $e = e^n \in G_n$ and $(a^n)^{-1} = (a^{-1})^n \in G_n$, so G_n is indeed a subgroup.

52. Let G be a group with no proper nontrivial subgroups. If $G = \{e\}$, then G is of course cyclic. If $G \neq \{e\}$, then let $a \in G$, $a \neq e$. We know that $\langle a \rangle$ is a subgroup of G and $\langle a \rangle \neq \{e\}$. Since G has no proper nontrivial subgroups, we must have $\langle a \rangle = G$ so G is indeed cyclic.

SECTION 1.4 - Cyclic Groups and Generators

1. $q = 4$, $r = 6$
2. $q = -5$, $r = 3$
3. $q = -7$, $r = 6$
4. $q = 6$, $r = 2$
5. 8
6. 8
7. 60
8. 4
9. 10
10. 0
11. 2
12. 4 (Generators are 1,2,3,4.)
13. 4 (Generators are 1,3,5,7.)
14. 4 (Generators are 1,5,7,11.)
15. 16 (Generators are 1,7,11,13,17,19,23,29,31,37,41,43,47,49, 53,59.)
16. 30/5 = 6
17. 42/6 = 7
18. 4
19. 8
20. An infinite cyclic group

21. (See the text answer.)

22.

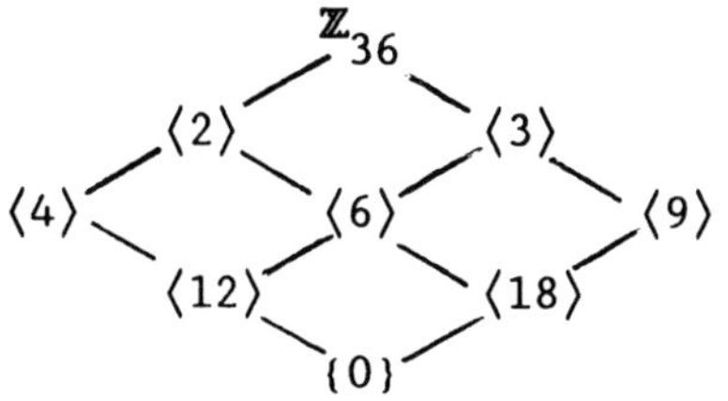

23. (See the text answer.)

24. 1,2,3,6

25. 1,2,4,8

26. 1,2,3,4,6,12

27. 1,2,3,4,5,6,10,12,15,20,30,60

28. 1,17

29. 0,1,2,3,4,5,6,7,8,9,10,11

30. 0,2,4,6,8,10

31. 0,2,4,6,8,10,12,14,16

32. 0,6,12,18,24,30

33. $\cdots,-24,-18,-12,-6,0,6,12,18,24,\cdots$

34. $\cdots,-15,-12,-9,-6,-3,0,3,6,9,12,15,\cdots$

35. T F F F T F F F T T f) The Klein 4-group is an example. g) 9 generates $\mathbb{Z}_{20}$.

36. The Klein 4-group

37. $\langle \mathbb{R}, +\rangle$

38. $\mathbb{Z}_2$

39. (See the text answer.)

40. $\mathbb{Z}_8$

41. i and $-i$

42. $\frac{1}{2}(1 + i\sqrt{3})$ and $\frac{1}{2}(1 - i\sqrt{3})$

43. $(1/\sqrt{2})(1 + i)$, $(1/\sqrt{2})(1 - i)$, $(1/\sqrt{2})(-1 + i)$, $(1/\sqrt{2})(-1 - i)$

44. $\frac{1}{2}(\sqrt{3} + i)$, $\frac{1}{2}(\sqrt{3} - i)$, $\frac{1}{2}(-\sqrt{3} + i)$, $\frac{1}{2}(-\sqrt{3} - i)$

45. The equation $(n_1r + m_1s) + (n_2r + m_2s) = (n_1 + n_2)r + (m_1 + m_2)s$ shows that the set is closed under addition. Since $0r + 0s = 0$, we see that 0 is in the set. Since $[-n)r + (-m)s] + (nr + ms) = 0$, we see that the set contains the inverse of each element. Thus it is a subgroup of $\mathbb{Z}$.

46. Let n be the order of ab so that $(ab)^n = e$. Multiplying this equation on the left by b and on the right by a, we find that $(ba)^{n+1} = bea = (ba)e$. Cancellation of the first factor ba from both sides shows that $(ba)^n = e$. Thus the order of ba is $\le n$. If the order of ba were less than n, a symmetric argument would show that the order of ab is less than n. Thus ba has order n.

Section 1.4

47. a) (See the text answer. Note that G can also be defined as $r\mathbb{Z} \cap s\mathbb{Z}$.)
    b) (See the text answer.)
    c) Let $d = ir + js$ be the gcd of r and s, and let $m = kr = qs$ be the least common multiple of r and s. Then $md = mir + mjs = qsir + krjs = (qi+kj)rs$, so rs is a divisor of md. Now let $r = ud$ and let $s = vd$. Then $rs = uvdd = (uvd)d$, and $uvd = rv = su$ is a common multiple of r and s, and hence $uvd = mt$. Thus $rs = mtd = (md)t$, so md is a divisor of rs. Hence $md = rs$.

48. Note that every group is the union of its cyclic subgroups, since every element of the group generates a cyclic subgroup that contains the element. Let G have only a finite number of subgroups, and hence only a finite number of cyclic subgroups. Now none of these cyclic subgroups can be infinite, for every infinite cyclic group is isomorphic to $\mathbb{Z}$ which has an infinite number of subgroups, namely $\mathbb{Z}$, $2\mathbb{Z}$, $3\mathbb{Z}$, $\cdots$. Such subgroups of an infinite cyclic subgroup of G would of course give an infinite number of subgroups of G, which is impossible. Thus G has only finite cyclic subgroups, and only a finite number if those. We see that the set G can be written as a finite union of finite sets, so G is itself a finite set.

49. S_3 gives a counterexample.

50. *Associativity:* Let $r, s, t \in \mathbb{Z}_n$. In accord with the division algorithm, write $s + t = nq_1 + r_1$ where $0 \le r_1 < n$ and write $r + r_1 = nq_2 + r_2$ where $0 \le r_2 < n$. Then $r +_n (s +_n t) = r +_n r_1 = r_2$. We also find that $r + s + t = r + nq_1 + r_1 = nq_1 + (r + r_1) = nq_1 + nq_2 + r_2 = n(q_1 + q_2) + r_2$. Thus $r_2 = r +_n (s +_n t)$ is indeed the remainder of $r + s + t$ when divided by n according to the division algorithm. A similar analysis shows that $(r +_n s) +_n t$ is this same remainder of $r + s + t$ when divided by n. Thus $+_n$ is associative.

    Identity: Clearly 0 acts as identity for $+_n$.

    Inverse: Clearly $n - r$ is the inverse of r in $\mathbb{Z}_n$.

51. Note that $xax^{-1} \ne e$ since $xax^{-1} = e$ would imply that $xa = x$ and $a = e$, and we are given that a has order 2. We have

$(xax^{-1})^2 = xax^{-1}xax^{-1} = xa^2x^{-1} = xex^{-1} = xx^{-1} = e$. Since a is given to be the *unique* element in G of order 2, we see that $xax^{-1} = a$, and we obtain $xa = ax$ for all $x \in G$ on multiplication on the right by x.

52. The positive integers less than pq and relatively prime to pq are those that are not multiples of p and are not multiples of q. There are $p - 1$ multiples of q and $q - 1$ multiples of p that are less than pq. Thus there are $(pq - 1) - (p - 1) - (q - 1) = pq - p - q + 1 = (p - 1)(q - 1)$ positive integers less than pq and relatively prime to pq.

53. The positive integers less than p^r *and relatively prime to* p^r are those that are not multiples of p. There are $p^{r-1} - 1$ multiples of p that are less than p^r. Thus there are $(p^r - 1) - (p^{r-1} - 1) = p^r - p^{r-1} = p^{r-1}(p - 1)$ positive integers less than p^r and relatively prime to p^r.

54. It is no loss of generality to suppose that $G = \mathbb{Z}_n$ and that we are considering the equation $mx = 0$ for a positive integer m dividing n. Clearly $0, n/m, 2n/m, \cdots, (m - 1)n/m$ are m solutions of $mx = 0$. If r is any solution in $\mathbb{Z}_n$ of $mx = 0$, then n is a divisor of mr, so that $mr = qn$. But then $r = qn/m < n$, so that q must be one of $0, 1, 2, \cdots m - 1$, and we see that the solutions exhibited above are indeed all the solutions.

55. There are exactly d solutions, where d is the gcd of m and n. Namely, working in $\mathbb{Z}_n$ again, it is clear that $0, n/d, 2n/d, \cdots, (d - 1)n/d$ are solutions of $mx = 0$. If r is any solution, then n divides mr so that $mr = nq$ and $r = nq/m$. Write $m = m_1d$ and $n = n_1d$ so that the gcd of m_1 and n_1 is 1. Then $r = nq/m$ can be written as $r = n_1dq/m_1d = n_1q/m_1$. Since m_1 and n_1 are relatively prime, we conclude that m_1 divides q ; let $q = m_1s$. Then $r = n_1q/m_1 = n_1m_1s/m_1 = n_1s_1 = (n/d)s$. Since $r < n$, we have $n_1s < n = n_1d$ so $s < d$. Consequently, s must be one of the numbers $0, 1, 2, \cdots, d - 1$ and we see that the solutions exhibited above are indeed all the solutions.

Section 1.4

56. All positive integers less than p are relatively prime to p since p is prime, and hence they all generate $\mathbb{Z}_p$. Thus $\mathbb{Z}_p$ has no proper cyclic subgroups, and thus no proper subgroups, since as a cyclic group, $\mathbb{Z}_p$ has only cyclic subgroups.

57. a) Let a be a generator of H and let b be a generator of K. Since G is abelian, we have $(ab)^{rs} = (a^r)^s(b^s)^r = e^r e^s = e$. We claim that no lower power of ab is equal to e. For suppose that $(ab)^n = a^n b^n = e$. Then $a^n = b^{-n} = c$ must be an element of both H and K, and thus generates a subgroup of H of order dividing r which must simultaneously be a subgroup of K of order dividing s. Since r and s are relatively prime, we see that we must have $c = e$, so $a^n = b^n = e$. But then n is divisible by both r and s, and since r and s are relatively prime, we have $n = rs$. Thus ab generates the desired cyclic subgroup of G of order rs.

    b) Let d be the gcd of r and s, and let $s = dq$ so that q and r are relatively prime and $rq = rs/d$ is the lcm of r and s (see Exercise 47c). Let a and b be generators of H and K respectively. Then $|\langle a\rangle| = r$ and $|\langle b^d\rangle| = q$ where r and q are relatively prime. Part (a) shows that the element ab generates a cyclic subgroup of order rq which is the lcm of r and s.

58. Choose a pair of generating directed arcs, call them *arc1* and *arc2*, start at any vertex of the digraph, and see if the sequences *arc1*, *arc2* and *arc2*, *arc1* lead to the same vertex. (This corresponds to asking if the two corresponding group generators commute.) The group is commutative if and only these two sequences lead to the same vertex for *every pair* of generating directed arcs.

59. It is not commutative.

60. It is not obvious, since a digraph of a cyclic group can be formed using a generating set of two or more elements, no one of which generates the group.

61. No. It does not contain the identity 0.

62.

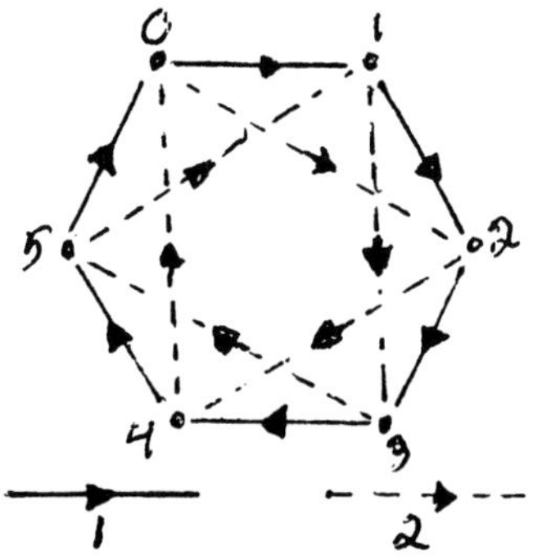

63. (See the text answer.)

64. a) Starting from any vertex, every path through the graph that terminates at the same vertex represents a product of generators or their inverses that is equal to the identity and thus gives a relation.

b) $a^4 = e$, $b^2 = e$, $(ab)^2 = e$

65. a) a^3b b) a^2 c) a^2

66.

	e	a	b	c
e	e	a	b	c
a	a	e	c	b
b	b	c	e	a
c	c	b	a	e

67. (See the text answer.)

68.

	e	a	b	c	d	f
e	e	a	b	c	d	f
a	a	c	f	e	b	d
b	b	d	e	f	a	c
c	c	e	d	a	f	b
d	d	f	c	b	e	a
f	f	b	a	d	c	e

69. (See the text answer.)

70. Generalizing Fig. 1.10c in the preceding exercise, consider two regular n-gons, one inside the other, with dashed lines joining corresponding vertices as shown in Fig. 1.10c for $n = 3$. Let the edges of the outer n-gon have clockwise arrows and the edges of the inner one have counterclockwise arrows as in the figure. The four properties listed after Example 12 in the text are satisfied, and the digraph represents a nonabelian group of order $2n$.

CHAPTER 2

MORE GROUPS AND COSETS

SECTION 2.1 - Groups of Permutations

1. $\begin{pmatrix} 1 & 2 & 3 & 4 & 5 & 6 \\ 1 & 2 & 3 & 6 & 5 & 4 \end{pmatrix}$

2. $\begin{pmatrix} 1 & 2 & 3 & 4 & 5 & 6 \\ 2 & 4 & 1 & 5 & 6 & 3 \end{pmatrix}$

3. $\begin{pmatrix} 1 & 2 & 3 & 4 & 5 & 6 \\ 3 & 4 & 1 & 6 & 2 & 5 \end{pmatrix}$

4. $\begin{pmatrix} 1 & 2 & 3 & 4 & 5 & 6 \\ 5 & 1 & 6 & 2 & 4 & 3 \end{pmatrix}$

5. $\begin{pmatrix} 1 & 2 & 3 & 4 & 5 & 6 \\ 2 & 6 & 1 & 5 & 4 & 3 \end{pmatrix}$

6. Starting with 1 and applying σ repeatedly, we see that σ takes 1 to 3 to 4 to 5 to 6 to 2 to 1, so σ^6 is the smallest possible power of σ that is the identity. It is easily checked that σ^6 carries 2, 3, 4, 5 and 6 to themselves also, so σ^6 is indeed the identity and $|\langle\sigma\rangle| = 6$.

7. $\tau^2 = \begin{pmatrix} 1 & 2 & 3 & 4 & 5 & 6 \\ 4 & 3 & 2 & 1 & 5 & 6 \end{pmatrix}$ and it is clear that $(\tau^2)^2$ is the identity. Thus we have $|\langle\tau^2\rangle| = 2$.

8. Since σ^6 is the identity (see Exercise 6), we have $\sigma^{100} = (\sigma^6)^{16}\sigma^4 = \sigma^4 = \begin{pmatrix} 1 & 2 & 3 & 4 & 5 & 6 \\ 6 & 5 & 2 & 1 & 3 & 4 \end{pmatrix}$.

9. We find that μ^2 is the identity, so $\mu^{100} = (\mu^2)^{50}$ is also the identity.

10. $\{1, 2, 3, 4, 5, 6\}$ 11. $\{1, 2, 3, 4\}$ 12. $\{1, 5\}$

13. (See the text answer.)

14. $\epsilon, \rho, \rho^2, \rho^3, \phi, \rho\phi, \rho^2\phi, \rho^3\phi$ where their ϕ is our μ_1. This gives our elements in the order
$$\rho_0, \rho_1, \rho_2, \rho_3, \mu_1, \delta_1, \mu_2, \delta_2.$$

15. σ may have the action of any of the six possible permutations on the set $\{1,2,4\}$, so there are six possibilities for σ.

16. There are 4 possibilities for $\sigma(1)$, then 3 possibilites for $\sigma(3)$, then 2 possibilities for $\sigma(4)$, and 1 possibility for $\sigma(5)$, for $4\cdot3\cdot2\cdot1 = 24$ possibilities in all.

17. (See the text answer.)

18. Referring to Table 2.2, we find that $\langle\rho_0\rangle = \{\rho_0\}$, $\langle\rho_1\rangle = \langle\rho_3\rangle = \{\rho_0, \rho_1, \rho_2, \rho_3\}$, $\langle\rho_2\rangle = \{\rho_0, \rho_2\}$, $\langle\mu_1\rangle = \{\rho_0, \mu_1\}$, $\langle\mu_2\rangle = \{\rho_0, \mu_2\}$, $\langle\delta_1\rangle = \{\rho_0, \delta_1\}$, and $\langle\delta_2\rangle = \{\rho_0, \delta_2\}$. These are all the cyclic subgroups. A subgroup containing one of the "turn the square over" permutations μ_1, μ_2, δ_1, or δ_2 and also containing ρ_1 or ρ_3 will describe all positions of the square so must be the entire group D_4. Checking the line of the table opposite μ_1, we see that the only other elements that can be in a proper subgroup with μ_1 are ρ_2, μ_2, and of course ρ_0. We check that that $\{\rho_0, \rho_2, \mu_1, \mu_2\}$ is closed under multiplication and is a subgroup. Checking the row of the table opposite μ_2 gives the same subgroup. Checking the rows opposite δ_1 and opposite δ_2 gives the subgroup $\{\rho_0, \rho_2, \delta_1, \delta_2\}$ as the only remaining possibility, using the same reasoning.

19.

	ρ^0	ρ	ρ^2	ρ^3	ρ^4	ρ^5
ρ^0	ρ^0	ρ	ρ^2	ρ^3	ρ^4	ρ^5
ρ	ρ	ρ^2	ρ^3	ρ^4	ρ^5	ρ^0
ρ^2	ρ^2	ρ^3	ρ^4	ρ^5	ρ^0	ρ
ρ^3	ρ^3	ρ^4	ρ^5	ρ^0	ρ	ρ^2
ρ^4	ρ^4	ρ^5	ρ^0	ρ	ρ^2	ρ^3
ρ^5	ρ^5	ρ^0	ρ	ρ^2	ρ^3	ρ^4

This group is not isomorphic to S_3 since it is an abelian group and S_3 is nonabelian.

20. (See the text answer.)

21.

$$\begin{bmatrix}1&0&0&0\\0&1&0&0\\0&0&1&0\\0&0&0&1\end{bmatrix} \quad \begin{bmatrix}0&0&0&1\\1&0&0&0\\0&1&0&0\\0&0&1&0\end{bmatrix} \quad \begin{bmatrix}0&0&1&0\\0&0&0&1\\1&0&0&0\\0&1&0&0\end{bmatrix} \quad \begin{bmatrix}0&1&0&0\\0&0&1&0\\0&0&0&1\\1&0&0&0\end{bmatrix}$$

$$\begin{bmatrix}0&1&0&0\\1&0&0&0\\0&0&0&1\\0&0&1&0\end{bmatrix} \quad \begin{bmatrix}0&0&0&1\\0&0&1&0\\0&1&0&0\\1&0&0&0\end{bmatrix} \quad \begin{bmatrix}0&0&1&0\\0&1&0&0\\1&0&0&0\\0&0&0&1\end{bmatrix} \quad \begin{bmatrix}1&0&0&0\\0&0&0&1\\0&0&1&0\\0&1&0&0\end{bmatrix}$$

Section 2.1

22. The identity and flipping over on the vertical axis that falls on the vertical line segment of the figure give the only symmetries; the symmetry group is isomophic to $\mathbb{Z}_2$.

23. As symmetries other than the identity, the figure admits a rotation through 180°, a horizontal flip, and a vertical flip. This group of four elements is isomorphic to the Klein 4-group.

24. If we join endpoints of the line segments, we have a square with the given lines as its diagonals. The symmetries of that square produce all symmetries of the given figure, so the group of symmetries is isomorphic to D_4.

25. The only symmetries are those obtained by sliding the figure to the left or to the right. We consider the vertical line segments to be one unit apart. For each integer n, we can slide the figure n units to the right if $n > 0$ and $|n|$ units to the left if $n < 0$, leaving the figure alone if $n = 0$. A moment of thought shows that performing the symmetry corresponding to an integer n and then the one corresponding to an integer m yields the symmetry corresponding to $n + m$. We see that the symmetry group isomorphic to $\mathbb{Z}$.

26. A permutation

27. Not a permutation. $f(1) = f(-1)$ so f is not one to one.

28. A permutation

29. Not a permutation. $f(x) \neq -1$ for any x so f is not onto $\mathbb{R}$.

30. Not a permutation. $f(2) = f(-1) = 0$ so f is not one to one.

31. The name *two-to-two function* suggests that such a function f should carry every pair of distinct points into two distinct points. Such a function is one-to-one in the conventional sense. (If the domain has only one element, a function cannot fail to be two-to-two, since the only way it can fail to be two-to-two is to carry two points into one point, and the set does not have two points.) Conversely, every function that is one-to-one in the conventional sense carries each pair of points into two distinct points. Thus the functions conventionally called one-to-one are precisely those that carry two points into two points, which is a much more intuitive unidirectional way of regarding them. Also, the standard way of trying to show a function is one-to-one is precisely to show that it is does not fail to be two-to-two. That is, proving that a

function is one-to-one becomes more natural in the two-to-two terminology.

32. T F T F T T F F F T

33. Every proper subgroup of S_3 is abelian, for such a subgroup has order either 1, 2, or 3 by Exercise 17b.

34. Composition of these functions is associative, and the identity map acts as identity for function composition. A function that is not one to one has no inverse, so we do not have inverses. Thus we have a *monoid*.

35. Yes, it is a subgroup. *Closure:* if $\sigma(b) = b$ and $\mu(b) = b$ then $(\sigma\mu)() = \sigma(\mu(b)) = \sigma(b) = b$. *Identity:* the identity carries every element into itself, and hence carries b into b. *Inverse:* if $\sigma(b) = b$, then $\sigma^{-1}(b) = b$.

36. No, the set need not be closed under the operation if B has more than one element. Suppose that $b, c \in B$ and $\sigma(b) = c$ and $\mu(b) = b$ while $\mu(c)$ is not in B. Then $(\mu\sigma)(b) = \mu(\sigma(b)) = \mu(c)$ is not in B, so $\mu\sigma$ is not in the given set.

37. No, an inverse need not exist. Suppose $A = \mathbb{Z}$ and $B = \mathbb{Z}^+$, and let $\sigma\colon A \longrightarrow A$ be defined by $\sigma(n) = n + 1$. Then σ is in the given set, but σ^{-1} is not since $\sigma^{-1}(1) = 0$.

38. Yes, it is a subgroup. Use the proof of Exercise 29, but replace b by B everywhere.

39. The order of D_n is $2n$ since the n-gon can be rotated to n possible positions, and then turned over and rotated to give another n positions. The rotations of the n-gon, without turning it over, clearly form a subgroup of order n.

40. The group has 24 elements, for any one of six faces can be on top, and for each such face on top, the cube can be rotated in four different positions leaving that face on top. One obtains a subgroup of order four by performing the four such rotations, leaving the top face and bottom face on the top and bottom. There are two more such rotation groups of order four, one corresponding to rotations leaving the front and back faces in those positions and one corresponding to rotations leaving the side faces in those positions. One obtains a subgroup of order three by taking hold of a pair of diagonally opposite vertices and rotating through the three possible positions, corresponding to the three edges

emanating from a vertex. There are four such diagonally opposite pairs of vertices, giving the desired four groups of order three.

41. Let $n \geq 3$, and let $\rho \in S_n$ be defined by $\rho(1) = 2$, $\rho(2) = 3$, $\rho(3) = 1$, and $\rho(m) = m$ for $3 < m \leq n$. Let $\mu \in S_n$ be defined by $\mu(1) = 1$, $\mu(2) = 3$, $\mu(3) = 2$, and $\mu(m) = m$ for $3 < m \leq n$. Then $\rho\mu \neq \mu\rho$ so S_n is not commutative. (Note that ρ coincides with our element ρ_1 and μ with our element μ_1 if $n = 3$.)

42. Suppose $\sigma(i) = m \neq i$. Find $\gamma \in S_n$ such that $\gamma(i) = i$ and and $\gamma(m) = r$ where $r \neq m$. Then $(\sigma\gamma)(i) = \sigma(\gamma(i)) = \sigma(i) = m$ while $(\gamma\sigma)(i) = \gamma(\sigma(i)) = \gamma(m) = r$, so $\sigma\gamma \neq \gamma\sigma$. Thus $\sigma\gamma = \gamma\sigma$ for all $\gamma \in S_n$ only if σ is the identity.

43. a) Let ϕ be one to one and onto B. For each $y \in B$, let $\phi^{-1}(y)$ be the $x \in A$ such that $\phi(x) = y$. Such an $x \in A$ exists since ϕ is onto B, and is unique since ϕ is one to one. This definition ensures that $\phi^{-1}(\phi(x)) = x$ and $\phi(\phi^{-1}(y)) = y$ for all $x \in A$, $y \in B$.

    Conversely, let such an inverse function exist. Let $y \in B$. Since $\phi(\phi^{-1}(y)) = y$, we see that ϕ is onto B. If $\phi(a) = \phi(b)$, then $a = \phi^{-1}(\phi(a)) = \phi^{-1}(\phi(b)) = b$ so we see that ϕ is one to one.

    b) Let $y \in B$, and let $x \in A$ be such that $\phi(x) = y$. We are told that an inverse ϕ^{-1} of ϕ exists, satisfying $\phi^{-1}(\phi(x)) = x$. Since $\phi^{-1}(\phi(x)) = \phi^{-1}(y)$, we see that $\phi^{-1}(y)$ must be the x chosen so that $\phi(x) = y$. There is a unique such x since ϕ is one to one.

44. Let c be an element in both $O_{a,\sigma}$ and $O_{b,\sigma}$. Then there exist integers r and s such that $\sigma^r(a) = c$ and $\sigma^s(b) = c$. Then $\sigma^{r-s}(a) = \sigma^{-s}(\sigma^r(a)) = \sigma^{-s}(c) = b$. Thus for each integer $n \in \mathbb{Z}$, we see that $\sigma^n(b) = \sigma^{n+r-s}(a)$. Clearly then $\{\sigma^n(b) \mid n \in \mathbb{Z}\} = \{\sigma^n(a) \mid n \in \mathbb{Z}\}$.

45. Let $A = \{a_1, a_2, \cdots, a_n\}$. Let $\sigma \in S_A$ be defined by $\sigma(a_i) = a_{i+1}$ for $1 \leq i < n$ and $\sigma(a_n) = a_1$. That is, σ essentially

performs a rotation if the elements of A are spaced evenly about a circle. It is clear that σ^n is the identity and $|\langle\sigma\rangle| = n = |A|$. We let $H = \langle\sigma\rangle$. Let a_i and a_j be given; suppose $i < j$. Then $\sigma^{j-i}(a_i) = a_j$ and $\sigma^{i-j}(a_j) = a_i$, so H is transitive on A.

46. Let $\langle\sigma\rangle$ be transitive on A and let $a \in A$. Then $\{\sigma^n(a) \mid n \in \mathbb{Z}\}$ must include all elements of A, that is, $O_{a,\sigma} = A$. Conversely, suppose that $O_{a,\sigma} = A$ for some $a \in A$. Then $\{\sigma^n(a) \mid n \in \mathbb{Z}\} = A$. Let $b, c \in A$ and let $b = \sigma^r(a)$ and $c = \sigma^s(a)$. Then $\sigma^{s-r}(b) = \sigma^s(\sigma^{-r}(b)) = \sigma^s(a) = c$, showing that $\langle\sigma\rangle$ is transitive on A.

47. a) (See the text answer.)
b) (Associativity) Let $a, b, c \in G'$. Then $a *' (b *' c) = a *' (c * b) = (c * b) * a = c * (b * a) = (b * a) *' c = (a *' b) *' c$, where we used the fact that G is a group and the definition of $*'$.
(Identity) We have $e *' a = a * e = a$ and $a *' e = e * a = a$ for all $a \in G'$.
(Inverse). Let $a \in G'$ and let a^{-1} be the inverse of a in G. Then $a^{-1} *' a = a * a^{-1} = e = a^{-1} * a = a *' a^{-1}$, so a^{-1} is also the inverse of a in G'.
Therefore G' is a group.

48. We use the notation in Table 2.2 of the text.

49. (See the text answer.)

50. a) $T_{0110}(s_0) = s_0$
$T_{0110}(s_1) = s_0$
$T_{0110}(s_2) = s_0$

b) $T_{0110111}(s_0) = s_2$
$T_{0110111}(s_1) = s_2$
$T_{0110111}(s_2) = s_2$

c) $T_{1101}(s_0) = s_1$
$T_{1101}(s_1) = s_1$
$T_{1101}(s_2) = s_1$

The transition function given by T_1 where $T_1(s_0) = s_1$, $T_1(s_1) = s_2$, and $T_1(s_2) = s_2$ is missing. This is the only one missing since a string ending with 0 carries all states to s_0, all strings ending with 01 have the same action as 1101, and all strings ending with 11 have the same action as 0110111. The answer does change if the empty string ϵ is is included since the identity transition function is not represented without it.

51. By having a sufficiently large input alphabet, conceivably every function mapping the set of states into itself could be a transition function. To count how many such functions there are, note that the state s_0 can be mapped into any of $n + 1$ choices, for each of these the state s_1 can be mapped into any of the $n + 1$ states, etc. Thus the number of functions appears as a product of $n + 1$ factors, each of which is $n + 1$, so the answer is $(n + 1)^{n+1}$.

52. The input string yx has as transition function $T_x \circ T_y$ since for any state s_i, we have $(T_x \circ T_y)(s_i) = T_x(T_y(s_i))$ which is the state attained when in state s_i by first applying the input y, and then the input x. Since our input strings are read from left to right, this amounts to the action yx.

53. (See the text answer.)

54.

$\circ$	T_ϵ	T_0	T_1	T_{01}	T_{11}
T_ϵ	T_ϵ	T_0	T_1	T_{01}	T_{11}
T_0	T_0	T_0	T_0	T_0	T_0
T_1	T_1	T_{01}	T_{11}	T_{11}	T_{11}
T_{01}	T_{01}	T_{01}	T_{01}	T_{01}	T_{01}
T_{11}	T_{11}	T_{11}	T_{11}	T_{11}	T_{11}

The monoid is not a group since T_0 has no inverse.

55. (See the text answer.)

56. 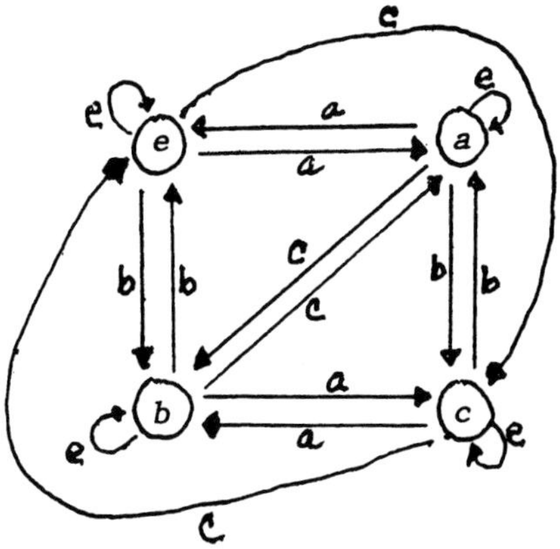

57. Let g_i be an element of G which we think of as an input. We must show that $\phi: G \to G$ where $\phi(x) = xg_i$ is a permutation of G. If $\phi(x) = \phi(y)$ then $xg_i = yg_i$ and $x = y$ by cancellation in G. Thus ϕ is one to one. Since $\phi(xg_i^{-1}) = xg_i^{-1}g_i = x$ for all s in G, we see that ϕ is onto G. Thus ϕ ϕ is a permutation of G. Since every transition function can be formed using function composition from the transition functions given by the input alphabet, and since a composition of permutations of a set is again a permutation of the set, we see that every transition function is a permutation.

58. a group that is isomorphic to G. (It is the right regular representation of G mentioned in the text after the proof of Thereom 25)

SECTION 2.2 - Orbits, Cycles, and the Alternating Groups

1. $\{1,2,5\}$, $\{3\}$, $\{4,6\}$
2. $\{1,5,7,8\}$, $\{2,3,6\}$, $\{4\}$
3. $\{1,2,3,4,5\}$, $\{6\}$, $\{7,8\}$
4. $\mathbb{Z}$
5. $\{2n \mid n \in \mathbb{Z}\}$, $\{2n + 1 \mid n \in \mathbb{Z}\}$
6. $(3n \mid n \in \mathbb{Z}\}$, $\{3n + 1 \mid n \in \mathbb{Z}\}$, $\{3n + 2 \mid n \in \mathbb{Z}\}$
7. $\begin{pmatrix} 1 & 2 & 3 & 4 & 5 & 6 & 7 & 8 \\ 4 & 1 & 3 & 5 & 8 & 6 & 2 & 7 \end{pmatrix}$
8. $\begin{pmatrix} 1 & 2 & 3 & 4 & 5 & 6 & 7 & 8 \\ 3 & 7 & 2 & 8 & 5 & 4 & 1 & 6 \end{pmatrix}$
9. $\begin{pmatrix} 1 & 2 & 3 & 4 & 5 & 6 & 7 & 8 \\ 5 & 4 & 3 & 7 & 8 & 6 & 2 & 1 \end{pmatrix}$
10. $(1,8)(3,6,4)(5,7)$ and $(1,8)(3,4)(3,6)(5,7)$
11. $(1,3,4)(2,6)(5,8,7)$ and $(1,4)(1,3)(2,6)(5,7)(5,8)$
12. $(1,3,4,7,8,6,5,2)$ and $(1,2)(1,5)(1,6)(1,8)(1,7)(1,4)(1,3)$
13. F T F F F F T T T F

Section 2.2

14. ρ_0, ρ_1, and ρ_2 are even.

Here is the table for A_3:

	ρ_0	ρ_1	ρ_2
ρ_0	ρ_0	ρ_1	ρ_2
ρ_1	ρ_1	ρ_2	ρ_0
ρ_2	ρ_2	ρ_0	ρ_1

15. (See the text answer.)

16. a) Since $(1,2,3,4,\cdots,n) = (1,n)(1,n-1)\cdots(1,3)(1,2)$, we see than a cycle of length n can be written as a product of $n - 1$ transpositions. Now a permutation in S_n can be be written as a product of disjoint cycles, the sum of whose lengths is $\le n$. If there are r disjoint cycles involved, then writing each as a product of transpositions, we see the permutation can be written as a product of at most $n - r$ transpositions. Since $r \ge 1$, we can always write the permutations as a product of at most $n - 1$ transpositions.

b) This follows from our proof of (a), since we must have $r \ge 2$.

c) Write the odd permutation σ as a product of s transpositions, where $s \le n - 1$ by part (a). Then s is an odd number and $2n + 3$ is an odd number, so $2n + 3 - s$ is an even number. Adjoin $2n + 3 - s$ transpositions $(1,2)$ as factors at the right of the product of the s transpositions that comprise σ. The same permutation σ results since the product of an even number of factors $(1,2)$ is the identity permutation. Thus σ can be written as a product of $2n + 3$ permutations.

If σ is even, we proceed in exactly the same way, but this time s is even so $2n + 8 - s$ is also even. We tack the identity permutation, written as a product of $2n + 8 - s$ factors $(1,2)$, onto the end of σ and obtain σ as a product of $2n + 8$ transpositions.

17. (See the text answer.)

18. Suppose $\sigma \in H$ is an odd permutation. Let $\phi: H \longrightarrow H$ be defined by $\phi(\mu) = \sigma\mu$ for $\mu \in H$. If $\phi(\mu_1) = \phi(\mu_2)$, then $\sigma\mu_1 = \sigma\mu_2$, so $\mu_1 = \mu_2$ by group cancellation. Also, for any any $\mu \in H$, we have $\phi(\sigma^{-1}\mu) = \sigma\sigma^{-1}\mu = \mu$. This shows that ϕ is a one-to-one map of H onto itself. Since σ is an odd permutation, we see that ϕ maps an even permutation into an odd one, and an odd permutation into an even one. Since ϕ maps the set of even permutations in H one to one onto the

set of odd permutations in H, it is immediate that H has the same number of even permutations as odd permutations. Thus we have shown that if H has one odd permutation, it has the same number of even permutations as odd permutations.

19. Since only the elements of the cycle are moved, there are just n elements moved by a cycle of length n.

20. Let σ, $\mu \in H$. If σ moves elements $s_1, s_2, \cdots, s_k$ of A and μ moves elements $r_1, r_2, \cdots, r_m$ of A, then $\sigma\mu$ cannot move any elements not in the list $s_1, s_2, \cdots, s_k, r_1, r_2, \cdots, r_m$ so $\sigma\mu$ moves at most a finite number of elements of A, and hence is in H. Thus H is closed under the operation of S_A. The identity permutation is in H since it moves no elements of A. Since the elements moved by $\sigma \in H$ are the same as the elements moved by σ^{-1}, we see that for each $\sigma \in H$, we have $\sigma^{-1} \in H$ also. Thus H is a subgroup of S_A.

21. No. If σ, $\mu \in H$ and σ moves 40 elements while μ moves 30 totally different elements, then $\sigma\mu$ moves 70 elements, and is thus not in H. Since H is not closed under permutation multiplication, it cannot be a subgroup of S_A.

22. Let μ be any odd permutation in S_n. Since σ is an odd permutation, so is σ^{-1}, and consequently $\sigma^{-1}\mu$ is an even permutation, and thus in A_n. Since $\mu = \sigma(\sigma^{-1}\mu)$, we see that μ is indeed a product of σ and a permutation in A_n.

23. It is no loss of generality to assume that $\sigma = (1,2,\cdots,m)$ where m is odd. Since m is odd, we easily compute that $\sigma^2 = (1,3,5,\cdots,m,2,4,6,\cdots,m - 1)$, which is again a cycle.

24. If σ is a cycle of length n, then σ^r is also a cycle if and only if n and r are relatively prime, that is, n and r have no common factors greater than 1.

25. We must show that λ_a is one to one and onto G. Suppose that $\lambda_a(g_1) = \lambda_a(g_2)$. Then $ag_1 = ag_2$. The group cancellation property then yields $g_1 = g_2$, so λ_a is one to one. Let $g \in G$. Then $\lambda_a(a^{-1}g) = a(a^{-1}g) = g$, so λ_a is onto G.

Section 2.2

26. Let $\lambda_a, \lambda_b \in H$. For $g \in G$, we have $\lambda_a(\lambda_b(g)) = \lambda_a(bg) = a(bg) = (ab)g = \lambda_{ab}(g)$. Thus $\lambda_a\lambda_b = \lambda_{ab} \in H$, so H is closed under permutation multiplication (function composition). Clearly $\lambda_e \in H$ is the identity permutation of G, and $\lambda_a^{-1} = \lambda_{a^{-1}} \in H$ for each $\lambda_a \in H$. Thus H is indeed a subgroup of S_G.

27. We must show that for each a, $b \in G$, there exists some $\lambda_c \in H$ such that $\lambda_c(a) = b$. We need only choose c such that $ca = b$. That is, we take $c = ba^{-1}$.

28. a) If σ could be written as a product of both an even number of transpositions and an odd number of transposition, then $\iota = \sigma\sigma^{-1}$ could written as a product of an odd number of transpositions, since the inverse of a product of transpositions is the product of the same transpositions in the opposite order.

    b) Thinking of the numbers 1, 2, 3, $\cdots$, n arranged in a row, keep switching i with the number to its right until it is finally switched with $i + k$, and occupies that position. This requires k switches, and leaves $i + k$ in the position formerly occupied by $i + k - 1$. Then keep switching $i + k$ with the number on its left until it reaches the position originally occupied by i. This requires $k - 1$ switches. The numbers between i and $i + k$ were moved once one position to the left and then once one position to the right, so they end in their original position.

    c) Since $2k - 1$ is an odd number, part (b) shows that if ι could be written as an odd number s of transpositions, then it could be written as a number of switches appearing as a sum of s odd numbers, which totals to an odd number since s is odd. Thus if ι can't be written as an odd number of switches, it can't be written as an odd number of transpositions.

    d) For $n = 2$, the only switch is the transposition (1, 2) and the product of an odd number of these tranpositions is equal to (1, 2) which is not the identity permutation.

    e) Thinking of the numbers 1, 2, 3, $\cdots$, n, $n + 1$ as arranged in a row, for each switch that moves $n + 1$ one place to the left, there must be another that moves it one place to the right in order to have it wind up in its original position at the end. Thus an even number of switches move

$n + 1$.

f) Note that a switch that involves $n + 1$ does not change the order of the other numbers among themselves. Thus if ι is written as a product of switches, deleting from the product those switches that move $n + 1$ gives the same permutation of the numbers from 1 to n, and thus gives the identity permutation of those numbers. By our induction hypothesis, there are thus an even number of switches that don't move $n + 1$.

g) Since the number of switches whose product is ι is the number that move $n + 1$ plus the number that don't move $n + 1$, we see from parts (e) and (f) that the total number is even.

29. (See the text answer.)

30. a) The only isometries of $\mathbb{R}$ leaving a number c fixed are the reflection through c that carries $c + x$ to $c - x$ for all $x \in \mathbb{R}$, and the identity map.

b) The isometries of $\mathbb{R}^2$ that leave a point P fixed are the rotations through any angle θ where $0 \le \theta < 360°$ and the reflections across any axis that passes through P.

c) The only isometries of $\mathbb{R}$ that carry a line segment into itself are the reflection through the midpoint of the line segment (see the answer to part (a)) and the identity map.

d) The isometries of $\mathbb{R}^2$ that carry a line segment into itself are a rotation of 180° about the midpoint of the line segment, a reflection in the axis containing the line segment, a reflection in the axis perpendicular to the line segment at its midpoint, and the identity map.

e) The isometries of $\mathbb{R}^3$ that carry a line segment into itself include rotations through any angle about an axis that contains the line segment, reflections across any plane plane that contains the line segment, and reflection across the plane perpendicular to the line segment at is midpoint.

31. (See the text answer.)

32.

	τ	ρ	μ	γ
τ	τ	ρ	$\mu\ \gamma$	$\mu\ \tau$
ρ	ρ	$\rho\ \tau$	$\mu\ \gamma$	$\mu\ \tau$
μ	$\mu\ \gamma$	$\mu\ \gamma$	$\tau\ \rho$	$\tau\ \rho$
γ	$\mu\ \gamma$	$\mu\ \gamma$	$\tau\ \rho$	$\tau\ \rho$

33. (See the text answer.)

34.

35. (See the text answer.)

36.

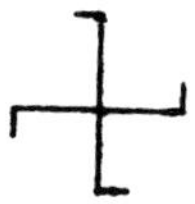

Section 2.2

37. (See the text answer.)

38. *Translation:* order ∞
Rotation: order any $n \geq 2$ or ∞
Reflection: order 2
Glide reflection: order ∞

39. Rotations, reflections

40. Rotations

41. None, since if P and Q are fixed point, then so are all the points on the line segment joining them.

42. Only the identity

43. If P, Q, and R are three non collinear points, then three circles with centers at P, Q, and R have at most one point in common. Namely, two circles intersect in two points, and if the center of the third circle does not lie on the line through the centers of the first two, then it can't pass through both points of intersection of the first two. An isometry ϕ that leaves P, Q, and R fixed must leave every other point S fixed since it must preserve its distance to P, Q, and R so that both S and $\phi(S)$ must be the unique point of intersection of three circles with P, Q, and R as centers and the appropriate radii.

44. If $\phi(P_i) = \psi(P_i)$ for $i = 1, 2, 3$ then $\phi^{-1}(\psi(P_i)) = P_1$ for $i = 1, 2, 3$. Thus by Exercise 43, $\phi^{-1}\psi = \iota$, the identity map, so $\psi = \phi$.

45. No. The product of two rotations (about different points) may be a translation, so the set of rotations is not closed under multiplication.

46. Yes. The product of two translations is a translation and the inverse of a translation is a translation.

47. (See the text answer.)

48. Yes. There is only one reflection μ across one particular line L, and μ^2 is the identity, so we have a group isomorphic to $\mathbb{Z}_2$.

49. No. The product of two glide reflections is orientation preserving, and hence is not a glide reflection.

50. Only reflections and rotations (and the identity) since translations and glide reflections do not have finite order in the group of all plane isometries.

SECTION 2.3 - Cosets and the Theorem of Lagrange

1. (See the text answer.)

2. $4\mathbb{Z} = \{\cdots, -8, -4, 0, 4, 8, \cdots\}$
   $2 + 4\mathbb{Z} = \{\cdots, -6, -2, 2, 6, 10, \cdots\}$

3. (See the text answer.)

4. $\langle 4\rangle = \{0,4,8\}$, $1 + \langle 4\rangle = \{1,5,9\}$, $2 + \langle 4\rangle = \{2,6,10\}$,
   $3 + \langle 4\rangle = \{3,7,11\}$

5. (See the text answer.)

6. $\{\rho_0, \mu_2\}$, $\{\rho_1, \delta_2\}$, $\{\rho_2, \mu_1\}$, $\{\rho_3, \delta_1\}$

7. $\{\rho_0, \mu_2\}$, $\{\rho_1, \delta_1\}$, $\{\rho_2, \mu_1\}$, $\{\rho_3, \delta_2\}$. The right cosets are not the same as the left cosets.

8.

	ρ_0	μ_2	ρ_1	δ_2	ρ_2	μ_1	ρ_3	δ_1
ρ_0	ρ_0	μ_2	ρ_1	δ_2	ρ_2	μ_1	ρ_3	δ_1
μ_2	μ_2	ρ_0	δ_1	ρ_3	μ_1	ρ_2	δ_2	ρ_1
ρ_1	ρ_1	δ_2	ρ_2	μ_1	ρ_3	δ_1	ρ_0	μ_2
δ_2	δ_2	ρ_1	μ_2	ρ_0	δ_1	ρ_3	μ_1	ρ_2
ρ_2	ρ_0	μ_1	ρ_3	δ_1	ρ_0	μ_2	ρ_1	δ_2
μ_1	μ_1	ρ_2	δ_2	ρ_1	μ_2	ρ_0	δ_1	ρ_3
3	ρ_3	δ_1	ρ_0	μ_2	ρ_1	δ_2	ρ_2	μ_1
δ_1	δ_1	ρ_3	μ_1	ρ_2	δ_2	ρ_1	μ_2	ρ_0

No, we do not get a coset group. The 2×2 blocks in the table do not all have elements of one coset.

9. (See the text answer.)

10. The same cosets are obtained as in Exercise 9, so the right cosets of $\{\rho_0, \rho_2\}$ are the same as the left cosets.

11. (See the text answer.)

12. $\langle 3\rangle = \{0,3,6,9,12,15,18,21\}$ has eight elements, so its index (the number of cosets) is $24/8 = 3$.

13. $\langle \mu_1\rangle = \{\rho_0, \mu_1\}$ has two elements, so its index (the number of left cosets) is $6/2 = 3$.

14. $\langle \mu_3\rangle = \{\rho_0, \mu_3\}$ has two elements, so its index (the number of left cosets) is $8/2 = 4$.

15. T T T F T F T T F T

Section 2.3

16. Impossible. For a subgroup H of an abelian group G, we have $a + H = H + a$ for all $a \in G$.

17. For any group G, just take the subgroup $H = G$.

18. The subgroup $\{0\}$ of $\mathbb{Z}_6$.

19. Impossible. Since the cells are disjoint and nonempty, their number cannot exceed the order of the group.

20. Impossible. The number of cells must divide the order of the group, and 4 does not divide 6.

21. *Reflexive:* Let $a \in G$. Then $aa^{-1} = e$ and $e \in H$ since H is a subgroup. Thus $a \sim_R a$.

Symmetric: Suppose $a \sim_R b$. Then $ab^{-1} \in G$. Since H is a subgroup, $(ab^{-1})^{-1} \in H$ and $(ab^{-1})^{-1} = ba^{-1}$, so $ba^{-1} \in H$ and $b \sim_R a$.

Transitive: Suppose $a \sim_R b$ and $b \sim_R c$. Then $ab^{-1} \in H$ and $bc^{-1} \in H$. Since H is a subgroup, $(ab^{-1})(bc^{-1}) = ac^{-1}$ is in H, so $a \sim_R c$.

22. Let $\phi_g\colon H \to Hg$ by $\phi_g(h) = hg$ for $h \in H$. If $\phi_g(h_1) = \phi_g(h_2)$ for $h_1, h_2 \in H$, then $h_1 g = h_2 g$ and $h_1 = h_2$ by group cancellation, so ϕ_g is one to one. Clearly ϕ_g is onto Hg, for if $hg \in Hg$, then $\phi_g(h) = hg$.

23. We show that $gH = Hg$ by showing each coset is a subset of the other. Let $gh \in gH$ where $g \in G$ and $h \in H$. Then $gh = ghg^{-1}g$ $= [(g^{-1})^{-1}hg^{-1}]g$ is in Hg since $(g^{-1})^{-1}hg^{-1}$ is in H by by hypothesis. Thus gH is a subset of Hg.
    Now let $hg \in Hg$ where $g \in G$ and $h \in H$. Then $hg = gg^{-1}hg$ $= g(g^{-1}hg)$ is in gH by hypothesis. Thus Hg is a subset of gH also, so $gH = Hg$.

24. Let $h \in H$ and $g \in G$. By hypothesis, $Hg = gH$. Thus $hg = gh_1$ for some $h_1 \in H$. Then $g^{-1}hg = h_1$, showing that $g^{-1}hg \in H$.

25. (See the text answer.)

26. True. $b = eb$ and $e \in H$ so $b \in Hb$. Since $Hb = Ha$, we have $b \in Ha$.

27. True. Since H is a subgroup, we have $\{h^{-1} \mid h \in H\} = H$. Thus $Ha^{-1} = \{ha^{-1} \mid h \in H\} = \{h^{-1}a^{-1} \mid h \in H\} = \{(ah)^{-1} \mid h \in H\}$. That is, Ha^{-1} consists of all inverses of elements in aH. Similarly, Hb^{-1} consists of all inverses of elements in bH. Since $aH = bH$, we must have $Ha^{-1} = Hb^{-1}$.

28. False. Let H be the subgroup $\{\rho_0, \mu_2\}$ of D_4 in Table 2.2. Then $\rho_1 H = \delta_2 H = \{\rho_1, \delta_2\}$, but $\rho_1{}^2 H = \rho_2 H = \{\rho_2, \mu_1\}$ while $\delta_2{}^2 H = \rho_0 H = H = \{\rho_1, \delta_2\}$.

29. The possible orders for a proper subgroup are p, q, and 1. Now p and q are primes and every group of prime order is cyclic, and of course every group of order 1 is cyclic. Thus every proper subgroup of a group of order pq must be cyclic.

30. From our proof in Exercise 27, we see that $Ha^{-1} = \{(ah)^{-1} \mid h \in H\}$. This shows that the map ϕ of the collection of left cosets into the collection of right cosets defined by $\phi(aH) = Ha^{-1}$ is well defined, for if $aH = bH$, then surely $\{(ah)^{-1} \mid h \in H\} = \{(bh)^{-1} \mid h \in H\}$. Since Ha^{-1} may be any right coset of H, the map is onto the collection of right cosets. Since elements in disjoint sets have disjoint inverses, we see that ϕ is one to one.

31. Let G be abelian of order $2n$ where n is odd. Suppose that G contains two elements, a and b, of order 2. Then $(ab)^2 = abab = aabb = ee = e$ and $ab \neq e$ since the inverse of a is a itself. Thus ab also has order 2. It is easily checked that then $\{e, a, b, ab\}$ is a subgroup of G of order 4. But this is impossible since n is odd and 4 does not divide $2n$. Thus there can't be two elements of order 2.

32. Let G be of order ≥ 2 but with no proper nontrivial subgroups. Let $a \in G$, $a \neq e$. Then $\langle a \rangle$ is a nontrivial subgroup of G, and thus must be G itself. Since every cyclic group not of prime order has proper subgroups, we see that G must be finite of prime order.

Section 2.3

33. Following the hint and using the notation there, it suffices to prove that $\{(a_ib_j)K \mid i = 1,\cdots,r,\ j = 1,\cdots,s\}$ is the collection of distinct left cosets of K in G. Let $g \in G$ and let g be in the left coset a_iH of H. Then $g = a_ih$ for $h \in H$. $h \in H$. Let h be in the left coset b_jK of K in H. Then $h = b_jk$ for $k \in K$, so $g = a_ib_jk$ and $g \in a_ib_jK$. This shows that the collection given in the hint includes all left cosets of K in G. It remains to show the cosets in the collection are distinct. Suppose that $a_ib_jK = a_pb_qK$, so that $a_ib_jk_1 = a_pb_qk_2$ for some k_1,k_2 in K. Now $b_jk_1 \in H$ and and $b_qk_2 \in H$. Thus a_i and a_p are in the same left coset of H, and therefore $i = p$ and $a_i = a_p$. Using group cancellation, we deduce that $b_jk_1 = b_qk_2$. But this means that b_j and b_q are in the same left coset of K, so $j = q$.

34. The partition into left cosets must be H and $G - H = \{g \in G \mid g \text{ not in } H\}$, since G has finite order and H must have half as many elements as G. For the same reason, this must be the partition into right cosets of H. Thus every left coset is also a right coset.

35. Let $a \in G$. Then $\langle a\rangle$ has order d that must divide the order of G, so that $n = dq$. We know that $a^d = e$. Thus $a^n = (a^d)^q = e^q = e$ also.

36. Let $r + \mathbf{Z}$ be a left coset of $\mathbf{Z}$ in $\mathbf{R}$, where $r \in \mathbf{R}$. Let $[r]$ be the greatest integer less than or equal to r. Then $0 \le r - [r] < 1$ and $r + (-[r])$ is in $r + \mathbf{Z}$. Since the difference of any two distinct elements in $r + \mathbf{Z}$ is at least 1, we see that $r - [r]$ must be the unique element x in $r + \mathbf{Z}$ satisfying $0 \le x < 1$.

37. Consider a left coset $r + \langle 2\pi\rangle$ of $\langle 2\pi\rangle$ in $\mathbf{R}$. Then every element of this coset is of the form $r + n(2\pi)$ for $n \in \mathbf{Z}$. We know that $\sin(r + n(2\pi)) = \sin r$ for all $n \in \mathbf{Z}$ since the function *sine* is periodic with period 2π. Thus *sine* has the same value at each element of the coset $r + \langle 2\pi\rangle$.

38. a) *Reflexive:* $a = eae$ where $e \in H$ and $e \in K$, so $a \sim a$.

    Symmetric: Let $a \sim b$ so $a = hbk$. Then $b = h^{-1}ak^{-1}$ and $h^{-1} \in H$ and $k^{-1} \in K$ since H and K are subgroups. Thus $b \sim a$.

Transitive: Let $a \sim b$ and $b \sim c$ so $a = hbk$ and $b = h_1ck_1$ where h, $h_1 \in H$ and k, $k_1 \in K$. Then $a = hh_1ck_1k$ and $hh_1 \in H$ and $k_1k \in K$ since H and K are subgroups. Thus $a \sim c$.

b) The equivalence class containing a is $HaK = \{hak \mid h \in H,\ k \in K\}$. It can be formed by taking the union of all right cosets of H that contain elements in the left coset aK.

39. a) If $\sigma(c) = c$ and $\mu(c) = c$, then $(\sigma\mu)(c) = \sigma(\mu(c)) = \sigma(c) = c$, so $S_{c,c}$ is closed under permutation multiplication. Clearly the identity is in $S_{c,c}$ and if σ leaves c fixed, then σ^{-1} does also. Thus $S_{c,c}$ is a subgroup of S_A.

b) (See the text answer.) c) (See the text answer.)

40. We may suppose the group is $\mathbb{Z}_n$. Let d be a divisor of n. Then $\langle n/d\rangle = \{0, n/d, 2n/d, \cdots, (d-1)n/d\}$ is a subgroup of $\mathbb{Z}_n$ of order d. It consists precisely of all elements x in $\mathbb{Z}_n$ satisfying $dx = 0$. Since an element x of any subgroup of order d of $\mathbb{Z}_n$ must satisfy $dx = 0$, we see that $\langle n/d\rangle$ is the only such subgroup. Since the order of a subgroup must divide the order of the whole group, we see these are the only subgroups that $\mathbb{Z}_n$ has.

41. Every element in $\mathbb{Z}_n$ of order d dividing n generates a cyclic subgroup of order d, and the number of generators of that subgroup is $\phi(d)$ by the corollary of Theorem 1.9. By the preceding exercise, there is a unique such subgroup of order d dividing n. Thus $\mathbb{Z}_n$ contains precisely $\phi(d)$ elements of each order d dividing n. Since the order of each elements of $\mathbb{Z}_n$ is a divisor of n, we obtain the desired result.

42. Let d be a divisor of $n = |G|$. Now if G contains a subgroup of order d, then each element of that subgroup satisfies the equation $x^d = e$. By the hypothesis that $x^m = e$ has as most m solutions in G, we see that there can be at most one subgroup of each order d dividing n. Now each element a of G has order d dividing n and $\langle a\rangle$ has exactly $\phi(d)$ generators. Since it must be the only subgroup of order d, we see that the number of elements of order d for each divisor d of n

cannot exceed $\phi(d)$. Thus we have

$$n = \sum_{d|n} (\text{number of elements of } G \text{ of order } d)$$
$$\leq \sum_{d|n} \phi(d) = n.$$

This shows that G must have exactly $\phi(d)$ elements of each order d dividing n, and thus must have $\phi(n) \geq 1$ elements of order n. Hence G is cyclic.

SECTION 2.4 - Direct Products and Finitely Generated Abelian Groups

1. (See the text answer.)

2.

Element	Order	Element	Order	Element	Order
(0,0)	1	(1,0)	3	(2,0)	3
(0,1)	4	(1,1)	12	(2,1)	12
(0,2)	2	(1,2)	6	(2,2)	6
(0,3)	4	(1,3)	12	(2,3)	12

3. 2 has order 2 in $\mathbb{Z}_4$ and 6 has order 2 in $\mathbb{Z}_{12}$. The lcm of 2 and 2 is 2.

4. 2 has order 3 in $\mathbb{Z}_6$ and 3 has order 5 in $\mathbb{Z}_{15}$. The lcm of 3 and 5 is 15.

5. 8 has order 3 in $\mathbb{Z}_{24}$ and 10 has order 9 in $\mathbb{Z}_{18}$. The lcm of 3 and 9 is 9.

6. 3 has order 4 in $\mathbb{Z}_4$ and 10 has order 6 in $\mathbb{Z}_{12}$ and 9 has order 5 in $\mathbb{Z}_{15}$. The lcm of 4 and 6 and 5 is 60.

7. 3 has order 4 in $\mathbb{Z}_4$ and 6 has order 2 in $\mathbb{Z}_{12}$ and 12 has order 5 in $\mathbb{Z}_{20}$ and 16 has order 3 in $\mathbb{Z}_{24}$. The lcm of 4 and 2 and 5 and 3 is 60.

8. For $\mathbb{Z}_6 \times \mathbb{Z}_8$: the lcm of 6 and 8 which is 24.
   For $\mathbb{Z}_{12} \times \mathbb{Z}_{15}$: the lcm of 12 and 15 which is 60.

9. (See the text answer.)

10. Seven order 2 subgroups: $\langle(1,0,0)\rangle$, $\langle(0,1,0)\rangle$, $\langle(0,0,1)\rangle$, $\langle(1,1,0)\rangle$, $\langle(1,0,1)\rangle$, $\langle(0,1,1)\rangle$, $\langle(1,1,1)\rangle$
Seven order 4 subgroups:
$\{(0,0,0), (1,0,0), (0,1,0), (1,1,0)\}$
$\{(0,0,0), (1,0,0), (0,0,1), (1,0,1)\}$
$\{(0,0,0), (1,0,0), (0,1,1), (1,1,1)\}$
$\{(0,0,0), (1,1,0), (0,0,1), (1,1,1)\}$
$\{(0,0,0), (1,1,0), (0,1,1), (1,0,1)\}$
$\{(0,0,0), (1,1,1), (0,1,0), (1,0,1)\}$
$\{(0,0,0), (0,1,1), (0,0,1), (0,1,0)\}$

11. (See the text answer.)

12. $\{(0,0,0), (1,0,0), (0,1,0), (1,1,0)\}$
$\{(0,0,0), (1,0,0), (0,0,2), (1,0,2)\}$
$\{(0,0,0), (1,0,0), (0,1,2), (1,1,2)\}$
$\{(0,0,0), (1,1,0), (0,0,2), (1,1,2)\}$
$\{(0,0,0), (1,1,0), (0,1,2), (1,0,2)\}$
$\{(0,0,0), (1,1,2), (0,1,0), (1,0,2)\}$
$\{(0,0,0), (0,1,2), (0,0,2), (0,1,0)\}$

13. (See the text answer.)

14. a) 4 b) 12 c) 12 d) 2, 2 e) 8

15. (See the text answer.)

16. $H = \{0, 2, 6, 8, 10\}$. Left cosets: H, $1 + H$

17. (See the text answer.)

18. $H = \{\rho_0, \rho_2, \mu_1, \mu_2\}$. Left cosets: H and $\{\rho_1, \rho_3, \delta_1, \delta_2\}$

19. D_4. Left cosets: D_4

20. $H = \{(0,0), (2,1), (4,2), (0,3), (2,0), (4,1), (0,2), (2,3), (4,0), (0,1), (2,2), (4,3)\}$.
Left cosets: H and the set of all group elements not in H.

21. (See the text answer.)

22. $\mathbb{Z}_{16}$, $\mathbb{Z}_2 \times \mathbb{Z}_8$, $\mathbb{Z}_4 \times \mathbb{Z}_4$, $\mathbb{Z}_2 \times \mathbb{Z}_2 \times \mathbb{Z}_4$, $\mathbb{Z}_2 \times \mathbb{Z}_2 \times \mathbb{Z}_2 \times \mathbb{Z}_2$

23. (See the text answer.)

24. $\mathbb{Z}_{16} \times \mathbb{Z}_9 \times \mathbb{Z}_5$, $\mathbb{Z}_2 \times \mathbb{Z}_8 \times \mathbb{Z}_9 \times \mathbb{Z}_5$
$\mathbb{Z}_4 \times \mathbb{Z}_4 \times \mathbb{Z}_9 \times \mathbb{Z}_5$, $\mathbb{Z}_2 \times \mathbb{Z}_2 \times \mathbb{Z}_4 \times \mathbb{Z}_9 \times \mathbb{Z}_5$
$\mathbb{Z}_2 \times \mathbb{Z}_2 \times \mathbb{Z}_2 \times \mathbb{Z}_2 \times \mathbb{Z}_9 \times \mathbb{Z}_5$, $\mathbb{Z}_{16} \times \mathbb{Z}_3 \times \mathbb{Z}_3 \times \mathbb{Z}_5$
$\mathbb{Z}_2 \times \mathbb{Z}_8 \times \mathbb{Z}_3 \times \mathbb{Z}_3 \times \mathbb{Z}_5$, $\mathbb{Z}_4 \times \mathbb{Z}_4 \times \mathbb{Z}_3 \times \mathbb{Z}_3 \times \mathbb{Z}_5$

$\mathbb{Z}_2 \times \mathbb{Z}_2 \times \mathbb{Z}_4 \times \mathbb{Z}_3 \times \mathbb{Z}_3 \times \mathbb{Z}_5$, $\mathbb{Z}_2 \times \mathbb{Z}_2 \times \mathbb{Z}_2 \times \mathbb{Z}_2 \times \mathbb{Z}_3 \times \mathbb{Z}_3 \times \mathbb{Z}_5$

25. (See the text answer.)

26. There are 3 of order 24. There are 2 of order 25. There are 6 of order (24)(25).

27. Since there are no primes that divide both m and n, any abelian group of order mn is isomorphic to a direct product of cyclic groups of prime-power order where all cyclic groups given by primes dividing m appear before the primes dividing n. Thus any abelian group of order mn is isomorphic to a direct product of a group of order m with a group of order n. Since there are r choices for the group of order m and s choices for the group of order n, there are rs choices in all.

28. $(10)^5 = 2^5 5^5$. There are 7 abelian groups of order 2^5, up to isomorphism, by Exercise 23. Replacing factors 2 by factors 5 in the answer to Exercise 23, we see that there are also 7 abelian groups of order 5^5, up to isomorphism. By Exercise 27, there are (7)(7) = 49 abelian groups of order 10^5, up to isomorphism.

29. T T F T F F F F F T

30. $\mathbb{Z}_2$ is an example.

31. a) The only subgroup of $\mathbb{Z}_5 \times \mathbb{Z}_6$ that is isomorphic to $\mathbb{Z}_5 \times \mathbb{Z}_6$ is $\mathbb{Z}_5 \times \mathbb{Z}_6$ itself.
    b) An infinite number of subgroups of $\mathbb{Z} \times \mathbb{Z}$ are isomorphic to $\mathbb{Z} \times \mathbb{Z}$, since every subgroup of the form $n\mathbb{Z} \times m\mathbb{Z}$ is isomorphic to $\mathbb{Z} \times \mathbb{Z}$ for all positive integers n and m.

32. S_3 is an example.

33. T F F T T F T F T T

34. The numbers are the same.

35. (See the text answer.)

36. Let G be abelian and let $a, b \in G$ have finite order. Then $a^r = b^s = e$ for some positive integers r and s. Since G is abelian, we see that $(ab)^{rs} = (a^r)^s(b^s)^r = e^s e^r = ee = e$ so ab has finite order. This shows the subset H of G of all elements is closed under the group operation. Of course $e \in H$ since e has order 1. If $a^r = e$, then $a^{-r} = (a^{-1})^r = e$ also, showing that $a \in H$ implies $a^{-1} \in H$ and completing the

demonstration that H is a subgroup.

37. (See the text answer.) 38. $\{1, -1\}$

39. (See the text answer.)

40. Using the notation of Theorem 2.11, we see that G is isomorphic to $T \times F$ where
$$T = \mathbb{Z}_{(p_1)^{r_1}} \times \mathbb{Z}_{(p_2)^{r_2}} \times \cdots \times \mathbb{Z}_{(p_n)^{r_n}}$$
is a torsion group and $F = \mathbb{Z} \times \mathbb{Z} \times \cdots \times \mathbb{Z}$ is torsion free.

41. (See the text answer.)

42. Computation in a direct product of n groups consists of computing using the individual groups operations in each of the n components. In a direct product of abelian groups the individual group operations are all commutative, and it follows at once that the direct product is an abelian group.

43. Let a, $b \in H$. Then $a^2 = b^2 = e$. Since G is abelian, we see that $(ab)^2 = abab = aabb = ee = e$, so $ab \in H$ also. Thus H is closed under the group operation. We are given that $e \in H$, and for all $a \in H$, the equation $a^2 = e$ means that $a^{-1} = a \in H$ so H is a subgroup.

44. Yes for order 3. No for order 4 ; the square of an element of order 4 has order 2, so H is not closed under the operation. For prime positive integers H will be a subgroup.

45. S_3 is a counterexample.

46. i) $(h,k) = (h,e)(e,k)$
ii) $(h,e)(e,k) = (h,k) = (e,k)(h,e)$
iii) The only element of $H \times K$ of the form (h,e) and also of the form (e,k) is $(e,e) = e$.

47. *Uniqueness:* Suppose $g = hk = h_1k_1$ for h, $h_1 \in H$ and k, $k_1 \in K$. Then $h_1^{-1}h = k_1k^{-1}$ is in both H and K, and we know $H \cap K = \{e\}$. Thus $h_1^{-1}h = k_1k^{-1} = e$, from which we see that $h = h_1$ and $k = k_1$.

Isomorphic: Suppose $g_1 = h_1k_1$ and $g_2 = h_2k_2$. Then $g_1g_2 = h_1k_1h_2k_2 = h_1h_2k_1k_2$ since elements of H and K commute by hypothesis (ii). Thus by uniqueness, g_1g_2 is renamed $(h_1h_2,\ k_1k_2) = (h_1,\ k_1)(h_2,\ k_2)$ in $H \times K$.

48. Recall that every subgroup of a cyclic group is cyclic. Thus if a finite abelian group G contains a subgroup isomorphic to $\mathbb{Z}_p \times \mathbb{Z}_p$, which is not cyclic, then G cannot be cyclic. Conversely, suppose that G is a finite abelian group that is not cyclic. By Theorem 2.11, G contains a subgroup isomorphic to $\mathbb{Z}_{p^r} \times \mathbb{Z}_{p^s}$ for the same prime p; if all components in the direct product corresponded to distinct primes, then G would have been cyclic. The subgroup $\langle p^{r-1}\rangle \times \langle p^{s-1}\rangle$ of $\mathbb{Z}_{p^r} \times \mathbb{Z}_{p^s}$ is clearly isomorphic to $\mathbb{Z}_p \times \mathbb{Z}_p$.

49. By the Theorem of Lagrange, the order of an element of a finite group (that is, the order of the cyclic subgroup it generates) divides the order of the group. Thus if G has prime-power order, then the order of every element is also a power of the prime. The hypothesis of commutativity was not used.

50. By Theorem 2.11, the groups in the decomposition of $G \times K$ and $H \times K$ are unique except for the order of the factors. Since $G \times K$ and $H \times K$ are isomorphic, these factors in the decomposition must be the same. Since $G \times K$ and $H \times K$ can both be written in the order with the factors from K last, we see that G and H must have the same factors in their expression in the decomposition described in Theorem 2.11. Thus G and H are isomorphic.

51. Computation shows that $(1,2,3,\cdots,n)^r(1,2)(1,2,3,\cdots,n)^{n-r} = (1,2)$ for $r = 0$, $(2,3)$ for $r = 1$, $(3,4)$ for $r = 1$, $\cdots$, $(n,1)$ for $r = n - 1$. To see this, note that any number not mapped into 1 or 2 by $(1,2,3,\cdots,n)^{n-r}$ is left fixed by the given product. For $r = i$, we see that $(1,2,3,\cdots,n)^{n-i}$ maps $i + 1$ into 1 which is mapped into 2 by $(1,2)$ which is mapped into $i + 2$ by $(1,2,3,\cdots,n)^i$.

Let (i,j) be any transposition, written with $i < j$. We easily compute that

$$(i,j) = (i,i+1)(i+1,i+2)\cdots(j-2,j-1)(j-1,j)(j-2,j-1)\cdots(i+1,i+2)(i,i+1).$$

By the corollary of Theorem 2.2, every permutation in S_n can be written as a product of transpositions, which we now see can each be written as a product of the special transpositions $(1,2)$, $(2,3)$, $\cdots$, $(n,1)$ which we now know can be expressed as products of $(1,2)$ and $(1,2,\cdots,n)$. This completes the proof.

52. 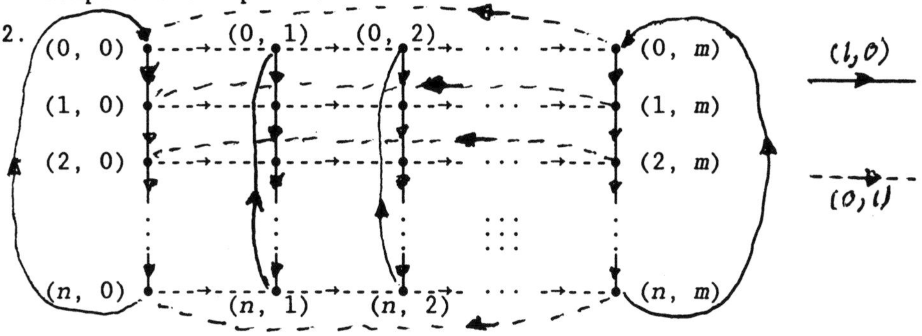

53. a) It is abelian if the two generators a and b representing the two arc types commute. From a diagram, we check that this is the case when the arrows on both n-gons have the same (clockwise or counterclockwise) direction.
b) $\mathbb{Z}_2 \times \mathbb{Z}_n$ c) $\mathbb{Z}_2 \times \mathbb{Z}_n$ is cyclic when n is odd.
d) The dihedral group D_n, for it is generated by an element ρ (a rotation) order n and an element μ (a reflections) of order 2 satisfying $\rho\mu = \mu\rho^{-1}$.

54. $\sin(2\pi x)$ 55. $\sin(2\pi x/\sqrt{3})$ 56. $\sin(2\pi x) + \cos(2\pi y)$

57. $\sin(2\pi x/3) + \cos(2\pi y/\sqrt{5})$

58. $\tau_1(x, y) = (x + \pi, y)$, $\tau_2(x, y) = (x, y + 2\pi/3)$

59. $x^2 + y^2$ 60. $(x - \sqrt{3})^2 + (y + 5)^2$

61. Since G is finite, it can contain no translations, so the orientation preserving isometries in G consist of the rotations in G and the identity map. Since the product of two orientation preserving isometries is orientation preserving, we see that the set H of all orientation preserving isometries in G is closed under multiplication (function composition). Since the inverse of a rotation is also a rotation, we see that H contains the inverse of each element, and is thus a subgroup of G. If $H \neq G$, let μ be an element of G that is not in H. If σ is another element of G

not in H, then $\mu^{-1}\sigma \in H$, since the product of two orientation reversing isometries is order preserving. Thus $\sigma \in \mu H$. This shows that the coset μH contains all elements of G that are not in H. Since $|\mu H| = |H|$, we see that in this case $|G| = 2|H|$.

62. We can consider all the rotations in G to be clockwise. Let ρ be the rotation in G which rotates the plane clockwise through the smallest positive angle. Such a rotation exists since G is a finite group. We claim that G is cyclic, generated by ρ. Let α be the angle of rotation for ρ. Let σ be another rotation in G with angle of rotation β. Write $\beta = q\alpha + \psi$, according to the division algorithm. Then $\psi = \beta - q\alpha$, and the isometry $\rho^{-q}\sigma$ rotates the plane through the angle ψ. By the division algorithm, either $\psi = 0$ or $0 < \psi < \alpha$. Since $0 < \psi < \alpha$ is impossible by our choice of α as the smallest nonzero angle of rotation, we see that $\psi = 0$. Hence $\beta = q\alpha$, so $\sigma = \rho^q$, showing that G is cyclic generated by ρ.

63. a) No b) No c) No d) No e) $\mathbb{Z}$

64. a) No b) No c) Yes d) No e) D_∞

65. a) No b) Yes c) No d) No e) $\mathbb{Z} \times \mathbb{Z}_2$

66. a) Yes b) No c) No d) No e) D_∞

67. a) Yes b) Yes c) Yes d) No e) $D_\infty \times \mathbb{Z}_2$

68. a) No b) No c) No d) Yes e) $\mathbb{Z}$

69. a) Yes b) No c) Yes d) Yes e) D_∞

70. a) Yes. 90°, 180° b) Yes c) No

71. a) Yes. 180° b) Yes c) No

72. a) No b) No c) No

73. a) No b) Yes c) No

74. a) Yes. 180° b) Yes c) No

75. a) Yes. 60°, 120°, 180° b) Yes c) No

76. a) Yes. 120° b) Yes c) No

77. a) No b) No c) Yes d) (1, 0) and (0, 1)

78. a) Yes. 90°, 180° b) Yes c) No d) (1, 0) and (1, 1)

79. a) Yes. 120° b) No c) No d) (0, 1) and $(\sqrt{3}, 1)$

80. a) Yes. 120° b) Yes c) No d) (0, 1) and $(\sqrt{3}, 1)$

81. Let us call the four diagonals of the cube through its center d_1, d_2, d_3, and d_4. By rotating the cube, any diagonal can be moved to fall on the line segment formerly occupied by any of the diagonals (including itself) in two ways. For example, if d_1 goes from point P to point Q and d_2 from point R to point S, then d_1 can be moved onto the segment from R to S with the vertex formerly at P falling on either point R or point S. Thus diagonal d_1 can be moved onto a diagonal (including itself) in $4 \cdot 2 = 8$ ways. Once diagonal d_1 is in position, we can keep the ends of d_1 fixed and rotate the cube through a total of three positions, giving a total of $3 \cdot 8 = 24$ possible rotations of the cube. But the set $\{d_1, d_2, d_3, d_4\}$ admits only $4! = 24$ permutations. Thus, identifying each rotation with one of these permutations of the four diagonals, we see that the group of rotations must be isomorphic to the full symmetric group S_4 on four letters.

SECTION 2.5 - Binary Linear Codes

1. 0001011 0000000 0111001 1111111 1111111 0100101 0000000 0100101 1111111 0111001
2. GONE HOME
3. $x_4 = x_1 + x_2$, $\quad x_5 = x_1 + x_3$, $\quad x_6 = x_2 + x_3$
4. 000000, 001011, 010101, 011110, 100110, 101101, 110011, 111000
5. Note that each x_i appears in at least two parity check equations and for each combination i,j of positions in a message word, some parity-check equation contains one of x_i, x_j but not the other. As explained after Eq.(2) in the text, this shows that the distance between code words is at least 3. Consequently, any two errors can be detected.
6. Since the minimum distance between code words is 3 as explained in the previous solution, any single error can be corrected.
7. The code C contains eight code words, so each coset contains

eight code words. Since the minimum weight of any nonzero word in C is 3, we see that the six cosets 100000 + C, 010000 + C, 001000 + C, 000100 + C, 000010 + C, and 000001 + C are all distinct. These six cosets and C account for $7 \cdot 8 = 56$ of the 64 elements of $\mathbf{B}$. Thus there is just one other coset. Since we see that 111111 is at least two units distance from every code word, the missing coset must be 111111 + C which is listed in the text answer.

8. a) 110 b) 001 c) 110 d) 100 or 110 or 111 e) 101

9. (See the first matrix in the next solution.)

10. We compute, using binary addition,

$$\begin{bmatrix} 1 & 1 & 0 & 1 & 0 & 0 \\ 1 & 0 & 1 & 0 & 1 & 0 \\ 0 & 1 & 1 & 0 & 0 & 1 \end{bmatrix} \begin{bmatrix} 1 & 0 & 1 & 1 & 1 \\ 1 & 0 & 1 & 0 & 0 \\ 0 & 1 & 1 & 1 & 0 \\ 1 & 0 & 0 & 0 & 1 \\ 1 & 1 & 1 & 1 & 0 \\ 1 & 1 & 1 & 0 & 1 \end{bmatrix} = \begin{bmatrix} 1 & 0 & 0 & 1 & 0 \\ 0 & 0 & 1 & 1 & 1 \\ 0 & 0 & 1 & 1 & 1 \end{bmatrix}$$

where the first matrix is the parity-check matrix and the second one has as *columns* the received words.

a) Since the first column of the product is the fourth column of the parity-check matrix, we change the fourth position of the received word 110111 to get the code word 110011 and decode as 110.

b) Since the second column of the product is the zero vector, the received word 001011 is a code word and we decode it as 001.

c) Since the third column of the product is the third column of the parity-check matrix, we change the third position of the received word 111011 to get the code word 110011 and decode as 110.

d) Since the fourth column of the product is not the zero vector and not a column of the parity-check matrix, there are at least two errors, and we ask for retransmission.

e) Since the fifth column of the product is the third column of the parity-check matrix, we change the third position of the received word 100101 to get the code word 101101 and decode as 101.

11. See the text answer for the table of words of weight at most 1 with their syndromes.

a) The syndrome of 110111 is 100 (see the first column of the product in the preceding solution) which is the syndrome of 000100, so we decode as the first three characters of 110111 + 000100 = 110011, namely 110.

b) The syndrome of 001011 is 000 so we decode as the first three characters of 001011 + 000000, namely 001.

c) The syndrome of 111011 is 011 so we decode as the first three characters of 111011 + 001000, namely 110.

d) The syndrome of 101010 is 111 which is not a syndrome in the table, so we ask for retransmission.

e) The syndrome of 100101 is 011 so we decode as the first three characters of 100101 + 001000, namely 101.

12. a) 7 b) 6 c) $\mathrm{wt}(u + v) = \mathrm{wt}(1010011001) = 5$
d) $\mathrm{wt}(u - v) = \mathrm{wt}(1010011001) = 5$

13. Since we add by components, this follows from the fact that using binary addition, $1 + 1 = 1 - 1 = 0$, $1 + 0 = 1 - 0 = 1$, $0 + 1 = 0 - 1 = 1$, and $0 + 0 = 0 - 0 = 0$.

14. If the transmitted word v and the received word w agree in component i, then the sum of the ith components is either $0 + 0 = 0$ or $1 + 1 = 0$. If they are different in their jth components, the sum of the jth components is $0 + 1 = 1$ or $1 + 0 = 1$. Thus $u + v$ has 1's in precisely the components where the error was made; hence the name *error pattern*.

15. From the solution to Exercises 13 and 14, we see that $u - v = u + v$ contains 1's in precisely the positions where the two words differ. The number of places where u and v differ is distance between them, and the number of 1's in $u - v$ is $\mathrm{wt}(u - v)$, so these numbers are equal.

16. a) By definition, $d(u, v)$ is the number of components where u and v differ, which is 0 if and only if u and v are the same word.

b) The number of components in which u and v differ is the same as the number of places where v and u differ.

c) Consider the contributions of the ith components of u, v, and w to each side of this equation. If u and w agree in their ith components, then these components contribute 0 to $d(u, v)$ but either 0 or 1 to both of $d(u, v)$ and $d(v, w)$. If u and w are different in their ith conponents, then these components contribute 1 to $d(u, w)$, and 1 to just one of $d(u, v)$ or $d(v, w)$. Thus the contribution of the ith component to $d(u, v) + d(v, w)$ is greater than or equal to its contribution to $d(u, w)$. Summing over all components, we obtain the desired inequality.

d) From Exercise 15, we obtain $d(u + w, v + w) = \mathrm{wt}((u + w) - (v + w)) = \mathrm{wt}(u - v) = d(u, v)$.

Section 2.5

17. Word addition is closed on $\mathbf{B}^n$ and is associative since binary addition, which is just addition in $\mathbb{Z}_2$, is associative. The word with all components 0 acts as additive identity, and the binary equation $1 + 1 = 0$ shows that every word in $\mathbf{B}^n$ is its own inverse. Thus we have a group, which we note is isomorphic to $\mathbb{Z}_2 \times \mathbb{Z}_2 \times \cdots \times \mathbb{Z}_2$ for n factors.

18. Let G be the generator matrix of the Hamming (7, 4) code. Let $u = (u_1, u_2, u_3, u_4)$ and $v = (v_1, v_2, v_3, v_4)$ be the row vectors whose components are the corresponding components of two words u and v in $\mathbf{B}^4$. The distributive property $(v + w)G = vG + wG$ for matrices shows that the sum of the code words associated with v and w is the code word associated with $v + w$, so addition is closed on the subset C of code words in $\mathbf{B}^7$. Addition of code words is associative since addition in $\mathbf{B}^7$ is associative (see Exercise 17). Since 0000000 is a code word and the inverse of every code word is itself, we see that C is a group code.

19. Suppose that $d(u, v)$ is minimum in C. By Exercise 15, $d(u, v) = \text{wt}(u - v)$, showing that the minimum weight of nonzero code words is less than or equal to the minimum distance between two of them. On the other hand, if w is a nonzero code word of minimum weight, then $\text{wt}(w)$ is the distance from w to the zero code word, showing the opposite inequality, so we have the equality stated in the exercise.

20. When the distance between code words is $m + 1$, then no received word that differs from the transmitted word in m or fewer components can be a code word, so the error in transmission will be detected.

21. The triangle inequality in Exercise 16 shows that if the distance from received word v to code word u is at most m and from code word w is at most m, then $d(u, v) \leq 2m$. Thus if the distance between code words is at least $2m + 1$, a received word v with at most m incorrect components has a *unique* nearest neighbor code word. This number $2m + 1$ can't be improved since if $d(u, v) = 2m$, then a possible received word w at distance m from both u and v can be constructed by having its components agree with those of u and v where the components of u and v agree, and making m of the 2m components of w in which u and v differ opposite from the components of u and the other m components of w opposite from

those of v.

22. By Exercise 19, the minimum nonzero weight of code words is equal to the minimum distance between them. By the answer to Exercise 20, if this distance is $m = 2t + 1$, then any $m - 1 = 2t$ or fewer errors can be detected. By the answer to Exercise 21, if the distance is $2m + 1 = 2t + 1$, then any $m = t$ or fewer errors can be corrected.

23. Let $e_1 = 1000\cdots0$, $e_2 = 0100\cdots$, $e_3 = 0010\cdots0$, $\cdots$, $e_n = 0000\cdots1$ in $\mathbf{B}^n$. If the distance between the 2^k words in C is at least 3, then the cosets $e_1 + C$, $e_2 + C$, $e_3 + C$, $\cdots$, $e_n + C$ are all distinct, since if w were in two of these cosets, so that $w = e_i + u = e_j + v$ for $i \neq j$ and $u, v \in C$, then the distance between u and v would be 2. Thus n must be large enough to have $\mathbf{B}^n$ contain at least the 2^k words in C and the $n\cdot 2^k$ words in all these cosets. Consequently, $2^n \geq 2^k + n\cdot 2^k$. Dividing by 2^k, we obtain $2^{n-k} \geq 1 + n$.

24. Since the minimum distance between words in C is at least 5, no word $w \in \mathbf{B}^n$ has distance ≤ 2 from two different words $u, v \in C$, for if $d(u, w) \leq 2$ and $d(v, w) \leq 2$, then $d(u, v) \leq 4$ by the triangle inequality in Exercise 16. Let e_i be the word in $\mathbf{B}^n$ with 1 in the ith component and 0's elsewhere, and let e_{ij} for $i \neq j$ be the word in $\mathbf{B}^n$ with 1's in components i and j and 0's elsewhere. By the first sentence, the cosets $e_i + C$ and $e_{ij} + C$ for $i, j = 1, \cdots, n$, $i \neq j$, are distinct from each other, and are distinct from C. Thus n must be large enough to have $\mathbf{B}^n$ contains all these words. Now C has 2^k elements, and the number of e_i is n while the the number of e_{ij} is $n(n - 1)/2$. Thus we must have

$$2^n \geq 2^k + n\cdot 2^k + \frac{n(n - 1)}{2}\cdot 2^k.$$

Dividing by $2k$, we obtain $2^{n-k} \geq 1 + n + n(n - 1)/2$.

25. a) 3 b) 3 c) 4 d) 5 e) 6 f) 7

26. $x_6 = x_1 + x_2 + x_3$, $x_7 = x_2 + x_4 + x_5$, $x_8 = x_1 + x_2 + x_4$,

$x_9 = x_1 + x_3 + x_5$ (Other answers are possible.)

27. $x_9 = x_1 + x_2 + x_3 + x_6 + x_7$, $x_{10} = x_5 + x_6 + x_7 + x_8$
$x_{11} = x_2 + x_3 + x_4 + x_6 + x_8$, $x_{12} = x_1 + x_3 + x_4 + x_5$
(Other answers are possible.)

28. a) Let u and v be two words in C of even weight; let wt(u) = n and let wt(v) = m. Then the total number of 1's in u and v is $n + m$, which is an even number. When u and v are added, the only time this total number of 1's is reduced is when an ith component of both words is 1, so that when u and v are added, the ith component of $u + v$ is 0. This addition in the ith component reduces the sum $n + m$ by 2 for each such i so $u + v$ still has an even number of 1's. This shows that H is closed under addition. Since the zero word has zero (an even number) 1's and each word is its own inverse, we see that the words of even weight form a subgroup H of C.

b) Suppose that $w \in C$ has odd weight. Part (a) shows that adding two words u and v of weights n and m respectively produces a word with weight $n + m - 2r$ where r is the number of components in which both u and v have 1's. Thus if H is the subgroup of words in C having even weight, then all words in the coset $w + H$ have odd weight. If v is any word of odd weight, then $v + w = v - w$ has even weight, so $v - w$ is in C. Hence v is in $w + C$. Thus $w + C$ contains all words of odd weight. Since $|H| = |w + H|$, we see that half the words of C have even weight and half have odd weight.

29. Note that rank(H) = $n - k$ since the last $n - k$ columns of H comprise the $(n - k)\times(n - k)$ identity matrix. The number of columns of H is n. Thus by the rank equation, nullity(H) = k, so the nullspace of H is a k-dimensional vector space. Since 0 and 1 are the only possible scalars, we see that there are precisely 2^k possible linear combinations of any k vectors that comprise a basis for the nullspace of H. Thus the nullspace of H contains 2^k words. Since C contains 2^k words and C is contained in the nullspace of H, we see that C is the nullspace of H. Thus if $H\mathbf{w} = \mathbf{0}$, then $w \in C$.

CHAPTER 3

HOMOMORPHISMS AND FACTOR GROUPS

SECTION 3.1 - Homomorphisms

1. Yes. $\phi(n + m) = m + n = \phi(m) + \phi(n)$ for $m, n \in \mathbf{Z}$.
2. No. $\phi(2.6 + 1.6) = \phi(4.2) = 4$ but $\phi(2.6) + \phi(1.6) = 2 + 1 = 3$.
3. Yes. $\phi(x,y) = |xy| = |x||y| = \phi(x)\phi(y)$ for $x, y \in \mathbf{R}^*$.
4. Yes. Let $m, n \in \mathbf{Z}_6$. In $\mathbf{Z}$, let $m + n = 6q + r$ by the division algorithm in $\mathbf{Z}$. Then $\phi(m +_6 n)$ is the remainder of r modulo 2. Since 2 divides 6, the remainder of $m + n$ in $\mathbf{Z}$ modulo 2 is also the remainder of r modulo 2. Not the map $\gamma: \mathbf{Z} \longrightarrow \mathbf{Z}_2$ of Example 5 is a homomorphism, and $\phi(m) = \gamma(m)$ and $\phi(n) = \gamma(m)$. Thus we have $\phi(m +_6 n) = \gamma(m + n) = \gamma(m) + \gamma(n) = \phi(m) + \phi(n)$, so ϕ is a homomorphism.
5. No. $\phi(5 +_9 7) = \phi(3) = 1$ but $\phi(5) +_2 \phi(7) = 1 +_2 1 = 0$.
6. Yes. $\phi(x + y) = 2^{x+y} = 2^x 2^y = \phi(x)\phi(y)$.
7. Yes. Let $a, b \in G_i$. Then
$$\begin{aligned}\phi(ab) &= (e_1, e_2, \cdots, ab, \cdots, e_r)\\ &= (e_1, e_2, \cdots, a, \cdots, e_r)(e_1, e_2, \cdots, b, \cdots, e_r)\\ &= \phi(a)\phi(b).\end{aligned}$$
8. No. Let $G = S_3$ with our usual notation. Then $\phi(\rho_1\mu_1) = \phi(\mu_3) = \mu_3^{-1} = \mu_3$ but $\phi(\rho_1)\phi(\mu_1) = \rho_1^{-1}\mu_1^{-1} = \rho_2\mu_1 = \mu_2$.
9. Yes. Let $f, g \in F$. By calculus, $(f + g)'' = f'' + g''$. Thus $\phi(f + g) = (f + g)'' = f'' + g'' = \phi(f) + \phi(g)$.
10. Yes. By calculus, $\int_a^b [f(x) + g(x)]dx = \int_a^b f(x)\,dx + \int_a^b g(x)dx$ so $\phi(f + g) = \int_0^4 [f(x) + g(x)]dx = \int_0^4 f(x)\,dx + \int_0^4 g(x)\,dx = \phi(f) + \phi(g)$.

Section 3.1

11. Yes. Let $f, g \in F$. By definition $3(f + g)(x) = 3[f(x) + g(x)] = 3f(x) + 3g(x) = (3f + 3g)(x)$. Thus $3(f + g)$ and $3f + 3g$ are the same function, showing that $\phi(f + g) = 3(f + g) = 3f + 3g = \phi(f) + \phi(g)$.

12. No. Let $n = 2$ and $A = \begin{bmatrix} 1 & 0 \\ 0 & 1 \end{bmatrix}$ and $B = \begin{bmatrix} 1 & 1 \\ 1 & 1 \end{bmatrix}$, so that

$A + B = \begin{bmatrix} 2 & 1 \\ 1 & 2 \end{bmatrix}$. Then $\phi(A + B) = \det(A + B) = 4 - 1 = 3$ but

$\phi(A) + \phi(B) = \det(A) + \det(B) = 1 + 0 = 1$.

13. Yes. Let $A = (a_{ij})$ and $B = (b_{ij})$ where the element with subscript ij is in the ith row and jth column. Then

$$\phi(A) = \mathrm{tr}(A + B) = \sum_{i=1}^{n} (a_{ii} + b_{ii}) = \sum_{i=1}^{n} a_{ii} + \sum_{i=1}^{n} b_{ii} =$$

$\mathrm{tr}(A) + \mathrm{tr}(B) = \phi(A) + \phi(B)$.

14. No. Let I_n be the $n \times n$ identity matrix. Then $\phi(I_n I_n) = \phi(I_n) = \mathrm{tr}(I_n) = n$ but $\phi(I_n) + \phi(I_n) = \mathrm{tr}(I_n) + \mathrm{tr}(I_n) = n + n = 2n$.

15. No. Let $f(x) = x$. Then $\phi(f \cdot f) = \int_0^1 x^2\, dx = 1/3$ but

$$\phi(f)\phi(f) = \left[\int_0^1 x\, dx\right]^2 = (1/2)^2 = 1/4.$$

16. $A_3 = \{\rho_0, \rho_1, \rho_2\}$

17. (See the text answer.)

18. 2. $\phi(1)$ must be a generator of $\mathbb{Z}$ if ϕ is to be onto $\mathbb{Z}$, so $\phi(1)$ must be either 1 or -1. If $\phi(1) = 1$, then $\phi(n) = n$ for all $n \in \mathbb{Z}$, while if $\phi(1) = -1$, then $\phi(n) = -n$ for all $n \in \mathbb{Z}$.

19. Infinitely many. For any $m \in \mathbb{Z}$, the map $\phi_m: \mathbb{Z} \to \mathbb{Z}$ defined by $\phi_m(n) = mn$ for all $n \in \mathbb{Z}$ is a homomorphism.

20. 2. Either $\phi(1) = 1$ and ϕ is the map γ of Example 5, or $\phi(1) = 0$ and $\phi(n) = 0$ for all $n \in \mathbb{Z}$.

21. Since $\phi_g(e) = e$ by Theorem 3.1, we must have $ge = e$, so $g = e$. Since $\phi_e(x) = ex = x$ is the identity map, it is

indeed a homomorphism.

22. $\phi_g(xy) = g(xy)g^{-1} = (gxg^{-1})(gyg^{-1}) = \phi_g(x)\phi_g(y)$ for any $x, y \in G$, so ϕ_g is a homomorphism for all $g \in G$.

23. T T F T F F T T F F

24. No nontrivial homomorphism. By Theorem 3.1, the image of ϕ would have to be a subgroup of $\mathbb{Z}_5$, and hence all of $\mathbb{Z}_5$ for a nontrivial ϕ. By Theorem 3.2, the number of cosets of Ker(ϕ) must then be 5. But the number of cosets of a subgroup of a finite group is a divisor of the order of the group, and 5 does not divide 12.

25. Let $\phi(n)$ be the remainder of n when divided by 4 for $n \in \mathbb{Z}_{12}$. Replacing 6 by 12 and 2 by 4 in the solution of Exercise 4 shows that ϕ is a homomorphism.

26. Let $\phi(m,n) = (m,0)$ for $(m,n) \in \mathbb{Z}_2 \times \mathbb{Z}_4$.

27. No nontrivial homomorphisms since $\mathbb{Z}$ has no finite subgroups other than $\{0\}$.

28. Let $\phi(n) = \rho_n$ for $n \in \mathbb{Z}_3$, using our notation in the text for elements of S_3.

29. (See the text answer.)

30. Let $\phi(m,n) = 2m$.

31. Let $\phi(2n) = (2n,0)$.

32. Viewing D_4 as a group of permutations, let $\phi(\sigma) = (1,2)$ for odd $\sigma \in D_4$ and $\phi(\sigma)$ be the identity for even $\sigma \in D_4$.

33. (See the text answer.)

34. Let $\phi(\sigma) = (1,2)$ for odd $\sigma \in S_3$ and $\phi(\sigma)$ be the identity for even $\sigma \in S_3$.

35. Since $\phi[G] = \{\phi(g) \mid g \in G\}$, we see that $|\phi[G]| \leq |G|$, so $|\phi[G]|$ must be finite also. By Theorem 3.2, each element of $\phi[G]$ corresponds to a distinct coset of Ker(ϕ) in G, and conversely, each coset corresponds to an element of $\phi[G]$. Thus $|\phi[G]| = |G|/|\text{Ker}(G)|$, so $|\phi(G)|$ divides $|G|$.

36. By Theorem 3.2, $\phi[G]$ is a subgroup of G' so if $|G'|$ is finite, then $|\phi[G]|$ is finite. By the Theorem of Lagrange, we see that $|\phi[G]|$ is then a divisor of $|G'|$.

Section 3.1

37. Let $g \in G$. By Theorem 1.11, we must have

$$g = a_{i_1} a_{i_2} \cdots a_{i_m} \quad \text{for } i_1, i_2, \cdots, i_m \in I.$$

By the homomorphism requirement, extended by induction to more than two factors, we must have

$$\phi(g) = \phi(a_{i_1})\phi(a_{i_2}) \cdots \phi(a_{i_m}).$$

Thus $\phi(g)$ is completely determined if $\phi(a_i)$ is known for generators a_i of G for $i \in I$.

38. By Theorem 3.1, Ker(ϕ) is a subgroup of G. By the Theorem of Lagrange, either Ker(ϕ) = $\{e\}$ or Ker(ϕ) = G since $|G|$ is a prime number. If Ker(ϕ) = $\{e\}$, then ϕ is one-to-one by the corollary of Theroem 2.2. If Ker(ϕ) = G, then ϕ is the trivial homomorphism, mapping everything into the identity element.

39. Ker(sgn_n) = A_n. The multiplicative group $\{-1, 1\}$ is isomorphic to the group $\mathbb{Z}_2$, and if $1 \in \{-1,1\}$ is renamed 0 and -1 is renamed 1, then this becomes the homomorphism of Example 2.

40. Let $a, b \in G$. For the composite homomorphism $\gamma\phi$, we have $\gamma\phi(ab) = \gamma(\phi(ab)) = \gamma(\phi(a)\phi(b)) = \gamma(\phi(a))\gamma(\phi(b)) = \gamma\phi(a)\gamma\phi(b)$ where the first equality uses the definition of the composite map $\gamma\phi$, the second equality uses the fact that ϕ is a homomorphism, the third uses the fact that γ is a homomorphism, and the last uses the definition of $\gamma\phi$ again. This shows that $\gamma\phi$ is indeed a homomorphism.

41. Let $m, n \in \mathbb{Z}$. We have $\phi(m + n) = a^{m+n} = a^m a^n = \phi(m)\phi(n)$, showing that ϕ is a homomorphism. The image of ϕ is the cyclic subgroup $\langle a \rangle$ of G, and Ker(ϕ) must be one of the subgroups of $\mathbb{Z}$, which must be cyclic and consist of all multiples of some integer $j \in \mathbb{Z}$.

42. We show each of $S = \{x \in G \mid \phi(x) = \phi(a)\}$ and Ha is a subset of the other. Let $s \in S$. Using Theorem 3.1 and the homomorphism property, we have $\phi(sa^{-1}) = \phi(s)\phi(a^{-1}) = \phi(a)\phi(a^{-1}) = \phi(a)\phi(a)^{-1} = e'$ so $sa^{-1} = h \in H$. Then $s = ha$ so $s \in Ha$. Thus S is a subset of Ha.

Let $h \in H$ so that $ha \in Ha$. Then $\phi(ha) = \phi(h)\phi(a) = e'\phi(a) = \phi(a)$, showing that $ha \in S$. Thus Ha is a subset of S, so

The possibilities for $\phi(1,1)$ are 1, 3, 7, 9 giving four possible isomorphisms.

4. Only one. A generator must be carried into a generator, and one is the only generator, so the identity map is the only possibility.

5. Two. A generator must be carried into a generator, so 1 must be carried into either 1 or 5. Each such choice completely determines the automorphism.

6. Four. A generator must be carried into a generator, so 1 must be carried into either 1, 3, 5 or 7. Each such choice completely determines the automorphism.

7. Two. A generator must be carried into a generator, so 1 must be carried into either 1 or -1. Each such choice completely determines the automorphism.

8. Four. A generator must be carried into a generator, so 1 must be carried into either 1, 5, 9 or 11. Each such choice completely determines the automorphism.

9. (See the text answer.)

10. 1) Every element of V is its own inverse, while $1 \in \mathbb{Z}_4$ is not its own inverse.
    2) $\mathbb{Z}_4$ is cyclic while V is not.

11. (See the text answer.)

12. T T F T F T T T F T

13. U has an element i of order 4, while $\langle \mathbb{R}, + \rangle$ does not.

14. U has an element i of order 4, while $\langle \mathbb{R}^*, \cdot \rangle$ does not.

15. Let $\iota(g) = g$ for all $g \in G$. Let $a, b \in G$. If $\iota(a) = \iota(b)$, then $a = b$ so ι is one to one. Clearly, ι is onto G. Finally, we have $\iota(ab) = ab = \iota(a)\iota(b)$, so the identity map ι is an isomorphism of G onto G.

16. Since $\phi: G \longrightarrow G'$ is one to one and onto G', for each $g' \in G'$ there is a *unique* $g \in G$ such that $\phi(g) = g'$. We define the inverse map $\phi^{-1}: G' \longrightarrow G$ by $\phi^{-1}(g') = g$. Let $a', b' \in G'$. If $\phi^{-1}(a') = \phi^{-1}(b')$, then applying ϕ to both sides of the equation, we find that $a' = b'$, so ϕ^{-1} is one to one. Given $g \in G$, we find that $\phi^{-1}(\phi(g)) = g$ by definition of ϕ^{-1}, so ϕ^{-1} is onto G. Finally, $\phi(\phi^{-1}(a'b')) = a'b'$ and since ϕ is an isomorphism, $\phi(\phi^{-1}(a')\phi^{-1}(b')) = \phi(\phi^{-1}(a'))\phi(\phi^{-1}(b') =$

$a'b'$ also. Since ϕ is a one to one map, we must have $\phi^{-1}(a'b') = \phi^{-1}(a')\phi^{-1}(b')$, showing that ϕ^{-1} is an isomorphism.

17. Let $\phi\colon G \longrightarrow G'$ and $\gamma\colon G' \longrightarrow G''$ be isomorphisms. We show that $\gamma\phi\colon G \longrightarrow G''$ is an isomorphism. Let $a, b \in G$. Suppose that $\gamma\phi(a) = \gamma\phi(b)$. Then $\gamma(\phi(a)) = \gamma(\phi(b))$ and since γ is one to one, we must have $\phi(a) = \phi(b)$. But then since ϕ is one to one, we must have $a = b$. This shows that $\gamma\phi$ is also one to one. Now let $g'' \in G''$. Since γ is onto G'', there exists $g' \in G$ such that $\gamma(g') = g''$. Since ϕ is onto G', there exists $g \in G$ such that $\phi(g) = g'$. Then $\gamma\phi(g) = \gamma(\phi(g)) = \gamma(g') = g''$, showing that $\gamma\phi$ is onto G''. Finally, since ϕ and γ are both isomorphisms, we have $\gamma\phi(ab) = \gamma(\phi(ab)) = \gamma(\phi(a)\phi(b)) = \gamma(\phi(a))\gamma(\phi(b)) = \gamma\phi(a)\gamma\phi(b)$, so $\gamma\phi$ is also an isomorphism.

18. This is actually a property of homomorphisms, the fact that ϕ is one to one and onto G' is not needed. By the property of homomorphisms, extended by induction, we have $\phi(a^n) = \phi(a)^n$ for all positive integers n. By Theorem 3.1, we know that $\phi(a^0) = \phi(e) = e'$, and $\phi(a^{-1}) = \phi(a)^{-1}$. Extending this last equation by induction, we see that $\phi(a^{-n}) = \phi(a)^{-n}$ for all negative integers $-n$. Thus $\phi(a^n)$ is known for all integers n if $\phi(a)$ is known. Since G is cyclic with generator a, this means that $\phi(g)$ is known for every $g \in G$ if $\phi(a)$ is known.

19. Let $\phi\colon G \longrightarrow G'$ be an isomorphism and suppose that G is abelian. Let $a', b' \in G'$. There exists unique $a, b \in G$ such that $\phi(a) = a'$ and $\phi(b) = b'$. By the homomorphism property, $\phi(ab) = \phi(a)\phi(b) = a'b'$, and $\phi(ba) = \phi(b)\phi(a) = b'a'$. Since G is commutative, $ab = ba$ so $\phi(ab) = \phi(ba)$ and we have $a'b' = b'a'$, showing that G' is commutative.

20. Let $\phi\colon G \longrightarrow G'$ be an isomorphism and suppose that G is cyclic with generator a, so that $G = \langle a \rangle$. Let $\phi(a) = a'$ and let $g' \in G'$. There exists a unique $g \in G$ such that $\phi(g) = g'$. Since $G = \langle a \rangle$, we have $g = a^n$ for some $n \in \mathbb{Z}$. Using Theorem 3.2 (in case $n \leq 0$) and the induction extension of the homomorphism property of ϕ, we have $g' = \phi(g) = \phi(a^n) = \phi(a)^n = (a')^n$, showing that G' is cyclic and generated by a'.

Section 3.2

21. Let $\phi(a) = a^{-1}$ for $a \in G$, where a^{-1} is the inverse of a in the group $\langle G, \cdot\rangle$. Uniqueness of inverses and the fact that $(a^{-1})^{-1} = a$ show at once that ϕ is one to one and onto G. Also, $\phi(a \cdot b) = (a \cdot b)^{-1} = b^{-1} \cdot a^{-1} = a^{-1} * b^{-1} =$ $\phi(a) * \phi(b)$, showing that ϕ is an isomorphism of $\langle G, \cdot\rangle$ onto $\langle G, *\rangle$, assuming that $\langle G, *\rangle$ is a group. It is trivial to show that $\langle G, *\rangle$ is a group, since the order of multiplication in G is simply reversed: e continues to act as identity, the inverse of each element is unchanged, and $(a * b) * c = a * (b * c)$ follows at once from $c \cdot (b \cdot a) = (c \cdot b) \cdot a$.

22. Let $G = \langle a\rangle$ be a cyclic group of order n, so that $G = \{a^i \mid i = 0, 1, 2, \cdots, n - 1\}$. Define $\phi: G \to \mathbb{Z}_n$ by $\phi(a^i) = i$ for $i = 0, 1, 2, \cdots, n\text{-}1$. Clearly ϕ is one to one and onto $\mathbb{Z}_n$. Let i and j be integers from 0 to $n - 1$, and let $i + j = qn + r$ for $0 \leq r < n$ by the division algorithm. Then $\phi(a^i a^j) = \phi(a^r) = r = i +_n j = \phi(a^i) +_n \phi(a^j)$, showing that ϕ satisfies the homomorphism property. Thus ϕ is an isomorphism.

23. Let $a, b \in G$. If $gag^{-1} = gbg^{-1}$, then $a = b$ by group cancellation, so i_g is a one-to-one map. Since $\phi(g^{-1}ag) = gg^{-1}agg^{-1} = a$, we see that ϕ is onto G. We have $i_g(ag) = gabg^{-1} = ga(g^{-1}g)bg^{-1} = (gag^{-1})(gbg^{-1}) = i_g(a)i_g(b)$, so i_g satisfies the homomorphism property also, and is thus an isomorphism.

24. Let $\phi(r) = r - 1$ for $r \in \mathbb{R}^*$. Clearly ϕ is one to one and onto S. Let $r, s \in \mathbb{R}^*$. Then $\phi(rs) = rs - 1 =$ $(r - 1) + (s - 1) + (r - 1)(s - 1) = \phi(r) * \phi(s)$, so ϕ satisfies the homomorphism property also, and is thus an isomorphism.

25. (See the text answer.)

26. Step 1: We show that ρ_a is a permutation of G. If $\rho_a(x) = \rho_a(y)$, then $xa = ya$ and $x = y$ by group cancellation,

so ρ_a is one to one. Since $\rho_a(xa^{-1}) = xa^{-1}a = x$, we see that see that ρ_a is onto G. Thus ρ_a is a permutation of the set G. Let $G'' = \{\rho_a \mid a \in G\}$.

For $a, b \in G$, we have $(\rho_a\rho_b)(x) = \rho_a(\rho_b(x)) = \rho_a(xb) = xba = \rho_{ba}(x)$, showing that G'' is closed under permutation multiplication. Since ρ_e is the identity permutation and $\rho_{a^{-1}}\rho_a = \rho_{aa^{-1}} = \rho_e$, we see that G'' is a subgroup of the group S of all permutations of G.

Let $\phi: G \longrightarrow G''$ be defined by $\phi(a) = \rho_{a^{-1}}$. Clearly ϕ is one to one and is onto G''. From the equation $\rho_a\rho_b = \rho_{ba}$ derived above, we have $\phi(ab) = \rho_{(ab)^{-1}} = \rho_{b^{-1}a^{-1}} = \rho_{a^{-1}}\rho_{b^{-1}} = \phi(a)\phi(b)$, which is the homomorphism property for ϕ. Therefore ϕ is an isomorphism of G onto G''.

SECTION 3.3 - Factor Groups

1. $\langle 3\rangle$ has 2 elements, so $\mathbb{Z}_6/\langle 3\rangle$ has $6/2 = 3$ elements.
2. $\langle 2\rangle \times \langle 2\rangle$ has $2\cdot 6 = 12$ elements, so the factor group has $48/12 = 4$ elements.
3. $\langle(2, 1)\rangle$ has 2 elements, so the factor group has $8/2 = 4$ elements.
4. $\{0\} \times \mathbb{Z}_5$ has 5 elements so the factor group has $15/5 = 3$ elements.
5. $\langle(1, 1)\rangle$ has 4 elements so the factor group has $8/4 = 2$ elements.
6. $\langle(4, 3)\rangle$ has 6 elements so the factor group has $(12 \cdot 18)/6 = 36$ elements.
7. $\langle(1, \rho_1)\rangle$ has 6 elements so the factor group has $(2 \cdot 6)/6 = 2$ elements.
8. $(1,1)$ generates the entire group, so the factor group has just one element.

Section 3.3

9. $\langle 4\rangle = \{0,4,8\}$. Since $5 + 5 + 5 + 5 = 4(5) = 0$ is the first repeated sum of 5 in $\langle 4\rangle$, we see that the coset $5 + \langle 4\rangle$ is of order 4 in this factor group.

10. $\langle 12\rangle = \{0,12,24,36,48\}$. Computing, $26+26 = 52 = -8$, $-8 + 26 = 18$, $18 + 26 = 44 = -16$, $-16 + 26 = 10$, $10 + 26 = 36 \in \langle 12\rangle$. Thus $25 + \langle 12\rangle$ has order 6 in the factor group.

11. $\langle(1,1)\rangle = \{(0,0), (1,1), (2,2), (0,3), (1,4), (2,5)\}$. Computing $(2,1) + (2,1) = (1,2)$, $(1,2) + (2,1) = (0,3)$ and $(0,3) \in \langle(1,1)\rangle$. Thus $(2,1) + \langle(1,1)\rangle$ has order 3 in the factor group.

12. $\langle(1,1)\rangle = \{(0,0), (1,1), (2,2), (3,3)\}$. Computing, $(3,1) + (3,1) = (2,2) \in \langle(1,1)\rangle$, so $(3,1) + \langle(1,1)\rangle$ has order 2 in the factor group.

13. $\langle(0,2)\rangle = \{(0,0), (0,2), (0,4), (0,6)\}$. Computing, $(3,1) + (3,1) = (2,2)$, $(2,2) + (3,1) = (1,3)$, $(1,3) + (3,1) = (0,4) \in \langle(0,2)\rangle$, so $(3,1) + \langle(0,2)\rangle$ has order 4 in the factor group.

14. $\langle(1,2)\rangle = \{(0,0), (1,2), (2,4), (3,6)\}$. Computing, $(3,3) + (3,3) = (2,6)$, $(2,6) + (3,3) = (1,1)$, $(1,1) + (3,3) = (0,4)$, $(0,4) + (3,3) = (3,7)$, $(3,7) + (3,3) = (2,2)$, $(2,2) + (3,3) = (1,5)$, $(1,5) + (3,3) = (0,0) \in \langle(1,2)\rangle$, so $(3,3) + \langle(1,2)\rangle$ has order 8 in the factor group. It generates the entire factor group.

15. $\langle(4,4)\rangle = \{(0,0), (4,4), (2,0), (0,4), (4,0), (2,4)\}$. Now $(2,0) \in \langle(4,4)\rangle$, so $(2,0) + \langle(4,4)\rangle$ has order 1 in the factor group.

16. Since $\rho_1\mu_1\rho_1^{-1} = \rho_1\mu_1\rho_2 = \mu_2$ and $\rho_1\rho_0\rho_1^{-1} = \rho_1\rho_0\rho_2 = \rho_0$, we see that $i_{\rho_1}(H) = \{\rho_0, \mu_2\}$.

17. (See the text answer for parts (a) and (b).)
    c) Taking a and b as representatives, we see that $(aH)(bH) = (ab)H$. Since G is abelian, $ab = ba$, so $(ab)H = (ba)H = (bH)(aH)$. Thus $(aH)(bH) = (bH)(aH)$ so G/H is abelian.

18. a) When working with a factor group G/H, one would let x be an element of G, not an element of G/H. The student probably does not understand what elements of G/H look like and can write nothing sensible concerning them.
    b) We must show that each element of G/H is of finite order. Let $xH \in G/H$.

c) Since G is a torsion group, we know that $x^m = e$ in G for some positive integer m. Computing $(xH)^m$ in G/H using the representative x, we have $(xH)^m = x^mH = eH = H$, so xH is of finite order. Since xH can be any element of G/H, we see that G/H is a torsion group.

19. (See the text answer.)

20. $|A_n| = |S_n|/2$, so the only cosets of A_n are A_n and the set $S_n - A_n$ of odd permutations. Thus the left and right cosets must be the same, and A_n is a normal subgroup of S_n. Since S_n/A_n has order 2, it is isomorphic to $\mathbb{Z}_2$.

21. By Exercise 36 of Section 3.4, we know the elements of G of finite order do form a subgroup T of G. Since G is abelian, every subgroup of G is a normal subgroup, so T is normal in G. Let $xT \in G/T$. Suppose that xT *is of finite order in* G/T; in particular, suppose $(xT)^m = T$. Then $x^m \in T$. Since T is a torsion group, we must have $(x^m)^r = x^{mr} = e$ in G for some positive integer r. Thus x is of finite order in G, so that $x \in T$. This means that $xT = T$. Thus the only element of finite order in G/T is the identity T, so G/T is a torsion group.

22. *Reflexive:* Since $i_e[H] = H$ for every subgroup H of G, we see that every subgroup is conjugate to itself.

Symmetric: Suppose $i_g[H] = K$, so that for each $k \in K$, we have $k = ghg^{-1}$ for exactly one $h \in H$. Then $h = (g^{-1})kg = (g^{-1})k(g^{-1})^{-1}$, and we see that $i_{g^{-1}}[K] = H$, so H is also conjugate to K.

Transitive: Suppose that $i_a[H] = K$ and $i_b[K] = S$ for elements a, b in G and subgroups H, K and S of G. Then each $s \in S$ can be written as $s = bkb^{-1}$ for a unique $k \in K$. But $k = aha^{-1}$ for a unique $h \in H$. Substituting, we have $s = b(aja^{-1})b^{-1} = (ba)h(a^{-1}b^{-1}) = (ba)h(ba)^{-1}$, so $i_{ba}[H] = S$ and S is conjugate to H.

Section 3.3

23. Since $gHg^{-1} = H$ for all $g \in G$ if H is a normal subgroup of G, we see that $\bar{H} = \{H\}$. Conversely, if $\bar{H} = \{H\}$, then $gHg^{-1} = H$ for all $g \in G$. Thus H must be a normal subgroup of G.

24. We see that $\rho_1\mu_2\rho_1^{-1} = \rho_1\mu_2\rho_2 = \mu_3$ and $\rho_2\mu_2\rho_2^{-1} = \rho_2\mu_2\rho_1 = \mu_1$ and that conjugation of μ_2 by other elements of S_3 again yield either μ_1, μ_2 or μ_3. Thus the subgroups of S_3 conjugate to $\{\rho_0, \mu_2\}$ are $\{\rho_0, \mu_2\}$, $\{\rho_0, \mu_3\}$, and $\{\rho_0, \mu_1\}$.

25. We have $|G/H| = m$. Since the order of an element divides the order of the group for a finite group, we see that $(aH)^m = H$ for all elements aH of G/H. Computing using the representative a of aH, we see that $a^m \in H$ for all $a \in G$.

26. Let $\{H_i \mid i \in I\}$ be a collection of normal subgroups of a group G. Let $K = \bigcap_{i \in I} H_i$. If $a, b \in K$, then $a, b \in H_i$ for each $i \in I$, and $ab \in H_i$ for each $i \in I$ since H_i is a subgroup of G. Thus $ab \in K$, and K is closed under the group operation of G. We see that $e \in K$ since $e \in H_i$ for each $i \in I$. Since $a^{-1} \in H_i$ for each $i \in I$ we see that $a^{-1} \in K$ also. Thus K is a subgroup of G. Let $g \in G$ and $k \in K$. Then $k \in H_i$ for $i \in I$ and $gkg^{-1} \in H_i$ for each $i \in I$ since each H_i is a normal subgroup of G. Hence $gkg^{-1} \in K$, and K is a normal subgroup of G.

27. Let $\{H_i \mid i \in I\}$ be the collection of all normal subgroups of G containing S. Note that G is such a subgroup of G, so I is nonempty. Let $K = \bigcap_{i \in I} H_i$. By Exercise 26, we know that K is a normal subgroup of G, and of course K contains S since H_i contains S for each $i \in I$. By our construction, we see that K is contained in *every* normal subgroup H_i of G containing S, so K must be the smallest normal subgroup of G containing S.

28. Consider two elements aC and bC in G/C. Now $(aC)^{-1} = a^{-1}C$ and $(bC)^{-1} = b^{-1}C$. Consequently, choosing representatives,

we see that $(aC)(bC)(aC)^{-1}(bC)^{-1} = aba^{-1}b^{-1}C$. However, $aba^{-1}b^{-1} \in C$ since C contains all commutators in G, so so $(aC)(bC)(aC)^{-1}(bC)^{-1} = C$. Thus $(aC)(bC) = C(bC)(aC) = (bC)(aC)$ which shows that G/C is abelian.

29. Let $g \in G$. Since the inner automorphism i_g: $G \longrightarrow G$ is a one-to-one map, we see that $i_g[H]$ has the same order as H. Since H is the only subgroup of G of that order, we find that $i_g[H] = H$ for all $g \in G$. Therefore H is invariant under all inner automorphisms of G, and hence is a normal subgroup of G.

30. By Exercise 39 of Section 1.3, we know that $H \cap N$ is a subgroup of G, and is contained in H, so it is a subgroup of H. Let $h \in H$ and $x \in H \cap N$. Then $x \in N$ and since N is normal in G, we find that $hxh^{-1} \in N$, and of course $hxh^{-1} \in H$ since $h, x \in H$. Thus $hxh^{-1} \in H \cap N$ so $H \cap N$ is a normal subgroup of H.

    Let $G = N = S_3$, and let $H = \{\rho_0, \mu_1\}$. Then N is normal in G, but $H \cap N = H$ is not normal in B.

31. Let H be the intersection of all subgroups of G of order s. We are told that this intersection is nonempty. By Exercise 26, H is a subgroup of G. Let $x \in H$ and $g \in G$. Let K be any subgroup of G of order s. To show that $gxg^{-1} \in H$, we must show that $gxg^{-1} \in K$. Now $g^{-1}Kg$ is a subgroup of G of order s, so $x \in g^{-1}Kg$. Let $x = g^{-1}kg$ for $k \in K$. Then $k = gxg^{-1}$, so gxg^{-1} is indeed in K. Since K can be any subgroup of G of order s, we see that $gxg^{-1} \in H$, so H is a normal subgroup of G.

32. a) By Exercise 40 of Section 3.1, the composition of two automorphisms of G is a homomorphism of G into G. Since each automorphism is a one-to-one map of G onto G, their composition also has this property, and is thus an automorphism of G. Thus composition gives a binary operation on the set of all automorphisms of G. The identity map acts as identity automorphism, and the inverse map of an automorphism of G is again an automorphism of G. Thus the automorphisms form a group under function composition.

b) For $a, b, x \in G$, we have $i_a(i_b(x)) = i_a(bxb^{-1}) = (aba^{-1})(axa^{-1})(ab^{-1}a^{-1}) = abxb^{-1}a^{-1} = (ab)x(ba)^{-1} = i_{ab}(x)$, so the composition of two inner automorphisms is again an inner automorphism. Clearly i_e acts as identity and the equation $i_a i_b = i_{ab}$ shows that $i_a i_{a^{-1}} = i_e$, so $i_{a^{-1}}$ is the inverse of a. Thus the set of inner automorphisms is a group under function composition.

Let $a \in G$ and let ϕ be any automorphism of G. We must show that $\phi i_a \phi^{-1}$ is an inner automorphism of G in order to show that the inner automorphisms are a normal subgroup of the entire automorphism group of G. For any $x \in G$, we have $(\phi i_a \phi^{-1})(x) = \phi(i_a(\phi^{-1}(x))) = \phi(a\phi^{-1}(x)a^{-1}) = \phi(a)\phi(\phi^{-1}(x))\phi(a^{-1}) = \phi(a)x(\phi(a))^{-1} = i_{\phi(a)}(x)$, so $\phi i_a \phi^{-1} = i_{\phi(a)}$ which is indeed an inner automorphism of G.

33. Let $H = \{g \in G \mid i_g = i_e\}$. Let $a \in H$ and $x \in G$. We must show that $xax^{-1} \in H$, that is, that $i_{xax^{-1}} = i_e$. For any $y \in G$, we have $i_{xax^{-1}}(y) = (xax^{-1})y(xax^{-1})^{-1} = x[a(x^{-1}yx)a^{-1}]x^{-1} = x(x^{-1}yx)x^{-1} = y = i_e(y)$, so $i_{xax^{-1}} = i_e$.

34. For $gH \in G/H$, let $\phi_*(gH) = \phi(g)H'$. Since we defined ϕ_* using the representative g of gH, we must show that ϕ_* is well defined. Let $h \in H$, so that gh is another representative of gH. Then $\phi(gh) = \phi(g)\phi(h)$. Since we are told that $\phi[H]$ is contained in H', we know that $\phi(h) = h' \in H'$, so $\phi(g)\phi(h) = \phi(g)h' \in \phi(g)H'$. This shows that ϕ_* is well defined, for the same coset $\phi(g)H'$ was obtained using the representatives g and gh.

For the homomorphism property, let $aH, bH \in G/H$. Then since ϕ is a homomorphism, we have $\phi_*((aH)(bH)) = \phi_*((ab)H) = \phi(ab)H' = \phi(a)\phi(b)H' = (\phi(a)H')(\phi(b)H') = \phi_*(aH)\phi_*(bH)$.

Thus ϕ_* is a homomorphism.

35. a) Let H be the subset of $GL(n, \mathbb{R})$ consisting of the $n \times n$ matrices with determinant 1. The property $\det(AB) = \det(A)\cdot\det(B)$ shows that H is closed under matrix multiplication. Now $\det(I_n) = 1$ and every matrix in $GL(n, \mathbb{R})$ has a nonzero determinant and is invertible. From $1 = \det(I_n) = \det(AA^{-1}) = \det(A)\cdot\det(A^{-1})$, it follows that $\det(A^{-1}) = 1/\det(A)$, so if $A \in H$, then $A^{-1} \in H$. Thus H is a subgroup of $GL(n, \mathbb{R})$. Let $A \in H$ and let $X \in GL(n, \mathbb{R})$; Since X is invertible, $\det(X) \neq 0$. Then $\det(XAX^{-1}) = \det(X)\cdot\det(A)\cdot\det(X^{-1}) = \det(X)\cdot\det(A)\cdot(1/\det(X)) = \det(A) = 1$, so $XAX^{-1} \in H$. Thus H is a normal subgroup of $GL(n, \mathbb{R})$.

b) Note from part (a) that if $\det(A) = -1$, then $\det(A^{-1}) = 1/(-1) = -1$. The same arguments as in part (a) then show that if K is the subset of $n \times n$ matrices with determinant ± 1, then K is a subgroup of $GL(n, \mathbb{R})$. Part (a) shows that if $A \in K$ and $X \in GL(n, \mathbb{R})$, then $\det(XAX^{-1}) = \det(A) = \pm 1$ so that again, $XAX^{-1} \in K$ and K is a normal subgroup of $GL(n, \mathbb{R})$.

SECTION 3.4 - Factor-Group Computations and Simple Groups

1. Since $\langle(0,1)\rangle$ has order 4, the factor group has order 2 and must be isomorphic to $\mathbb{Z}_2$. This is also obvious since this factor group essentially collapses everything in the second factor of $\mathbb{Z}_2 \times \mathbb{Z}_4$ to the identity, leaving just the first factor.

2. In this factor group, the first factor is not touched, but in the second factor, the element 2 is collapsed to 0. Since $\mathbb{Z}_4/\langle 2\rangle$ is isomorphic to $\mathbb{Z}_2$, we see the factor group is isomorphic to $\mathbb{Z}_2 \times \mathbb{Z}_2$. Alternatively, we can argue that the factor group has order four but no element of order greater than two.

3. We have $\langle(1,2)\rangle = \{(0,0), (1,2)\}$, so the factor group is of

order $8/2 = 4$. We easily see that $(1,1) + \langle(1,2)\rangle$ has order 4 in this factor group, which must then be isomorphic to $\mathbb{Z}_4$.

4. We have $\langle(1,2)\rangle = \{(0,0), (1,2), (2,4), (3,6)\}$, so the factor group has order $32/4 = 8$. Since $(0,1)$ must be added to itself eight times for the sum to lie in $\langle(1,2)\rangle$, we see that $(0,1) + \langle(1,2)\rangle$ is of order 8 in this factor group, which is is thus cyclic and isomorphic to $\mathbb{Z}_8$.

5. $\langle(1,2,4)\rangle = \{(0,0,0), (1,2,4), (2,0,0), (3,2,4)\}$; we see the factor group has order $(4)(4)(8)/4 = 32$. The factor group can have no element of order greater than 8 since $\mathbb{Z}_4 \times \mathbb{Z}_4 \times \mathbb{Z}_8$ has no elements of order greater than 8. Since $(0,0,1)$ must be added to itself eight times for the sum to lie in $\langle(1,2,4)\rangle$, we see the factor group has an element $(0,0,1) + \langle(1,2,4)\rangle$ of order 8, and is thus either isomorphic to $\mathbb{Z}_4 \times \mathbb{Z}_8$ or to $\mathbb{Z}_2 \times \mathbb{Z}_2 \times \mathbb{Z}_8$ The first group has only three elements of order 2, while the second one has seven elements of order 2. We count the elements of $\mathbb{Z}_4 \times \mathbb{Z}_4 \times \mathbb{Z}_8$ not in $\langle(1,2,4)\rangle$ but which when added to themselves yield an element of $\langle(1,2,4)\rangle$. No element added to itself yields $(1,2,4)$ or $(3,2,4)$. There are six elements that yield $(0,0,0)$ when added to themselves. There are another six elements that yield $(2,0,0)$ when added to themselves. These twelve elements are only enough to form three 4-element cosets in the factor group, which must be isomorphic to $\mathbb{Z}_4 \times \mathbb{Z}_8$.

6. Factoring out by $\langle(0,1)\rangle$ collapses the second factor of $\mathbb{Z} \times \mathbb{Z}$ to zero without touching the first factor, so the factor group is isomophic to $\mathbb{Z}$.

7. The 1 in the generator $(1,2)$ of $\langle(1,2)\rangle$ shows that each coset of $\langle(1,2)\rangle$ contains a unique element of the form $(0,m)$, and of course, every such element of $\mathbb{Z} \times \mathbb{Z}$ is in some coset of $\langle(1,2)\rangle$. We can choose these representatives $(0,m)$ to compute in the factor group, which must therefore be isomorphic to $\mathbb{Z}$.

8. The first 1 in the generator $(1,1,1)$ of $\langle(1,1,1)\rangle$ shows that each coset of $\langle(1,1,1)\rangle$ contains a unique element of the form $(0,m,n)$, and of course every such element $(0,m,n)$ of $\mathbb{Z} \times \mathbb{Z} \times \mathbb{Z}$ is in some coset of $\langle(1,1,1)\rangle$. We can choose these representatives $(0,m,n)$ to compute in the factor group, which must therefore be isomorphic to $\mathbb{Z} \times \mathbb{Z}$.

9. We conjecture that $(\mathbb{Z} \times \mathbb{Z} \times \mathbb{Z}_4)/\langle(3, 0, 0)\rangle$ is isomorphic to $\mathbb{Z}_3 \times \mathbb{Z} \times \mathbb{Z}_4$, since only the multiples of 3 in the first factor are collapsed to zero. It is easy to check that $\phi: \mathbb{Z} \times \mathbb{Z} \times \mathbb{Z}_4 \to \mathbb{Z}_3 \times \mathbb{Z} \times \mathbb{Z}_4$ defined by $\phi(n, m, s) = (r, m, s)$, where r is the remainder of n when divided by 3 in the division algorithm, is an onto homomorphism with kernel $\langle(3, 0, 0)\rangle$. By Theorem 3.9, such a check proves our conjecture.

10. We conjecture that $(\mathbb{Z} \times \mathbb{Z} \times \mathbb{Z}_8)/\langle(0, 4, 0)\rangle$ is isomorphic to $\mathbb{Z} \times \mathbb{Z}_4 \times \mathbb{Z}_8$, since only the multiples of 4 in the second factor are collapsed to zero. It is easy to check that $\phi: \mathbb{Z} \times \mathbb{Z} \times \mathbb{Z}_8 \to \mathbb{Z} \times \mathbb{Z}_4 \times \mathbb{Z}_8$ defined by $\phi(n, m, s) = (n, r, s)$, where r is the remainder of m when divided by 4 in in the division algorithm, is an onto homomorphism with kernel $\langle(0, 4, 0)\rangle$. By Theorem 3.9, such a check proves our conjecture.

11. Note that $(1,1) + \langle(2,2)\rangle$ is of order 2 in the factor group group and $(0,1) + \langle(2,2)\rangle$ generates an infinite cyclic subgroup of the factor group. This suggests that the factor group is isomorphic to $\mathbb{Z}_2 \times \mathbb{Z}$. We construct a homomorphism homomorphism ϕ of $\mathbb{Z} \times \mathbb{Z}$ onto $\mathbb{Z}_2 \times \mathbb{Z}$ having kernel $\langle(2,2)\rangle$. By Theorem 3.9, we will then know that $(\mathbb{Z} \times \mathbb{Z})/\langle(2,2)\rangle$ is isomorphic to $\mathbb{Z}_2 \times \mathbb{Z}$.

We want to have $\phi(1, 1) = (1, 0)$ and $\phi(0, 1) = (0, 1)$. Since $(m, n) = m(1, 1) + (n - m)(0,1)$, we try to define ϕ by

$$\phi(m, n) = m(1, 0) + (n - m)(0,1) = (m\cdot 1, n - m).$$

Here $m\cdot 1$ means $1 + 1 + \cdots + 1$ for m summands in $\mathbb{Z}_2$, in other words, the remainder of m modulo 2. Since $\phi[(m, n) + (r, s)] = \phi(m + r, n + s) = ((m + r)\cdot 1, n + s - m - r) = (m\cdot 1, n - m) + (r\cdot 1, s - r) = \phi(m, n) + \phi(r, s)$, we see that ϕ is indeed a homomorphism. For $(r, s) \in (\mathbb{Z}_2 \times \mathbb{Z})$, we see that $\phi(r, s + r) = (r, s)$, so ϕ is onto $\mathbb{Z}_2 \times \mathbb{Z}$. If $\phi(m, n) = (0, 0)$, then $m\cdot 1 = 0$ in $\mathbb{Z}_2$ and $n - m = 0$ in $\mathbb{Z}$. Thus $m = n$ and m is even, so $(m,n) = (m,m)$ lies in $\langle(2,2)\rangle$. Thus $\text{Ker}(\phi)$ is contained in $\langle(2,2)\rangle$. It is easy to see that $\langle(2,2)\rangle$ lies in $\text{Ker}(\phi)$, so $\text{Ker}(\phi) = \langle(2,2)\rangle$. As we observed above, Theorem 3.9 shows that our factor group is isomorphic to $\mathbb{Z}_2 \times \mathbb{Z}$.

Section 3.4

12. Clearly $(1,1,1) + \langle(3,3,3)\rangle$ is of order 3 in the factor group while $(0,1,0) + \langle(3,3,3)\rangle$ and $(0,0,1) + \langle(3,3,3)\rangle$ both generate infinite subgroups of the factor group. We conjecture that the factor group is isomorphic to $\mathbb{Z}_3 \times \mathbb{Z} \times \mathbb{Z}$. As in the solution to Exercise 11, we show this by defining a homomorphism ϕ of $\mathbb{Z} \times \mathbb{Z} \times \mathbb{Z}$ onto $\mathbb{Z}_3 \times \mathbb{Z} \times \mathbb{Z}$ having kernel $\langle(3,3,3)\rangle$. Just as in Exercise 11, we are motivated to let

$$\phi(m, n, s) = (m\cdot 1, n - m, s - m).$$

We easily check that ϕ is a homomorphism with the onto property and kernel we desire, completing the proof. Just follow the arguments in the solution of Exercise 11.

13. Checking the table for D_4, we find that only ρ_0 and ρ_2 commute with every element of D_4. Thus $Z(D_4) = \{\rho_0, \rho_2\}$. It follows that $\{\rho_0, \rho_2\}$ is a normal subgroup of D_4. Now $D_4/Z(D_4)$ has order 4 and is hence abelian. Therefore the commutator subgroup C is contained in $Z(D_4)$. Since D_4 is not abelian, we see that $C \neq \{\rho_0\}$, so $C = Z(D_4) = \{\rho_0, \rho_2\}$.

14.

Subgroup	Factor group (up to isomorphism)
$\langle(1, 0)\rangle$	$\mathbb{Z}_4$
$\langle(0, 1)\rangle$	$\mathbb{Z}_4$
$\langle(1, 1)\rangle$	$\mathbb{Z}_4$
$\langle(1, 2)\rangle$	$\mathbb{Z}_4$
$\langle(2, 1)\rangle$	$\mathbb{Z}_4$
$\langle(1, 3)\rangle$	$\mathbb{Z}_4$
$\langle 2\rangle \times \langle 2\rangle$	$\mathbb{Z}_2 \times \mathbb{Z}_2$
$\langle(2, 0)\rangle$	$\mathbb{Z}_2 \times \mathbb{Z}_4$
$\langle(0, 2)\rangle$	$\mathbb{Z}_4 \times \mathbb{Z}_2$
$\langle(2, 2)\rangle$	$\mathbb{Z}_2 \times \mathbb{Z}_4$
$\langle(0, 0)\rangle$	$\mathbb{Z}_4 \times \mathbb{Z}_4$

15. T F F T F T F F T F

16. $H = \{f \in F \mid f(0) = 0\}$. Also $H_a = \{f \in F \mid f(a) = 0\}$ for any $a \in \mathbb{R}$.

17. $H^* = \{f \in F^* \mid f(0) = 1\}$. Also $H_a^* = \{f \in F^* \mid f(a) = 1\}$ for any $a \in \mathbb{R}$.

18. No. If $f + f = g \in K$, then $f = g/2$ which is continuous also. The coset $f + K$ in the factor group then has order 1 rather rather than 2.

19. (See the text answer.)

20. The even integers in $\mathbb{Z}^+$. If n is odd and f is continuous, then for each $a \in \mathbb{R}$ there is a unique $b \in \mathbb{Z}$ such that $b^n = a$. Thus there is a unique function g, namely $g = f^{1/n}$, such that $g^n = f$, and g is continuous so that $gK^* = K^*$ and has order 1 rather than n. For even positive integers n, we can imitate the construction in the solution to Exercise 19.

21. U/z_0U is isomorphic to $\{e\}$, for $z_0U = U$.

22. It is isomorphic to U, for the map $\phi: U \longrightarrow U$ given by $\phi(z) = z^2$ is a homomorphism of U onto U with kernel $\{-1, 1\}$. By Theorem 3.9, $U/\langle -1\rangle$ is isomorphic to U.

23. It is isomorphic to U, for the map $\phi: U \longrightarrow U$ given by $\phi(z) = z^n$ is a homomorphism of U onto U with kernel $\langle z_n\rangle$. By Theorem 3.9, $U/\langle z_n\rangle$ is isomorphic to U.

24. It is isomorphic to U, the multiplicative group of complex numbers of absolute value 1. The map $\phi: \mathbb{R} \longrightarrow U$ given by $\phi(r) = \cos(2\pi r) + i\sin(2\pi r)$ is a homomorphism of $\mathbb{R}$ onto U with kernel $\mathbb{Z}$. By Theorem 3.9, $\mathbb{R}/\mathbb{Z}$ is isomorphic to U.

25. $\mathbb{Z}$. $\mathbb{Z}/\langle 2\rangle$ has only elements of finite order.

26. Let $G = \mathbb{Z}_2 \times \mathbb{Z}_4$. Then $H = \langle(1,0)\rangle$ is isomorphic to $K = \langle(0,2)\rangle$, but G/H is isomorphic to $\mathbb{Z}_4$ while G/K is isomorphic to $\mathbb{Z}_2 \times \mathbb{Z}_2$.

27. a) The center is the whole group.
b) $\{e\}$, since the center is a normal subgroup.

28. a) The commutator subgroup of an abelian group is $\{e\}$.
b) $\{e\}$, since the commutator subgroup is a normal subgroup.

29. Every subgroup H of index 2 is normal, since both left and right cosets of H are H itself and the elements not in H.

Thus G cannot be simple if it has a subgroup H of index 2.

30. We know that $\phi[N]$ is a subgroup of $\phi[G]$ by Theorem 3.1. We need only show that $\phi[N]$ is normal in $\phi[G]$. Let $g \in G$ and $x \in N$. Since ϕ is a homomorphism, Theorem 3.1 tells us that $\phi(g)\phi(x)\phi(g)^{-1} = \phi(g)\phi(x)\phi(g^{-1}) = \phi(4gxg^{-1})$. Since N is normal, we know that $gxg^{-1} \in N$, so $\phi(gxg^{-1})$ is in $\phi[N]$, and we are done.

31. We know that $\phi^{-1}(N')$ is a subgroup of G by Theorem 3.1. We need only show that $\phi^{-1}[N']$ is normal in G. Let $x \in \phi^{-1}[N']$, so that $\phi(x) = x'$. For each $g \in G$, we have $\phi(gxg^{-1}) = \phi(g)\phi(x)\phi(g^{-1}) = \phi(g)\phi(x)\phi(g)^{-1} \in N'$ since N' is a normal normal subgroup of G'. Thus $gxg^{-1} \in \phi^{-1}[N']$, showing that that $\phi^{-1}[N']$ is a normal subgroup of G. Note that N' only need to be normal in $\phi[G]$ for the conclusion to hold.

32. Suppose that $G/Z(G)$ is cyclic and is generated by the coset $aZ(G)$. Let $x, y \in G$. Then x is a member of a coset $a^mZ(G)$ and y is a member of a coset $a^nZ(G)$ for some $m, n \in \mathbb{Z}$. We can thus write $x = a^m z_1$ and $y = a^n z_2$ where $z_1, z_2 \in Z(G)$. Since z_1 and z_2 commute with every element of G, we have $xy = a^m z_1 a^n z_2 = a^{m+n} z_1 z_2 = a^n z_2 a^m z_1 = yx$, showing that G is abelian. Therefore if G is not abelian, then $G/Z(G)$ is not cyclic.

33. Let G be nonabelian of order pq. Suppose that $Z(G) \neq \{e\}$. Then $|Z(G)|$ is a divisor of pq greater than 1 but less than pq, and hence $|Z(G)|$ is either p or q. But then $|G/Z(G)|$ is either q or p, and hence is cyclic, which contradicts Exercise 32. Therefore $Z(G) = \{e\}$.

34. a) Since $(a, b, c) = (a, c)(a, b)$, we see that every 3-cycle is an even permutation, and hence is in A_n.

    b) Let $\sigma \in A_n$ and write σ as a product of transpositions. The number of transpositions in the product will be even by definition of A_n. The product of the first two transpositions will be either of the form $(a, b)(c, d)$ or of the form $(a, b)(a, c)$ or of the form $(a, b)(a, b)$

depending on repetition of letters in the transpositions. If the form is $(a, b)(a, b)$, it can be deleted from the product altogether. As the hint shows, either of the other two forms can be expressed as a product of 3-cycles. We then proceed with the next pair of transpositions in the product, and continue until we have expressed σ as a product of 3-cycles. Thus the 3-cycles generate A_n.

c) Following the hint, we find that $(r,s,i)^2 = (r,i,s)$, $(r,s,j)(r,s,i)^2 = (r,s,j)(r,i,s) = (r,i,j)$, $(r,s,j)^2(r,s,i) = (r,j,s)(r,s,i) = (s,i,j)$, $(r,s,i)^2(r,s,k)(r,s,j)^2(r,s,i) = (r,i,s)(r,s,k)(s,i,j) = (i,j,k)$. Now every 3-cycle either contains neither r nor s and is of the form (i,j,k), or one of r or s and is of the form (r,i,j) or (s,i,j), or both r and s and is of the form (r,s,i) or $(r,i,s) = (s,r,i)$. Since all of these forms can be obtained from our special 3-cycles, we see that the special 3-cycles generate A_n.

d) Following the hint, we find that $((r,s)(i,j))(r,s,i)^2((r,s)(i,j))^{-1} = (r,s)(i,j)(r,i,s)(i,j)(r,s) = (r,s,j)$. Thus if N is a normal subgroup of A_n and contains a 3-cycle, which we may consider to be (r,s,i) since r and s could be any two numbers from 1 to n in part (c), we see that N must contain all the special 3-cycles and hence be all of A_n by part (c).

e) Before making the computations in the hints of the five cases, we observe that one of the cases must hold. If Case 1 is not true, and Case 2 is not true, then when elements of N are written as a product of disjoint cycles, no cycle of length greater than 3 occurs, and no element of N is a single 3-cycle. The remaining cases cover the possibilities that at least one of the products of disjoint cycles involves two cycles of length 3, involves one cycle of length 3, or involves no cycle of length 3. Thus all possibilities are covered, and we turn to the computations in the hints.

Case 1. By part (d), if N contains a 3-cycle, then $N = A_n$ and we are done.

Case 2. Note that the a_1 through a_r do not appear in μ. We have $\sigma^{-1}[(a_1,a_2,a_3)\sigma(a_1,a_2,a_3)^{-1}] =$

$(a_r, \cdots, a_2, a_1)\mu(a_1, a_2, a_3)\mu(a_1, a_2, \cdots, a_r)(a_1, a_3, a_2) = (a_1, a_3, a_r)$, and this element is in N since it is the product of σ^{-1} and a conjugate of σ by an element of A_n. Thus in this case, N contains a 3-cycle and is equal to A_n by part (d).

Case 3. Note that a_1 through a_6 do not appear in μ. We see that $\sigma^{-1}[(a_1, a_2, a_4)\sigma(a_1, a_2, a_4)^{-1}] = (a_1, a_3, a_2)(a_4, a_6, a_5)\mu(a_1, a_2, a_4)\mu(a_4, a_5, a_6)(a_1, a_2, a_3)(a_1, a_4, a_2) = (a_1, a_4, a_2, a_6, a_3)$ is in N as explained in the last sentence of our treatement of Case 2. Thus N contains a cycle of length greater than 3, and $N = A_n$ by Case 2.

Case 4. Note that a_1 through a_3 do not appear in μ. Of course $\sigma^2 \in N$ since $\sigma \in N$, and we have $\sigma^2 = \mu(a_1, a_2, a_3)\mu(a_1, a_2, a_3) = (a_1, a_3, a_2) \in N$, so N contains a 3-cycle and hence $N = A_n$ as shown by part (d).

Case 5. Note that a_1 through a_4 do not appear in μ. We see that $\sigma^{-1}[(a_1, a_2, a_3)\sigma(a_1, a_2, a_3)^{-1}] = (a_1, a_2)(a_3, a_4)\mu(a_1, a_2, a_3)\mu(a_1, a_2)(a_3, a_4)(a_1, a_3, a_2) = (a_1, a_3)(a_2, a_4)$ is in N as explained in the last sentence of our treatment of Case 2. Continuing with the hint given, we find that $[\beta^{-1}\alpha\beta]\alpha = (a_1, i, a_3)(a_1, a_3)(a_2, a_4)(a_1, a_3, i)(a_1, a_3)(a_2, a_4) = (a_1, a_3, i)$ which is in N by the now familiar argument. Thus $N = A_n$ in this case also, by part (d).

35. Let $h_1n_1, h_2n_2 \in HN$ where $h_1, h_2 \in H$ and $n_1, n_2 \in N$. Since N is a normal subgroup, left cosets are right cosets so so $Nh_2 = h_2N$; in particular $n_1h_2 = h_2n_3$ for some $n_3 \in N$. Then $h_1n_1h_2n_2 = h_1h_2n_3n_2 \in HN$, so HN is closed under the group operation. Since $e \in H$ and $e \in N$, we see that $e = ee$ is in HN. Now $(h_1n_1)^{-1} = n_1^{-1}h_1^{-1} \in Nh_1^{-1}$ and $Nh_1^{-1} = h_1^{-1}N$

since N is normal. Thus $n_1^{-1}h_1^{-1} = h_1^{-1}n_4$ form some $n_4 \in N$, so $(h_1n_1)^{-1} \in HN$, and we see that HN is a subgroup of G.

Clearly it is the smallest subgroup of G containing both H and N, since any such subgroup must contain all the products hn for $h \in H$ and $n \in N$.

36. Exercise 35 shows that NM is a subgroup of G. We must show that $gnmg^{-1} \in NM$ for all $g \in G$, $n \in N$, $m \in M$. Since N is normal, we know that $gN = Ng$, so $gn = n_1g$ for some $n_1 \in N$, and $gnmg^{-1} = n_1gmg^{-1}$. *Since* M is normal, we know that $gM = Mg$, so $gm = m_1g$ for some $m \in M$. Then $gnmg^{-1} = n_1gmg^{-1} = n_1m_1gg^{-1} = n_1m_1 \in NM$. This concludes the proof that NM is is normal in G.

37. The fact that K is normal shows that $hkh^{-1} \in K$, so $(hkh^{-1})k^{-1} \in K$. The fact that H is normal shows that $kh^{-1}k^{-1} \in H$, so $h(kh^{-1}k^{-1}) \in H$. Thus $hkh^{-1}k^{-1} \in H \cap K$, so $hkh^{-1}k^{-1} = e$. It follows that $hk = kh$.

SECTION 3.5 - Series of Groups

1. (See the text answer.)
2. The refinement $\{0\} < 14700\mathbb{Z} < 300\mathbb{Z} < 60\mathbb{Z} < 20\mathbb{Z} < \mathbb{Z}$ of of $\{0\} < 10\mathbb{Z} < 20\mathbb{Z} < \mathbb{Z}$ and the refinement $\{0\} < 14700\mathbb{Z} < 4900\mathbb{Z} < 245\mathbb{Z} < 49\mathbb{Z} < \mathbb{Z}$ of $\{0\} < 245\mathbb{Z} < 49\mathbb{Z} < \mathbb{Z}$ are isomorphic.
3. The give series are already isomorphic.
4. The refinement $\{0\} < \langle 36\rangle < \langle 18\rangle < \langle 3\rangle < \mathbb{Z}_{72}$ of $\{0\} < \langle 18\rangle < \langle 3\rangle < \mathbb{Z}_{72}$ and the refinement $\{0\} < \langle 24\rangle < \langle 12\rangle < \langle 6\rangle < \mathbb{Z}_{72}$ of $\{0\} < \langle 24\rangle < \langle 12\rangle < \mathbb{Z}_{72}$ are isomorphic.
5. (See the text answer.)

Section 3.5

6. $\{0\} < \langle 30\rangle < \langle 15\rangle < \langle 5\rangle < \mathbb{Z}_{60}$,
$\{0\} < \langle 30\rangle < \langle 15\rangle < \langle 3\rangle < \mathbb{Z}_{60}$,
$\{0\} < \langle 30\rangle < \langle 10\rangle < \langle 5\rangle < \mathbb{Z}_{60}$,
$\{0\} < \langle 30\rangle < \langle 10\rangle < \langle 2\rangle < \mathbb{Z}_{60}$,
$\{0\} < \langle 30\rangle < \langle 6\rangle < \langle 3\rangle < \mathbb{Z}_{60}$,
$\{0\} < \langle 30\rangle < \langle 6\rangle < \langle 2\rangle < \mathbb{Z}_{60}$,
$\{0\} < \langle 20\rangle < \langle 10\rangle < \langle 5\rangle < \mathbb{Z}_{60}$,
$\{0\} < \langle 20\rangle < \langle 10\rangle < \langle 2\rangle < \mathbb{Z}_{60}$,
$\{0\} < \langle 20\rangle < \langle 4\rangle < \langle 2\rangle < \mathbb{Z}_{60}$,
$\{0\} < \langle 12\rangle < \langle 6\rangle < \langle 3\rangle < \mathbb{Z}_{60}$,
$\{0\} < \langle 12\rangle < \langle 6\rangle < \langle 2\rangle < \mathbb{Z}_{60}$,
$\{0\} < \langle 12\rangle < \langle 4\rangle < \langle 2\rangle < \mathbb{Z}_{60}$.
For each series the factor groups are isomorphic to $\mathbb{Z}_2$ $\mathbb{Z}_2$, $\mathbb{Z}_3$ and $\mathbb{Z}_5$ in some order.

7. (See the text answer for the series.) For each series, the factor groups are isomorphic to $\mathbb{Z}_2$, $\mathbb{Z}_2$, $\mathbb{Z}_2$, $\mathbb{Z}_2$ and $\mathbb{Z}_3$ in some order.

8. $\{0\}\times\{0\} < \mathbb{Z}_5 \times \{0\} < \mathbb{Z}_5 \times \mathbb{Z}_5$ and
$\{0\} \times \{0\} < \{0\} \times \mathbb{Z}_5 < \mathbb{Z}_5 \times \mathbb{Z}_5$.

9. (See the text answer.)

10. $\{0\} \times \{0\} \times \{0\} < \mathbb{Z}_2\times \{0\} \times \{0\} < \mathbb{Z}_2\times \mathbb{Z}_5\times \{0\} < \mathbb{Z}_2\times \mathbb{Z}_5\times \mathbb{Z}_7$,
$\{0\} \times \{0\} \times \{0\} < \mathbb{Z}_2\times \{0\} \times \{0\} < \mathbb{Z}_2\times \{0\} \times \mathbb{Z}_7 < \mathbb{Z}_2\times \mathbb{Z}_5\times \mathbb{Z}_7$,
$\{0\} \times \{0\} \times \{0\} < \{0\} \times \mathbb{Z}_5\times \{0\} < \{0\} \times \mathbb{Z}_5\times \mathbb{Z}_7 < \mathbb{Z}_2\times \mathbb{Z}_5\times \mathbb{Z}_7$,
$\{0\} \times \{0\} \times \{0\} < \{0\} \times \mathbb{Z}_5\times \{0\} < \mathbb{Z}_2\times \mathbb{Z}_5\times \{0\} < \mathbb{Z}_2\times \mathbb{Z}_5\times \mathbb{Z}_7$,
$\{0\} \times \{0\} \times \{0\} < \{0\} \times \{0\} \times \mathbb{Z}_7 < \{0\} \times \mathbb{Z}_5\times \mathbb{Z}_7 < \mathbb{Z}_2\times \mathbb{Z}_5\times \mathbb{Z}_7$,
$\{0\} \times \{0\} \times \{0\} < \{0\} \times \{0\} \times \mathbb{Z}_7 < \mathbb{Z}_2\times \{0\} \times \mathbb{Z}_7 < \mathbb{Z}_2\times \mathbb{Z}_5\times \mathbb{Z}_7$.

11. $\{\rho_0\} \times \mathbb{Z}_4$

12. $\{\rho_0\} \times \{\rho_0, \rho_2\}$

13. (See the text answer.)

14. $\{\rho_0\} \times \{\rho_0, \rho_2\} \leq \{\rho_0\} \times D_4 \leq \{\rho_0\} \times D_4 \leq \cdots$

15. T F T F F T F F T T i) The Jordan-Holder theorem applied

to the groups $\mathbb{Z}_n$ implies the Fundamental Theorem of Arithmetic.

16. $\{\rho_0\} \times \{\rho_0\} \leq A_3 \times \{\rho_0\} \leq S_3 \times \{\rho_0\} \leq S_3 \times A_3 \leq S_3 \times S_3$. Yes, $S_3 \times S_3$ is solvable since all factor groups in this series are of order either 2 or 3 and hence are abelian.

17. Yes. $\{\rho_0\} \leq \{\rho_0, \rho_2\} \leq \{\rho_0, \rho_1, \rho_2, \rho_3\} \leq D_4$ is a composition series with all factor groups of order 2 and hence abelian.

18. We use induction to show that $|H_i| = s_1 \cdots s_i$ for $i = 1, \cdots, n$. For $n = 1$, $s_1 = |H_1/H_0| = |H_1/\{e\}| = |H_1|$ because each coset in $H_1/\{e\}$ has only one element. Now suppose that $|H_k| = s_1 \cdots s_k$ for $k < i \leq n$. Now H_i/H_{i-1} consists of s_i cosets of H_{i-1}, each having $|H_{i-1}|$ elements. By our induction assumption, $|H_{i-1}| = s_1 \cdots s_{i-1}$ Thus $|H_i| = |H_{i-1}|s_i = s_1 \cdots s_{i-1}s_i$ Our induction is complete and the desired assertion follows by taking $i = n$.

19. By Definition 3.11, a composition series for G contains a *finite* number of subgroups of G. If G is infinite and abelian, and $\{e\} = H_0 < H_1 < H_2 < \cdots < H_n = G$ is a subnormal series, the factor groups H_i/H_{i-1} cannot all be of finite order for $i = 1, 2, \cdots, n$, or $|G|$ would be finite by Exercise 19. Suppose $|H_k/H_{k-1}|$ is infinite. Now every infinite abelian group has a proper nontrivial subgroup and hence a normal subgroup. To see this we need only consider the cyclic subgroup $\langle a \rangle$ for some $a \neq e$ in the group. If $\langle a \rangle$ is not the whole group, we are done. If $\langle a \rangle$ is the whole group, then $\langle a^2 \rangle$ is a proper subgroup. Thus we see that H_k/H_{k-1} has a proper nontrivial subgroup, so it is not simple and our series is not a composition series.

20. Let $G = G_1 \times G_2 \times \cdots \times G_m$ and suppose that G_i is solvable for $i = 1, 2, \cdots, m$. We form a composition series for G as follows: Start with $H_0 = \{e_1\} \times \{e_2\} \times \cdots \times \{e_m\}$ where e_i is the identity of G_i. Let $H_1 = H_{11} \times \{e_2\} \times \cdots \times \{e_m\}$ where H_{11} is the smallest nontrivial subgroup of G_1 in a

composition series $\{e_1\} \le H_{11} \le H_{12} \le \cdots \le H_{1n_1} = G_1$ for G_1. Continue to build the composition series for G by putting these subgroups H_{1i} in sequence in the first factor of the direct product series until you arrive at $G_1 \times \{e_2\} \times \cdots \times \{e_m\}$. Then start putting the sequence of subgroups $H_{21}, H_{22}, \cdots, H_{2n_2}$ in a composition series for G_2 into the second factor until you arrive at $G_1 \times G_2 \times \{e_3\} \times \cdots \times \{e_m\}$. Continue in this way across the factors in the direct product G until you arrive at

$$G = H_{1n_1} \times H_{2n_2} \times \cdots \times H_{mn_m} = G_1 \times G_2 \times \cdots \times G_m.$$

The factor group formed from two consecutive terms of this series for G is naturally isomorphic to one of the factor groups in a composition series for one of the groups G_i by our construction. Thus these factor groups are all simple so we have indeed constructed a composition series for G . Since all the factor groups in the composition series for G_i are abelian for $i = 1, 2, \cdots, m$ we see the factor groups of the composition series for G are abelian, so G is a solvable group.

SECTION 3.6 - Group Action on a Set

1. (See the text answer.)
2. $G_1 = G_3 = \{\rho_0, \delta_2\}$, $G_2 = G_4 = \{\rho_0, \delta_1\}$,
   $G_{s_1} = G_{s_3} = \{\rho_0, \mu_1\}$, $G_{s_2} = G_{s_4} = \{\rho_0, \mu_2\}$,
   $G_{m_1} = G_{m_2} = \{\rho_0, \rho_2, \mu_1, \mu_2\}$, $G_{d_1} = G_{d_2} = \{\rho_0, \rho_2, \delta_1, \delta_2\}$,
   $G_C = G$, $G_{P_1} = G_{P_3} = \{\rho_0, \mu_1\}$, $G_{P_2} = G_{P_4} = \{\rho_0, \mu_2\}$
3. (See the text answer.)
4. Every sub-G-set of a G-set X consists of a union of orbits in X under G .
5. A G-set is transitive if it has only one orbit.
6. F T F F F T T F F T

7. a) $\{P_1, P_2, P_3, P_4\}$ and $\{s_1, s_2, s_3, s_4\}$
   b) δ_1 leaves two elements, 1 and 3, of $\{1, 2, 3, 4\}$ fixed, but δ_1 leaves no elements of $\{s_1, s_2, s_3, s_4\}$ fixed.
   c) Yes. From the condition $g\phi(x) = \phi(gx)$, *it follows that if* g leaves x fixed, then g must leave $\phi(x)$ fixed. The only other conceivable candidates for isomorphic sub-D_4-sets, in view of part (b), are $\{m_1, m_2\}$ and $\{d_1, d_2\}$. Since μ_1 leaves m_1 and m_2 fixed but moves both d_1 and d_2, we see these are not isomorphic sub-D_4-sets.

8. a) Yes
   b) $\{1, 2, 3, 4\}$, $\{s_1, s_2, s_3, s_4\}$, and $\{P_1, P_2, P_3, P_4\}$

9. Suppose that G acts faithfully on X, and let g_1 and g_2 be two elements of G. Since G acts faithfully, there exists $a \in X$ such that $(g_1^{-1}g_2)a \neq a$. It follows that $g_2a \neq g_1a$, for if $g_2a = g_1a$, then $g_1^{-1}(g_2a) = (g_1^{-1}g_2)a = g_1^{-1}(g_1a) = (g_1^{-1}g_1)a = ea = a$, contrary to our choice of $a \in X$. Thus g_1 and g_2 have different action on a.

   Conversely, suppose that no two distinct elements of G have the same action on each element of X. Then G acts faithfully on X, for if $g \in G$ leaves each element of X fixed, then g has the same action as the identity e, so it must be the identity.

10. Let $g_1, g_2 \in G_Y$. Then for each $y \in Y$, we have $(g_1g_2)y = g_1(g_2y) = g_1y = y$, so $g_1g_2 \in G_Y$ and G_Y is closed under the group operation. Since $ey = y$ for all $y \in Y$, we see that $e \in G_Y$. From $y = g_1y$ for all $y \in Y$, it follows that that $g_1^{-1}y = g_1^{-1}(g_1y) = (g_1^{-1}g_1)y = ey = y$ for all $y \in Y$ so $g_1^{-1} \in G_Y$ also, and consequently $G_Y \leq G$.

11. a) Since rotation through 0 radians leaves each point of the plane fixed, the first requirement of Definition 3.17 is satisfied. The second requirement $(\theta_1 + \theta_2)P = \theta_1(\theta_2P)$ is also valid, because a rotation counterclockwise through

$\theta_1 + \theta_2$ radians can be achieved by sequential rotations through θ_2 radians and then through θ_1 radians.

b) It is the circle with center at the origin (0,0) and radius the distance from P to the origin.

c) The group G_P is the cyclic subgroup $\langle 2\pi \rangle$ of G.

12. a) Let $X = \bigcup_{i \in I} X_i$ and let $x \in X$. Then $x \in X_i$ for precisely one index $i \in I$ since the sets are disjoint, and we define gx for each $g \in G$ to be the value given by the action of G on X_i. Conditions (1) and (2) in Definition 3.17 are satisfied since X_i is a G-set by assumption.

b) We have seen that each orbit in X is a sub-G-set. The G-set X can be regarded as the union of these sub-G-sets since the action gx of $g \in G$ on $x \in X$ concides with the the sub-G-set action gx of $g \in G$ on the same element x viewed as an element of its orbit.

13. Let $\phi: X \longrightarrow G_{x_0}$ be defined by $\phi(x) = gG_{x_0}$ where $gx_0 = x$. Since G acts faithfully on X, we know that such a g exists. We must show that ϕ is well defined. Suppose that $g_1x_0 = x$ and $g_2x_0 = x$. Then $g_2x_0 = g_1x_0$, and as in the answer to Exercise 9, we can deduce that $(g_1^{-1}g_2)x_0 = x_0$. But then $g_1^{-1}g_2 \in G_{x_0}$ so $g_2 \in g_1G_{x_0}$ and $g_1G_{x_0} = g_2G_{x_0}$. This shows that the definition of $\phi(x)$ is independent of the choice of g such that $gx_0 = x_0$, that is, that ϕ is well defined.

It remains to show that ϕ is one to one and that $g\phi(x) = \phi(gx)$ for all $x \in X$ and $g \in G$. Suppose that $\phi(x_1) = \phi(x_2)$ for $x_1, x_2 \in X$, and let $g_1x_0 = x_1$ and $g_2x_0 = x_2$ Then $\phi(x_1) = \phi(x_2)$ implies that $g_1G_{x_0} = g_2G_{x_0}$ so $g_2 = g_1g_0$ for some $g_0 \in G_{x_0}$. The equation $g_2x_0 = x_2$ then yields $g_1g_0x_0 = x_2$. Since $g_0 \in G_{x_0}$, we then obtain $g_1x_0 = x_2$ so $x_1 = x_2$ and ϕ is one to one. To show that $g\phi(x) = \phi(gx)$, let $x = g_1x_0$. Then $gx = g(g_1x_0) = (gg_1)x_0$ so $\phi(x) = (gg_1)G_{x_0} = g(g_1G_{x_0}) = g\phi(x)$.

14. By Exercise 12, each G-set X is the union of its G-set orbits X_i for $i \in I$. By Exercise 13, each orbit X_i is isomorphic to a G-set consisting of left cosets of $G_{x_{i,0}}$ where $s_{i,0}$ is any point of X_i. It is possible that the group $G_{x_{i,0}}$ may be the same as the group $G_{x_{j,0}}$ for some $j \neq i$ in I, but by attaching the index i to each coset of $G_{x_{i,0}}$ and j to each coset of $G_{x_{j,0}}$ as indicated in the statement of the exercise, we can consider these ith and jth coset G-sets to be disjoint. Identifying X_i with this isomorphic ith coset G-set, we see X is isomorphic to a disjoint union of left coset G-sets.

15. a) If $g(g_0x_0) = g_0x_0$, then $g_0^{-1}(g(g_0x_0)) = (g_0^{-1}gg_0)x_0 = g_0^{-1}(g_0x_0) = (g_0^{-1}g_0)x_0 = ex_0 = x_0$, so $g_0^{-1}gg_0 \in H$. Thus $g \in g_0Hg_0^{-1}$. Since g may be any element of K, this shows that K is a subset of $g_0Hg_0^{-1}$. Since $g_0^{-1}(g_0x_0) = x_0$ this same argument, regarding g_0x_0 as the original base point, shows that H is a subset of $g_0^{-1}Kg_0$. These two subset relations show that $K = g_0Hg_0^{-1}$.

b) *Conjecture:* The G-set of left cosets of H is isomorphic to the G-set of left cosets of K if and only if H and K are conjugate subgroups of G.

c) We first show that if H and K are conjugate, then the G-set L_H of left cosets of H is isomorphic to the G-set L_K of left cosets of K. Let $g_0 \in G$ be chosen such that $K = g_0Hg_0^{-1}$. Note that for $aH \in L_H$, we have $aHg_0^{-1} = ag_0^{-1}g_0Hg_0^{-1} = ag_0^{-1}K \in L_K$. We define $\phi: L_H \longrightarrow L_K$ by $\phi(aH) = ag_0^{-1}K$. We just saw that $ag_0^{-1}K = (aH)g_0^{-1}$, so ϕ is independent of the choice of $a \in H$, that is, ϕ is well defined. Since ag_0^{-1} assumes all values in G as a varies through G we see that ϕ is onto L_K. If $\phi(aH) = \phi(bH)$, then $(aH)g_0^{-1} = ag_0^{-1}K = bg_0^{-1}K = (bH)g_0^{-1}$, so $aH = bH$ and ϕ is one to one. To show ϕ is an isomorphism of G-sets, it only remains to show that $\phi(g(aH)) = g\phi(aH)$ for all $g \in G$ and $aH \in L$. But $\phi(g(aH)) = \phi((ga)H) = (ga)g_0^{-1}K = g(ag_0^{-1}K) = g\phi(aH)$, and we are done.

Conversely, suppose that $\phi\colon L_H \longrightarrow L_K$ is an isomorphism of the G-set of left cosets of H onto the G-set of left cosets of K. Since ϕ is an onto map, there exists $g_0 \in G$ such that $\phi(g_0H) = K$. Since ϕ commutes with the action of G, we have $(g_0hg_0^{-1})K = (g_0hg_0^{-1})\phi(g_0H) = \phi(g_0hg_0^{-1}g_0H) = \phi(g_0H) = K$, so $g_0hg_0^{-1} \in K$ for all $h \in H$, that is, $g_0Hg_0^{-1}$ is a subset of K. From $\phi(g_0H) = K$, we we easily see that $\phi^{-1}(g_0^{-1}K) = H$, and an argument similar to the one just made then shows that $g_0^{-1}Kg_0$ is a subset of H. Thus $g_0Hg_0^{-1} = K$, that is, the subgroups are indeed conjugate.

16. There are three of them; call them X, Y, and $\mathbb{Z}_4$ corresponding to the three subgroups $\{0, 1, 2, 3\}$, $\{0, 2\}$, and $\{0\}$ of $\mathbb{Z}_4$, no two of which are conjugate.

	X	Y	
	a	a	b
0	a	a	b
1	a	b	a
2	a	a	b
3	a	b	a

and $\mathbb{Z}_4$ itself corresponding to $\{0\}$.

17. There are four of them; call them X, Y, Z, and $\mathbb{Z}_6$ corresponding respectively to the subgroups $\mathbb{Z}_6$, $\langle 2\rangle$, $\langle 3\rangle$ and $\{0\}$. See the text answer for the action tables for X, Y, and Z.

18. There are four of them; call them X, Y, Z and S_3 corresponding respectively to the subgroup S_3, the subgroup

$\{\rho_0, \rho_1, \rho_2\}$, the three conjugate subgroups $\{\rho_0, \mu_1\}$, $\{\rho_0, \mu_2\}$, and $\{\rho_0, \mu_3\}$, and the trivial subgroup $\{\rho_0\}$. (We have used the notation for S_3 given in Section 2.1.) The action table for X, Y, and Z is shown at the right.

	X	Y		Z		
	a	a	b	a	b	c
ρ_0	a	a	b	a	b	c
ρ_1	a	a	b	b	c	a
ρ_2	a	a	b	c	a	b
μ_1	a	b	a	a	c	b
μ_2	a	b	a	c	b	a
μ_3	a	b	a	b	a	c

SECTION 3.7 - Applications of G-Sets to Counting

1. $G = \langle(1\ 3\ 5\ 6)\rangle$ has order 4. Let $X = \{1,2,3,4,5,6,7,8\}$.
$|X_{(1)}| = 8$, $|X_{(1\ 3\ 4\ 5)}| = |\{2,4,7,8\}| = 4$
$|X_{(1\ 5)(3\ 6)}| = |\{2,4,7,8\}| = 4$
$|X_{(1\ 6\ 5\ 3)}| = |\{2,4,7,8\}| = 4$.
$\sum_{g\in G} |X_g| = 8 + 4 + 4 + 4 = 20$. The number of orbits under G is then $(1/4)(20) = 5$.

2. Let G be the subgroup, and let $X = \{1,2,3,4,5,6,7,8\}$.
$|X_{(1)}| = 8$, $|X_{(1\ 3)}| = 6$, $|X_{(2\ 4\ 7)}| = 5$, $|X_{(2\ 4\ 7)}| = 5$,
$|X_{(1\ 3)(2\ 4\ 7)}| = 3$, $|X_{(1\ 3)(2\ 7\ 4)}| = 3$.
$\sum_{g\in G} |X_g| = 8 + 6 + 5 + 5 + 3 + 3 = 30$. The number of orbits under G is then $(1/6)(30) = 5$.

3. The group of rigid motions of the tetrahedron has 12 elements since any one of the four triangles can be on the bottom and the die can then be rotated through three positions, keeping the bottom face the same. We see that $|X_g| = 0$ unless g is the identity ι of this group G, and $|X_\iota| = 4! = 24$. Thus there are $(1/12)(24) = 2$ distinguishable dice.

4. The total number of ways such a block can be painted with different colors on each face is $8\cdot7\cdot6\cdot5\cdot4\cdot3$. The group of rigid motions of the cube has 24 elements. The only rigid motion leaving unchanged a block with different colors on all faces is the identity, which leaves all such blocks fixed.

Thus the number of distinguishable blocks is $(1/24)(8\cdot7\cdot6\cdot5\cdot4\cdot3) = 8\cdot7\cdot5\cdot3 = 40\cdot21 = 840$.

5. There are 8^6 ways of painting the faces of a block, allowing for repetition of colors. Following the breakdown of the group G of rotations given in the hint, and using sublabels to suggest the categories in this breakdown, we have

   $|X_\iota| = 8^6$ where ι is the identity,

   $|X_{\text{opp face, }90°\text{ or }270°\text{ rotation}}| = 8\cdot8\cdot8$, there are 6 such,

   $|X_{\text{opp face, }180°\text{ rotation}}| = 8\cdot8\cdot8\cdot8$, there are 3 such,

   $|X_{\text{opp vert}}| = 8\cdot8$, there are 8 such,

   $|X_{\text{opp edges}}| = 8\cdot8\cdot8$, there are 6 such.

   Thus $\sum_{g\in G}|X_g| = 8^6 + 6\cdot8^3 + 3\cdot8^4 + 8\cdot8^2 + 6\cdot8^3 = 8^3(8^3 + 37)$.

   The number of distinguishable blocks is thus

   $(1/24)[8^3(8^3 + 37)] = 11{,}712$.

6. Proceding as in Exercise 5 using the same group G acting on the set X of 4^8 ways of coloring the eight vertices, we obtain

   $|X_\iota| = 4^8$ where ι is the identity,

   $|X_{\text{opp face, }90°\text{ or }270°\text{ rotation}}| = 4\cdot4$, there are 6 such,

   $|X_{\text{opp face, }180°\text{ rotation}}| = 4\cdot4\cdot4\cdot4$, there are 3 such,

   $|X_{\text{opp vert}}| = 4\cdot4\cdot4\cdot4$, there are 8 such,

   $|X_{\text{opp edges}}| = 4\cdot4\cdot4\cdot4$, there are 6 such.

   Thus $\sum_{g\in G}|X_g| = 4^8 + 6\cdot4^2 + 3\cdot4^4 + 8\cdot4^4 + 6\cdot4^4 = 4^4(273) + 96$

   The number of distinguishable blocks is thus

   $(1/24)[4^4(273) + 96] = 2{,}916$.

7. a) The group is $G = D_4$ and has eight elements. We label them as in Section 2.1. There are $6\cdot5\cdot4\cdot3$ ways of painting the edges of a square, and we let X be this set of 360 elements. We have

   $$|X_{\rho_0}| = 6\cdot5\cdot4\cdot3 \text{ and } |X_g| = 0 \text{ for } g \in D_4,\ g \neq \rho_0.$$

   Thus the number of distinguishable such painted squares is $(1/8)(360) = 45$.

b) We let G be as in part (a), and let X be the 6^4 ways of painting the edges of the squares, allowing repetition of colors. This time, we have

$$|X_{\rho_0}| = 6^4, \quad |X_{\rho_1}| = |X_{\rho_3}| = 6, \quad |X_{\rho_2}| = 6\cdot 6,$$
$$|X_{\mu_1}| = |X_{\mu_2}| = 6\cdot 6\cdot 6, \text{ and } |X_{\delta_1}| = |X_{\delta_2}| = 6\cdot 6.$$

Thus $\sum_{g\in G} |X_g| = 6^4 + 2\cdot 6 + 6^2 + 2\cdot 6^3 + 2\cdot 6^2 = 6^2(51) + 12.$

The number of distinguishable squares is thus

$(1/8)[6^2(51) + 12] = 231.$

8. The group of rigid motions of the tetrahedron is a subgroup G of the group of permutations of its vertices. The order of G is 12 since, viewing the tetrahedron as sitting on a table, any of the four faces may be on the bottom, and then the base can be rotated repeatedly through 120° to give three possible positions. If we call the vertex at the top of the tetrahedron number 1 and number the vertices on the table as 2, 3, and 4 counterclockwise when viewed from above, we can write our 12 group elements in cyclic notation as

#1 on top	#2 on top	#3 on top	#4 on top
(1)	(1 2)(3 4)	(1 3)(2 4)	(1 4)(2 3)
(2 3 4)	(1 3 2)	(1 2 3)	(1 2 4)
(2 4 3)	(1 4 2)	(1 4 3)	(1 3 4) .

Let X be the 2^6 ways of placing either a 50-ohm resistor or 100-ohm resistor in each edge of the tetrahedron. Now the the elements of G that are 3-cycles correspond to rotating holding a single vertex fixed. These carry the three edges from that vertex cyclically into themselves and the three edges of the triangle opposite that vertex cyclically into themselves. Thus $|X_{3\text{-cycle}}| = 2\cdot 2$. The element (1, 2)(3, 4) of G carries the edge joining vertices 1 and 2 the edge joining vertices 2 and 4 into themselves, swaps the edge joining vertices 1 and 3 with the one joining vertices 2 and 4 and swaps the edge joining 1 and 4 with the one joining 2 and 3. Thus we see that $|X_{(1,2)(3,4)}| = 2\cdot 2\cdot 2\cdot 2$, and of course an analogous count can be made for the group elements (1, 3)(2, 4) and (1, 4)(2, 3). Thus we obtain

$$|X_\iota| = 2^6, \quad |X_{3\text{-cycle}}| = 2^2, \quad \text{and} \quad |X_{\text{other type}}| = 2^4.$$

Consequently $\sum_{g \in G} |X_g| = 2^6 + 8 \cdot 2^2 + 3 \cdot 2^4 = 144$.

The number of distinguishable wirings is $(1/12)(144) = 12$.

9. a) The group G of rigid motions of the prism has order 8, four positions leaving the end faces in the same position, and four positions with the end faces swapped. There are $6 \cdot 5 \cdot 4 \cdot 3 \cdot 2 \cdot 1$ ways of painting the faces different colors. We let X be the set of these $6!$ possibilities. Then $|X_\iota| = 6$ and $|X_{\text{other}}| = 0$. Thus there are $(1/8)(6!) = 6 \cdot 5 \cdot 3 = 90$ distinguishable painted prisms using six different colors.

   b) This time the set X of possible ways of painting the prism has 6^6 elements. We have

   $|X_\iota| = 6^6$ where ι is the identity,

   $|X_{\text{same ends, rotate 90° or 270°}}| = 6 \cdot 6 \cdot 6$,

   $|X_{\text{same ends, rotate 180°}}| = 6 \cdot 6 \cdot 6 \cdot 6$,

   $|X_{\text{swap ends keeping top face on top}}| = 6 \cdot 6 \cdot 6 \cdot 6$,

   $|X_{\text{swap ends as above, rotate 90° or 270°}}| = 6 \cdot 6 \cdot 6$,

   $|X_{\text{swap ends as above, rotate 180°}}| = 6 \cdot 6 \cdot 6 \cdot 6$.

   Then $\sum_{g \in G} |X_g| = 6^6 + 2 \cdot 6^3 + 6^4 + 6^4 + 2 \cdot 6^3 + 6^4 =$

   $6^3(6^3 + 2 + 6 + 6 + 2 + 6) = 6^3(6^3 + 22)$. The number of distinguishable prisms is then $(1/8)[6^3(6^3 + 22)] = 6{,}426$.

CHAPTER 4

ADVANCED GROUP THEORY

SECTION 4.1 - Isomorphism Theorems; Proof of the Jordan-Holder Theorem

1. (See the text answer.)

2. a) $K = \{0,6,12\}$
   b) $0 + K = \{0,6,12\}$, $1 + K = \{1,7,13\}$, $2 + K = \{2,8,14\}$, $3 + K = \{3,9,15\}$, $4 + K = \{4,10,16\}$, $5 + K = \{5,11,17\}$
   c) $\phi[\mathbb{Z}_{18}]$ is the subgroup $\{0,2,4,6,8,10\}$ of $\mathbb{Z}_{12}$.
   d) $\psi(0 + K) = 0$, $\psi(1 + K) = 10$, $\psi(2 + K) = 8$, $\psi(3 + K) = 6$, $\psi(4 + K) = 4$, $\psi(5 + K) = 2$

3. (See the text answer.)

4. a) $HN = \{0,3,6,9,12,15,18,21,24,27,30,33\}$
   b) $0 + N = \{0,9,18,27\}$, $3 + N = \{3,12,21,30\}$, $6 + N = \{6,15,24,33\}$
   c) $0 + (H \cap N) = \{0,18\}$, $6 + (H \cap N) = \{6,24\}$, $12 + (H \cap N) = \{12,30\}$
   d) $\psi(0 + N) = 0 + (H \cap N)$, $\psi(3 + N) = 12 + (H \cap N)$, $\psi(6 + N) = 6 + (H \cap N)$

5. (See the text answer.)

6. a) $0 + H = \{0,9,18,27\}$, $1 + H = \{1,10,19,28\}$, $2 + H = \{2,11,20,29\}$, $3 + H = \{3,12,21,30\}$, $4 + H = \{4,13,22,31\}$, $5 + H = \{5,14,23,32\}$, $6 + H = \{6,15,24,33\}$, $7 + H = \{7,16,25,34\}$, $8 + H = \{8,17,26,35\}$
   b) $0 + K = \{0,18\}$, $1 + K = \{1,19\}$, $2 + K = \{2,20\}$, $3 + K = \{3,21\}$, $4 + K = \{4,22\}$, $5 + K = \{5,23\}$, $6 + K = \{6,24\}$, $7 + K = \{7,25\}$, $8 + K = \{8,26\}$, $9 + K = \{9,27\}$, $10 + K = \{10,28\}$, $11 + K = \{11,29\}$, $12 + K = \{12,30\}$, $13 + K = \{13,31\}$, $14 + K = \{14,32\}$, $15 + K = \{15,33\}$, $16 + K = \{16,34\}$, $17 + K = \{17,35\}$
   c) $0 + K = \{0,18\}$, $9 + K = \{9,27\}$
   d) $(0 + K) + H/K = H/K = \{0 + K,\ 9 + K\} = \{\{0,18\},\ \{9,27\}\}$
   $(1 + K) + H/K = \{1 + K,\ 10 + K\} = \{\{1,19\},\ \{10,28\}\}$
   $(2 + K) + H/K = \{2 + K,\ 11 + K\} = \{\{2,20\},\ \{11,29\}\}$
   $(3 + K) + H/K = \{3 + K,\ 12 + K\} = \{\{3,21\},\ \{12,30\}\}$
   $(4 + K) + H/K = \{4 + K,\ 13 + K\} = \{\{4,22\},\ \{13,31\}\}$
   $(5 + K) + H/K = \{5 + K,\ 14 + K\} = \{\{5,23\},\ \{14,32\}\}$
   $(6 + K) + H/K = \{6 + K,\ 15 + K\} = \{\{6,24\},\ \{15,33\}\}$
   $(7 + K) + H/K = \{7 + K,\ 16 + K\} = \{\{7,25\},\ \{16,34\}\}$
   $(8 + K) + H/K = \{8 + K,\ 17 + K\} = \{\{8,26\},\ \{17,35\}\}$

e) $\psi(0 + H) = (0 + K) + H/K$, $\psi(1 + H) = (1 + K) + H/K$,
$\psi(2 + H) = (2 + K) + H/K$, $\psi(3 + H) = (3 + K) + H/K$,
$\psi(4 + H) = (4 + K) + H/K$, $\psi(5 + H) = (5 + K) + H/K$,
$\psi(6 + H) = (6 + K) + H/K$, $\psi(7 + H) = (7 + K) + H/K$,
$\psi(8 + H) = (8 + K) + H/K$

7. (See the text answer.)

8. *Chain (3)*

$\{0\} \leq \langle 12\rangle \leq \langle 12\rangle \leq \langle 12\rangle$
$\leq \langle 12\rangle \leq \langle 12\rangle \leq \langle 4\rangle$
$\leq \langle 2\rangle \leq \mathbb{Z}_{24} \leq \mathbb{Z}_{24}$

Chain (4)

$\{0\} \leq \langle 12\rangle < \langle 12\rangle \leq \langle 6\rangle$
$\leq \langle 6\rangle \leq \langle 6\rangle \leq \langle 3\rangle$
$\leq \langle 3\rangle \leq \mathbb{Z}_{24} \leq \mathbb{Z}_{24}$

Isomorphisms

$\langle 12\rangle/\{0\} \simeq \langle 12\rangle/\{0\} \simeq \mathbb{Z}_2$, $\langle 12\rangle/\langle 12\rangle \simeq \langle 6\rangle/\langle 6\rangle \simeq \{0\}$,

$\langle 12\rangle/\langle 12\rangle \simeq \langle 3\rangle/\langle 3\rangle \simeq \{0\}$, $\langle 12\rangle/\langle 12\rangle \simeq \langle 12\rangle/\langle 12\rangle \simeq \{0\}$,
$\langle 12\rangle/\langle 12\rangle \simeq \langle 6\rangle/\langle 6\rangle \simeq \{0\}$, $\langle 4\rangle/\langle 12\rangle \simeq \mathbb{Z}_{24}/\langle 3\rangle \simeq \mathbb{Z}_3$

$\langle 2\rangle/\langle 4\rangle \simeq \langle 6\rangle/\langle 12\rangle \simeq \mathbb{Z}_2$, $\mathbb{Z}_{24}/\langle 2\rangle \simeq \langle 3\rangle/\langle 6\rangle \simeq \mathbb{Z}_2$

$\mathbb{Z}_{24}/\mathbb{Z}_{24} \simeq \mathbb{Z}_{24}/\mathbb{Z}_{24} \simeq \{0\}$

9. Let $x \in H \cap N$ and let $h \in H$. Since $x \in H$ and H is a subgroup, we know that $hxh^{-1} \in H$. Since $x \in N$ and N is normal in G, we also know that $hxh^{-1} \in N$. Thus $hxh^{-1} \in H \cap N$ so $H \cap N$ is a normal subgroup of H.

10. Let $a \in H^* \cap K$ and let $b \in H \cap K$. Then $b \in H$ and $a \in H^*$ so $bab^{-1} \in H^*$ since H^* is normal in H. Also, $b \in K$ and $a \in K$ so $bab^{-1} \in K$. Thus $bab^{-1} \in H^* \cap K$, so $H^* \cap K$ is a normal subgroup of $H \cap K$.

11. a) Let $\gamma: G \longrightarrow G/H$ be the natural homomorphism of a group onto a factor group. Then $\gamma[K] = K/H = B$ is a normal subgroup of $A = G/H$ by Theorem 3.13. Similarly, $\gamma[L] = L/H = C$ is a normal subgroup of A. It is clear that $B = K/H$ is a subgroup of $C = L/H$ since K is a subgroup of L.
b) We know that $(A/B)/(C/B) \simeq A/C$ by Theorem 4.3. Now let $\phi: A \longrightarrow A/C$ be the natural homomorphism. Since $C = L/H$ we see that the composite homomorphism $\phi\gamma: G \longrightarrow A/C$ has kernel L. Thus $A/C \simeq G/L$, so $(A/B)/(C/B) \simeq (G/L$.

12. By Lemma 4.1, we know that $K \vee L = KL = LK$, so $G = KL = LK$. By Theorem 4.2, $G/L = KL/L \simeq K/(K \cap L) = K/\{e\} \simeq K$. Similarly, $G/K = LK/K \simeq L/(L \cap K) = L/\{e\} \simeq L$.

13. Following the hint, Exercise 10 shows that $K \cap H_i$ is a normal subgroup of $K \cap H_{i+1}$ for $i = 0, 1, \cdots, n-1$, so the subgroups $K \cap H_{i+1}$ form a subnormal series for K. Taking $N = H_{i-1}$ and $H = K \cap H_i$ as subgroups of H_i and applying Theorem 4.2, we see that $HN/N = [(K \cap H_i)H_{i-1}]/H_{i-1} \simeq H/(H \cap N) = (K \cap H_i)/(K \cap H_{i-1})$. Now $(K \cap H_i)H_{i-1} \leq (K \cap H_i)H_i = H_i$ so $[(K \cap H_i)H_{i-1}]/H_{i-1}$ can be viewed as a subgroup of H_i/H_{i-1}. Since H_i/H_{i-1} is a simple group, we see that $[(K \cap H_i)H_{i-1}]/H_{i-1}$ is either the trivial group or is isomorphic to H_i/H_{i-1}, and hence is simple and abelian since G is solvable. Thus the distinct groups among the $K \cap H_i$ form a composition series for K with abelian factor groups, and consequently K is solvable.

14. Following the hint, we show that $H_{i-1}N$ is normal in H_iN. Let $h_{i-1}n_1 \in H_{i-1}N$ and $h_in_2 \in H_iN$ where the elements belong to the obvious sets. Using the fact that H_{i-1} is normal in H_i and that N is normal in G, we obtain $h_in_2h_{i-1}n_1(h_in_2)^{-1} = h_in_2h_{i-1}n_1n_2^{-1}h_i^{-1} = h_ih_{i-1}n_3n_1n_2^{-1}h_i^{-1} = h'_{i-1}h_in_4h_i^{-1} = h'_{i-1}h_ih_i^{-1}n_5 = h'_{i-1}n_5 \in H_{i-1}N$. Thus $H_{i-1}N$ is a normal subgroup of H_iN.

The hint does the rest of the work for us, except to observe at the end that H_i/H_{i-1} being simple implies that $[H_i \cap (H_{i-1}N)]/H_{i-1}$ is either trivial or isomorphic to H_i/H_{i-1}. Thus $(H_iN)/(H_{i-1}N)$ is either isomorphic to H_i/H_{i-1} or is trivial. Since $N = H_0N$ is itself simple, it follows at once that the distinct groups among the H_iN for $i = 0, 1, 2, \cdots, n$ form a composition series for G.

15. First we show that ψ is well defined. Let h_in_1 and h'_in_2 be the same element of H_iN. Then $h_in_1 = h'_in_2$ so $h_i = h'_in_2n_1^{-1} = h'_in_3$. Since γ has kernel N, we see that $\psi(h_in_1) =$

$\gamma(h_i n_1)\gamma[H_{i-1}] = (h_i N)\gamma[H_{i-1}] = (h_1' n_3 N)\gamma[H_{i-1}] = (h_1' N)\gamma[H_{i-1}]$
$= \gamma(h_1' n_2)\gamma[H_{i-1}]$, so ψ is well defined.

We show that ψ is a homomorphism. This follows from the fact that $\gamma(h_i n_1 h_i' n_2) = \gamma(h_i n_1)\gamma(h' n_2)$ since γ is s homomorphism.

The kernel of ψ consists of all $x \in H_i N$ such that $\gamma(x) \in \gamma[H_{i-1}] = H_{i-1}N$ and ψ is clearly an onto map. By Theorem 4.1, $\gamma[H_i]/\gamma[H_{i-1}]$ is isomorphic to $(H_i N)/(H_{i-1}N)$. Exercise 14 shows that these factor groups are simple, and the desired result follows immediately.

16. Let $H_0 = \{e\} < H_1 < H_2 < \cdots < H_n = G$ be a composition series for G, and let ϕ: $G \longrightarrow G'$ be a group homomorphism of G onto G' with kernel N. Then $G' \simeq G/N$. Exercise 15 shows that the distinct groups among the groups $H_i N$ for $i = 0, 1, \cdots, n$ form a composition series for G/N, and Exercise 14 shows that a factor group of this composition series is isomorphic to one of the factor groups in the composition series of groups H_i for G. Since G is a solvable group, it follows at once that all the factor groups in this composition series for G/N composed of some of the groups $H_i N$ are also abelian, so that G/N is solvable.

SECTION 4.2 - Sylow Theorems

1. 3 2. 27 3. 1 or 3 4. 1, 85, 1, 51

5. Since $|S_4| = 24$, the Sylow 3-subgroups have order 3 and are thus cyclic and generated by a single 3-cycle. The possibilities are
$\langle(1,2,3)\rangle$,
$\langle(1,2,4)\rangle = \langle(3,4)(1,2,3)(3,4)\rangle$,
$\langle(1,3,4)\rangle = \langle(2,4)(1,2,3)(2,4)\rangle$, and
$\langle(2,3,4)\rangle = \langle(1,4)(1,2,3)(1,4)\rangle$.

6. A Sylow 2-subgroup of S_4 must have order 8 and by Theorem 4.9 there must be either 1 or 3 of them. The group of symmetries of the square has order 8 and can viewed as a subgroup of S_4 if we number the vertices 1, 2, 3, and 4. Numbering vertices

in order 1,2,3,4 counterclockwise, we obtain the group $H = \{(1), (1,2,3,4), (1,3)(2,4), (1,4,3,2), (1,3), (2,4), (1,2)(3,4), (1,4)(2,3)\}$.
Numbering vertices in order 1,3,2,4 counterclockwise, we obtain the group $K = \{(1), (1,3,2,4), (1,2)(3,4), (1,4,2,3), (1,2), (3,4), (1,4)(2,3), (1,3)(2,4)\}$.
Numbering vertices in order 1,3,4,2 counterclockwise, we obtain the group $L = \{(1), (1,3,4,2), (1,4)(2,3), (1,2,4,3), (1,4), (2,3), (1,3)(2,4), (1,2)(3,4)\}$.
Since there can be at most three of them, we have found them all. We see that $K = (2,3)H(2,3)$ and $L = (2,3,4)H(2,4,3)$.

7. T T T F T F T T F F

8. Let H be a Sylow p-subgroup of G. For each $g \in G$, the conjugate group gHg^{-1} is also a Sylow p-subgroup of G. Since G has only one Sylow p-subgroup, it must be that $gHg^{-1} = H$ for all $g \in G$, so that H is a proper normal subgroup of G, and G is thus not a simple group.

9. The divisors of 45 are 1, 3, 5, 9, 15, and 45. Of these, only 1 is congruent to 1 modulo 3, so by Theorem 4.9, there is only one Sylow 3-subgroup of a group of order 45. By the argument in Exercise 8, this subgroup must be a normal subgroup.

10. Let G be a p-group, so that by definition, every element of G has order a power of p. If a prime q divides $|G|$ then G has an element of order q by Cauchy's theorem. Thus the order of G must be a power of p.

    Conversely, if the order of G is a power of p, then the order of each element of G is also a power of p by the Theorem of Lagrange. Thus G is a p-group.

11. Since $N[P]$ is a normal subgroup of $N[N[P]]$, conjugation of P by an element of $N[N[P]]$ yields a subgroup of $N[P]$ that is a Sylow p-subgroup of G, and also of $N[P]$. Such a Sylow p-subgroup must be conjugate to P under conjugation by an element of $N[P]$ (by Theorem 4.8), and thus must be P since P is a normal subgroup of $N[P]$. Thus P is invariant under conjugation by every element of $N[N[P]]$, so $N[N[P]]$ is contained in $N[P]$. Since $N[P]$ is contained in $N[N[P]]$ by definition, we see that $N[N[P]] = N[P]$.

Section 4.2

12. By Theorem 4.7, H is contained in a Sylow p-subgroup K of G. By Theorem 4.8, there exists $g \in G$ such that $gKg^{-1} = P$. Consequently, $gHg^{-1} \leq P$.

13. The divisors of $(35)^3$ that are not divisible by 5 are 1, 7, 49, and 343, which are congruent to 1, 2, 4, and 3 respectively modulo 5. By Theorem 4.9, there is only one Sylow 5-subgroup of a group of order $(35)^3$, and it must be a normal subgroup by the argument in Exercise 8.

14. The divisors of 255 that are not multiples of 17 are 1, 3, 5, 15. Of these, only 1 is congruent to 1 modulo 17. By Theorem 4.9, there is only one Sylow 17-subgroup of a group of order 255, and it must be a normal subgroup by the argument in Exercise 8.

15. The divisors of $p^r m$ that are not divisible by p are 1 and m. Since $m < p$, of these two divisors of $p^r m$, only 1 is congruent to 1 modulo p. By Theorem 4.9, there is a unique Sylow p-subgroup of a group of order $p^r m$, and this must be a normal subgroup by the argument in Exercise 8.

16. a) $G_G = \{g \in G \mid gxg^{-1} = x \text{ for all } x \in G\}$
$= \{g \in G \mid gx = xg \text{ for all } x \in G\}$
$= Z(G)$ by definition of $Z(G)$.

b) Let G be a nontrivial p-group, so that $|G| = p$ for $r \geq 1$. By Theorem 4.5, we see that $|G| \equiv |G_G| \pmod p$. Thus p is a divisor of $|G_G|$, and hence is a divisor of $|Z(G)|$ by part (a). Thus $Z(G)$ is nontrivial.

17. We proceed by induction on n. If $n = 1$, the statement is obviously true, for $H_0 = \{e\}$ and the entire group $G = H_1$ are the required subgroups. If $n = 2$, the subgroups are furnished by Theorem 4.7. Suppose the statement is true for $n = k$, and let G have order p^{k+1}. Let $H_j = Z(G)$ have order p^j where $j \geq 1$ by Exercise 16. If $j = k + 1$, then G is abelian and the subgroups provided by Theorem 4.7 are all normal subgroups of G and we are done. If $j < k + 1$, apply the induction hypothesis to H_j to find the desired subgroups $H_0, H_1, \cdots, H_{j-1}$. Then form the factor group G/H_j which has which has order p^{k+1-j} and apply the induction hypothesis to

find normal subgroups $K_1 \le K_2 \le \cdots \le K_{k+1-j}$ of G/H_j where the order of K_i is p^i. If $\gamma: G \longrightarrow G/H_j$ is the canonical homomorphism, then $H_{j+i} = \gamma^{-1}[K_i]$ is a normal subgroup of G of order p^{i+j}, and all these subgroups H_k form the desired chain of normal subgroups of G.

18. Let H be a normal p-subgroup of G. By Theorem 4.7, there exists a Sylow p-subgroup P of G containing H. By Theorem 4.8, every Sylow p-subgroup of G is of the form gPg^{-1} for some $g \in G$. Since $gHg^{-1} = H$, we see that H is contained in every Sylow p-subgroup of G.

SECTION 4.3 - Applications of the Sylow Theory

1. (See the text answer.)

2. As p-groups, groups of these orders are not simple by Theorem 4.7: 4, 8, 9, 16, 25, 27, 32, 49.

   As groups of order pq, groups of these orders are not simple by Theorem 4.13: 6, 10, 14, 15, 21, 22, 26, 33, 34, 35, 38, 39, 46, 51, 55, 57, 58.

   As groups of order $p^r m$, $m < p$, groups of these orders are not simple by Exercise 15 of Section 4.2: 18, 20, 21, 28, 42, 44, 50, 52, 54.

   The text showed that groups of these orders are not simple: 30, 36, 48.

   Order 12: Such a group has either 1 or 3 Sylow 2-subgroups of order 4 and either 1 or 4 Sylow 3-subgroups of order 3. To have 3 subgroups of order 4 and 4 subgroups of order 3 would require at least 4 elements of order divisible by 2 and at least 8 elements of order 3, which would require 12 elements other than the identity. Thus there is either only one subgroup of order 4 or only one of order 3, which must be normal.

   Order 24: Such a group has either 1 or 3 subgroups of order 8 and either 1 or 4 subgroups of order 3. If there is unique subgroup of either order, we are done. Suppose that H and K are two different subgroups of order 8. By Lemma 4.5 $H \cap K$ must have order 4, and is normal in both H and K, being of index 2. Thus $N[H \cap K]$ contains both H and K so has order a multiple > 1 of 8 and a divisor of 24. Hence $N[H \cap K]$ is of order 24 and $H \cap K$ is a normal subgroup.

Order 40: Theorem 4.9 shows there is a unique subgroup of order 5, which must be normal.

Order 45: Theorem 4.9 shows there is a unique subgroup of order 9, which must be normal.

Order 56: Such a group has either 1 or 7 subgroups of order 8 and either 1 or 8 subgroups of order 7. If there is a unique subgroup of either order, we are done. Eight subgroups of order 7 require 48 elements of order 7, and 7 subgroups of order 8 require at least 8 elements of order divisible by 2, which is impossible in a group of 56 elements.

Order 60: Theorem 4.9 shows there is a unique subgroup of order 5, which must be normal.

All orders from 2 to 60 that are not prime have been considered.

3. T T F T T T T T F F

4. By Theorem 4.9, a group G of order (5)(7)(47) contains a unique subgroup H of order 47, which must normal in G. By the same arguments, there exist unique normal subgroups K and L of orders 7 and 5 respectively. By Lemma 4.5, LK has order 35 since $L \cap K = \{e\}$. By the proof of Lemma 4.4, we see that $LK = L \vee K$ and is isomorphic to $L \times K$. By Lemma 4.5, $(LK)H$ has order (35)(47), so $(LK)H = G$. Now LK must be the unique subgroup of G of order 35, since another subgroup would lead to subgroups of orders 7 and 5 other than L and K, which is impossible. By Lemma 4.4, G is isomorphic to $LK \times H$ and consequently to $L \times K \times H$ which is abelian and cyclic.

5. A group of order 96 has either 1 or 3 subgroups of order 32. If there is only one such subgroup, it is normal and we are done. If not, let H and K be distinct subgroups of order 32. By Lemma 4.5, $H \cap K$ must have order 16, and is normal in both H and K, being of index 2. Thus $N[H \cap K]$ has order a multiple > 1 of 32 and a divisor of 96, so the order must be 96. Thus $H \cap K$ is a normal in the whole group.

6. A group G of order 160 has either 1 or 5 subgroups of order 32. If there is only one such subgroup, it is normal and we are done. If not, let H and K be distinct subgroups of order 32. By Lemma 4.4, $H \cap K$ must have order either 16 or 8. If $|H \cap K| = 16$, then $H \cap K$ is normal in G by an analogous argument to that in Exercise 5. If $|H \cap K| = 8$, then find

groups H_1 and K_1 of order 16 such that $H \cap K < H_1 < H$ and $H \cap K < K_1 < K$ by Theorem 4.7. Then $H \cap K$ is normal in both H_1 and K_1, being of index 2. Thus $N[H \cap K]$ has order a multiple > 1 of 16. If $|N[H \cap K]| = 32$, then $N[H \cap K] \cap H = H_1$ and H_1 is normal in G by an analogous argument to that in Exercise 5. Otherwise $N[H \cap K] = G$ so $H \cap K$ is normal in G.

7. a) We have $\tau\sigma\tau^{-1}(\tau a_i) = \tau\sigma(a_i) = \begin{cases} \tau(a_{i+1}) & \text{if } i < m, \\ \tau(a_1) & \text{if } i = m. \end{cases}$

For any element b that is not of the form $\tau(a_i)$, we have $\tau\sigma\tau^{-1}(b) = \tau\tau^{-1}(b) = b$. Thus $\tau\sigma\tau^{-1}$ has the desired action on each element of $\{1, 2, \cdots, n\}$.

b) Let $\sigma = (a_1, a_2, \cdots, a_m)$ and $\mu = (b_1, b_2, \cdots, b_m)$ be cycles of length m in S_n. Let τ be any permutation in S_n such that $\tau(a_i) = b_i$. Part (a) then shows that $\tau\sigma\tau^{-1} = \mu$.

c) Let σ and μ be products of s disjoint cycles with the ith cycle in each product of length r_i. Let τ be any permutation in S_n that carries the jth element of the ith cycle of σ into the jth element of the ith cycle of μ for $1 \le j \le r_i$ for each i where $1 \le i \le s$. (That is, if we write μ directly under σ and erase all the parentheses at the ends of cycles, we get a 2-rowed notation for the action of τ on the elements moved by σ. Fill in the action of τ on other elements in any way to yield an element of S_n.) Repetition of the computation in (a) shows that $\tau\sigma\tau^{-1} = \mu$.

d) Let σ be any permutation in S_n. Express σ as a product of disjoint cycles, supplying cycles of length 1 for all elements not moved by σ. The sum of the lengths of these cycles is then n, and the sum yields a partition of n. By part (c), σ is conjugate to any other permutation that can be expressed similarly as a product of disjoint cycles and yielding the same partition of n. To show the one-to-one correspondence between conjugate classes and partitions of n, it remains to show that conjugate permutations always do correspond to the same partition of n. The computations in parts (a) and (c) show the effect of conjugation on a product of disjoint cycles, and we see that this is indeed the case.

e) $1 = 1$, $p(1) = 1$

$2 = 2 = 1 + 1$, $p(2) = 2$

$3 = 3 = 1 + 2 = 1 + 1 + 1$, $p(3) = 3$

$4 = 4 = 1 + 3 = 2 + 2 = 1 + 1 + 2 = 1 + 1 + 1 + 1$, $p(4) = 5$

$5 = 5 = 1 + 4 = 2 + 3 = 1 + 1 + 3 = 1 + 2 + 2 = 1 + 1 + 1 + 2 = 1 + 1 + 1 + 1 + 1$, $p(5) = 7$

$6 = 6 = 1 + 5 = 2 + 4 = 3 + 3 = 1 + 1 + 4 = 1 + 2 + 3 = 2 + 2 + 2 = 1 + 1 + 1 + 3 = 1 + 1 + 2 + 2 = 1 + 1 + 1 + 1 + 2 = 1 + 1 + 1 + 1 + 1 + 1$, $p(6) = 11$

$7 = 7 = 1 + 6 = 2 + 5 = 3 + 4 = 1 + 1 + 5 = 1 + 2 + 4 = 1 + 3 + 3 = 2 + 2 + 3 = 1 + 1 + 1 + 4 = 1 + 1 + 2 + 3 = 1 + 2 + 2 + 2 = 1 + 1 + 1 + 1 + 3 = 1 + 1 + 1 + 2 + 2 = 1 + 1 + 1 + 1 + 1 + 2 = 1 + 1 + 1 + 1 + 1 + 1 + 1$, $p(7) = 15$

8. Using Exercise 7c, we see the conjugate classes are
$\{\iota\}$,
$\{(1,2), (1,3), (1,4), (2,3), (2,4), (3,4)\}$,
$\{(1,2)(3,4), (1,3)(2,4), (1,4)(2,3)\}$,
$\{(1,2,3), (1,2,4), (1,3,4), (1,3,2), (1,4,2), (1,4,3), (2,3,4), (2,4,3)\}$,
$\{(1,2,3,4), (1,2,4,3), (1,3,2,4), (1,3,4,2), (1,4,2,3), (1,4,3,2)\}$.
The class equation is $24 = 1 + 6 + 3 + 8 + 6$.

9. For S_5: There are $\binom{5}{2} = 10$ transpositions,

$\binom{5}{2}\binom{3}{2}/2 = (10)(3)/2 = 15$ products of two disjoint transpositions,

$\binom{5}{3}\cdot 2 = 10\cdot 2 = 20$ cycles of length 3,

$20\cdot 1 = 20$ products of disjoint cycles of lengths 3 and 2,
$5\cdot 6 = 30$ cycles of length 4 (see the last problem answer),
$4\cdot 3\cdot 2\cdot 1 = 24$ cycles of length 5.
The class equation is $120 = 1 + 10 + 15 + 20 + 20 + 30 + 24$.

For S_6: There are $\binom{6}{2} = 15$ transpositions,

$\binom{6}{2}\binom{4}{2}/2 = (15)(6)/2 = 45$ products of two disjoint 2-cycles,

$\binom{6}{2}\binom{4}{2}\binom{2}{2}/6 = (15)(6)(1)/6 = 15$ products of three disjoint 2-cycles,

$\binom{6}{3}\cdot 2 = 20\cdot 2 = 40$ cycles of length 3,

$40\cdot\binom{3}{2} = 120$ products of a 3-cycle and a disjoint 2-cycle,

$40\cdot 2/2 = 40$ products of two disjoint 3-cycles,

$\binom{6}{4}\cdot 6 = 15\cdot 6 = 90$ 4-cycles,

$90\cdot 1 = 90$ products of a 4-cycle and a disjoint 2-cycle,

$\binom{6}{5}\cdot 4\cdot 3\cdot 2\cdot 1 = 6\cdot 24 = 144$ 5-cycles,

$5\cdot 4\cdot 3\cdot 2\cdot 1 = 120$ 6-cycles.

The class equation is

$720 = 1 + 15 + 45 + 15 + 40 + 120 + 40 + 90 + 90 + 144 + 120.$

10. Exercise 7 shows that the number of conjugate classes in S_n is the number $p(n)$ of partitions of n. By Theorem 2.11, the number of abelian groups of order p^n (up to isomorphism) is also the number of partitions of n.

11. Each element of the center of a group G gives rise to a 1-element conjugate class of G. It is clear from Exercise 7a that for $n > 2$, every permutation in S_n having an orbit with more than one element is conjugate to some other permutation in S_n. Thus the identity in S_n is the only element that is conjugate only to itself, if $n > 2$.

SECTION 4.4 - Free Abelian Groups

1. (See the text answer.)

2. Yes. $(1,0) = (3,1) + (-1)(2,1)$ and $(0,1) = 3(2,1) + (-2)(3,1)$ so $\{(2,1), (3,1)\}$ generates $\mathbb{Z} \times \mathbb{Z}$. If $m(2,1) + n(3,1) = 0$, then $2m + 3n = 0$ and $m + n = 0$. Then $m = -n$ so $2(-n) + 3n = 0$, $n = 0$, and $m = 0$. Thus the conditions for a basis in Theorem 4.14 are satisfied.

3. (See the text answer.)

Section 4.4

4. By Cramer's rule, the equations

$$ax + cy = e$$
$$bx + dy = f$$

have a unique solution in $\mathbb{R}$ if and only if $ad - bc \neq 0$. The solution is then

$$x = \frac{ed - fc}{ad - bc} \quad \text{and} \quad y = \frac{af - be}{ad - bc}.$$

These values x and y are integers for *all* choices of e and f if and only if $D = ad - bc$ divides each of a, b, c, and d. Let $a = r_aD$, $b = r_bD$, $c = r_cD$, and $d = r_dD$. Then $D = ad - bc = (r_ar_d - r_br_c)D^2$ so that D is an integer which is an integer multiple of its square. The only possible values for D are ± 1, so we obtain the condition $|ad - bc| = 1$.

5. $2\mathbb{Z}$ is a proper subgroup of rank $r = 1$ of the free abelian group $\mathbb{Z}$ of rank $r = 1$.

6. T T T T T F F T T F

7. Let $\phi: G \to \underbrace{\mathbb{Z} \times \mathbb{Z} \times \cdots \times \mathbb{Z}}_{r \text{ factors}}$ be the map described before the statement of Theorem 4.15. Note that ϕ is well defined since each $a \in G$ has a *unique* expression in the form $n_1x_1 + n_2x_2 + \cdots + n_rx_r$ where each $n_i \in \mathbb{Z}$. Suppose $b \in G$, and $b = m_1x_1 + m_2x_2 + \cdots + m_rx_r$. Then

$$\begin{aligned}
\phi(a + b) &= \phi((n_1 + m_1)x_1 + (n_2 + m_2)x_2 + \cdots + (n_r + m_r)x_r) \\
&= (n_1 + m_1, n_2 + m_2, \cdots, n_r + m_r) \\
&= (n_1, n_2, \cdots, n_r) + (m_1, m_2, \cdots, m_r) \\
&= \phi(a) + \phi(b)
\end{aligned}$$

so ϕ is a homomorphism. If $\phi(a) = \phi(b)$ then $n_i = m_i$ for $i = 1, 2, \cdots, r$ so $a = b$; this shows that ϕ is one to one. Clearly ϕ is an onto map since $n_1x_1 + n_2x_2 + \cdots + n_rx_r$ is in G for all integer choices of the coefficients n_i, for $i = 1, 2, \cdots, r$. Thus ϕ is an isomorphism.

8. Let G be free abelian generated by X. Let $a \neq 0$ in G be given by $a = n_1x_1 + n_2x_2 + \cdots + n_rx_r$ where $x_i \in X$ and $n_i \in \mathbb{Z}$ for $i = 1, 2, \cdots, r$. If a has finite order $m > 0$, then $ma =$

$mn_1x_1 + mn_2x_2 + \cdots + mn_rx_r = 0$. Since G is free abelian, we deduce that $mn_1 = mn_2 = \cdots = mn_r = 0$, so $n_1 = n_2 = \cdots = n_r = 0$ and $a = 0$, contradicting our choice of a. Thus no element of G has finite order > 0.

9. Let G and G' be free abelian with bases X and X' respectively. Let $\overline{X} = \{(x,0) \mid x \in X\}$ and $\overline{X}' = \{(0,x') \mid x' \in X'\}$. We claim that $Y = \overline{X} \cup \overline{X}'$ is a basis for $G \times G'$. Let $(g,g') \in G \times G'$. Then

$$g = n_1x_1 + \cdots + n_rx_r \quad \text{and} \quad g' = m_1x_1' + \cdots + m_sx_s'$$

for *unique* choices of the n_i and m_j, except for possible zero coefficients. Thus

$$(g,g') = n_1(x_1,0)+\cdots+n_r(x_r,0) + m_1(0, x_1')+\cdots+m_s(0, x_s')$$

for *unique* choices of the n_i and m_j, except for possible zero coefficients. This shows that Y is a basis for $G \times G'$, which is thus free abelian.

10. If G is free abelian of finite rank, then G is of course finitely generated, and by Exercise 8, G has no elements of finite order. Conversely, if G is a finitely generated torsion-free abelian group, then Theorem 2.11 shows that G is isomorphic to a direct product of the group $\mathbb{Z}$ with itself a finite number of times, so G is free abelian of finite rank.

11. Since $\mathbb{Q}$ is not cyclic, any basis for $\mathbb{Q}$ must contain at least two elements. Suppose n/m and r/s are in a basis for $\mathbb{Q}$ where n, m, r, and s are nonzero integers. Then $mr(\frac{n}{m}) + (-ns)(\frac{r}{s}) = rn - nr = 0$, which is an impossible relation in a basis. Thus $\mathbb{Q}$ has no basis, so it is not a free abelian group.

12. Suppose $p^r a = 0$ and $p^s b = 0$. Then $p^{r+s}(a + b) = p^s(p^r a) + p^r(p^s b) = p^s 0 + p^r 0 = 0 + 0 = 0$, so $a + b$ is also of p-power order. Also $0 = p^r 0 = p^r[a + (-a)] = p^r a + p^r(-a) = 0 + p^r(-a)$, so $-a$ has p-power order also. Thus all elements of T of p-power order, together with zero, form a subgroup T_p of T.

Section 4.4

13. It is clear from Theorem 2.11 that the elements of T of p-power order are precisely those having 0 in all components except those of the form $\mathbb{Z}_{p^r}$ in a prime-power decomposition. (Recall that the order of an element in a direct product is the lcm of the orders of its components in the individual groups.) Thus T_p is isomorphic to the direct product of those factors having p-power order.

14. Suppose $na = nb = 0$ for $a, b \in G$. Then $n(a + b) = n(a + b) = na + nb = 0 + 0 = 0$. This shows that $G[n]$ is closed under the group addition. If $na = 0$, then $0 = n0 = n[a + (-a)] = na + n(-a) = 0 + n(-a)$, so $n(-a) = 0$ also. Of course $n0 = 0$. Thus $G[n]$ is a subgroup of G.

15. Let $x \in \mathbb{Z}_{p^r}$. If $px = 0$, then px, computed in $\mathbb{Z}$, is a multiple of p^r. The possibilities for x are then $0, 1p^{r-2}, 2p^{r-1}, 3p^{r-1}, \cdots, (p - 1)p^{r-1}$. Clearly these elements form subgroup of $\mathbb{Z}_{p^r}$ that is isomorphic to $\mathbb{Z}_p$.

16. This follows at once from Exercise 15 and the fact that for abelian groups G_i with subgroups H_i, we have

$$(G_1 \times G_2 \times \cdots \times G_m)[p] = G_1[p] \times G_2[p] \times \cdots \times G_m[p]$$

and

$$(G_1 \times G_2 \times \cdots \times G_m)/(H_1 \times H_2 \times \cdots \times H_m) \simeq (G_1/H_1) \times (G_2/H_2) \times \cdots \times (G_m/H_m).$$

Both of these relations follow at once from the fact the computation in a direct product is performed in the component groups.

17. a) By Exercise 16, both m and n are $\log_p|T_p/T_p[p]|$.

b) Suppose that $r_1 < s_1$. Then the prime-power decomposition of the subgroup $p^{r_1}T_p$ computed in the first decomposition of T_p would have less than m factors while the decomposition of the same subgroup computed using the second decomposition of T_p would still have $m = n$ factors. But applying part (a) to this subgroup, we see that this is an impossible situation; the number of factors in the prime-power decomposition of an abelian p-power group is well defined. Thus $r_1 = s_1$.

Proceeding by induction, suppose that $r_i = s_i$ for all $i < j$, and suppose $r_j < s_j$. Multiplication of elements of $T_p^{r_j}$ by p annihilates all components in the first decomposition given of T_p through at least component i, while the component $\mathbb{Z}_{p^{s_i}}$ of the second decomposition given is not annihilated. This would contradict the fact that by part (a), the number of factors in the prime-power decomposition of any abelian p-group, in particular of $p^{r_j}T_p$, is well defined. Thus $r_j = s_j$ and our induction proof is complete.

18. If $m = p_1^{r_1} p_2^{r_2} \cdots p_k^{r_k}$, then we know that $\mathbb{Z}_m \simeq \mathbb{Z}_{p_1^{r_1}} \times \mathbb{Z}_{p_2^{r_2}} \times \cdots \times \mathbb{Z}_{p_k^{r_k}}$ from Section 2.4. If we form this decomposition of each factor in a torsion-coefficient decomposition, we obtain the unique prime-power decomposition.

19. Following the notation defined in the exercise, we know that if the unique prime-power decomposition is formed from a torsion-coefficient decomposition, as described in Exercise 18, then cyclic factors of order $p_i^{h_i}$ must appear for each $i = 1, \cdots, t$. Since each torsion coefficient except the final one must divide the following one, the final one must contain as factors all these prime powers $p_i^{h_i}$ for $i = 1, \cdots, t$. Since the p_i for $i = 1, \cdots, t$ are the only primes dividing $|T|$, we see that m_r and n_r must both be equal to the product of these highest prime powers for $i = 1, \cdots, t$ as asserted.

20. If we cross off the last factors $\mathbb{Z}_{m_r}$ and $\mathbb{Z}_{n_r}$ in the two given torsion-coefficient decompositions of T, we obtain torsion-coefficient decompositions of isomorphic groups, since both decompositions must, by Exercises 17 and 18, be isomorphic to the group obtained by crossing off from the prime-power decomposition of T one factor of order $p_i^{h_i}$ for $i = 1, \cdots, t$. (We are using the notation of Exercise 19 here.) We now apply the argument of Exercise 19 to these torsion-coefficient decompositions of this smaller group, and deduce that $m_{r-1} = n_{r-1}$. Continuing to cross off identical final factors, we see that we must have the same number of factors, that is, $r = s$, and $m_{r-i} = n_{r-i}$ for $i = 1, \cdots, r - 1$.

SECTION 4.5 - Free Groups

1. (See the text answer.)

2. a) a^5c^3, $a^{-5}c^{-3}$ b) $a^2b^3c^6$, $a^{-2}b^{-3}c^{-6}$

3. a) 16 since each generator can be mapped into any one of 4 elements by Theorem 4.24
   b) 36 by analogous reasoning to part (a)
   c) 36 by analogous reasoning to part (a)

4. a) Let the free group have generators x and y. By Theorem 4.24, x and y can be mapped into any elements to give a homomorphism. The homomorphism will be onto $\mathbb{Z}_4$ if and only if not both x and y are mapped into the subgroup $\{0, 2\}$. Since 4 of the 16 possible homomorphisms map x and y into $\{0, 2\}$, there are $16 - 4 = 12$ homomorphisms onto $\mathbb{Z}_4$.
   b) Arguing as in part (a), we eliminate the 4 homomorphisms that map x and y into $\{0, 3\}$ and the 9 that map x and y into $\{0, 2, 4\}$. The homomorphism mapping both x and y into 0 is counted in both cases, so there are a total of 12 of the possible homomorphisms to eliminate, leaving $36 - 12 = 24$.
   c) Arguing as in part (a), we eliminate the 4 homomorphisms that map x and y into $\{\rho_0, \mu_1\}$, the 4 that map x and y

into $\{\rho_0, \mu_2\}$, and the 4 that map x and y into $\{\rho_0, \mu_3\}$. Then we eliminate the 9 homomorphisms mapping x and y into $\{\rho_0, \rho_1, \rho_2\}$. The homomorphism that maps x and y into $\{0\}$ is counted four times, so we have a total of $12 + 9 - 3 = 18$ homomorphisms to eliminate, leaving $36 - 18 = 18$.

5. a) 16 by the same count as in Exercise 3a.
   b) 36 by the same count as in Exercise 3b.
   c) Since the homomorphic image of an abelian group must be abelian, the image must be either $\{0\}$, $\{\rho_0, \mu_1\}$, $\{\rho_0, \mu_2\}$, $\{\rho_0, \mu_3\}$, or $\{\rho_0, \rho_1, \rho_2\}$. The count made in Exercise 4c shows that there are 18 such homomorphisms.

6. a) 12 by the same count as in Exercise 4a.
   b) 24 by the same count as in Exercise 4b.
   c) 0, since the homomorphic image of an abelian group must be abelian, and S_3 is not abelian.

7. Our reaction to these instances was given in the text.

8. T F F T F F F T F T

9. a) $3(2) + 2(3) = 0$ but $3(2) \neq 0$ and $2(3) \neq 0$. A basis for $\mathbb{Z}_4$ is $\{1\}$.
   b) $\{1\}$ is a basis for $\mathbb{Z}_6$ since the group is cyclic with generator 1, and since $m1 = 0$ if and only if $m1 = 0$.
   c) It is clear that $\{2,3\}$ generates $\mathbb{Z}_6$ since $1 = 2(2) + 1(3)$. If $m_1 2 + m_2 3 = 0$, then in $\mathbb{Z}$, we know that 6 divides $m_1 2 + m_2 3$. Thus 3 divides $m_1 2 + m_2 3$, and hence 3 divides $m_1 2$. Since 3 is prime and does not divide 2, it must be that 3 divides m_1. Thus 6 divides $m_1 2$ in $\mathbb{Z}$, so $m_1 2 = 0$ in $\mathbb{Z}_6$. A similar argument starting with the fact that 2 divides $m_1 2 + m_2 3$ shows that $m_2 3 = 0$ in $\mathbb{Z}_6$. Thus $\{2, 3\}$ is a basis for $\mathbb{Z}_6$.
   c) Yes
   d) By Theorem 4.20, a finite abelian group G is isomorphic to a direct product $\mathbb{Z}_{m_1} \times \mathbb{Z}_{m_2} \times \cdots \times \mathbb{Z}_{m_r}$ where m_i divides divides m_{i+1} for $i = 1, 2, \cdots, r - 1$. Let b_i be the element of this direct product having 1 in the ith component and 0 in the other components. The computation

by components in a direct product shows at once that $\{b_1, b_2, \cdots, b_r\}$ is a basis, and since the order of each b_i is m_i, we see that the orders have the desired divisibility property.

10. a) Let $x \in G_1^*$. Since ϕ_1 is onto, we have $x = \phi_1(y)$ for some $y \in G$. Then $\theta_2(x) = \theta_2\phi_1(y) = \phi_2(y)$ and $\theta_1\theta_2(x) = \theta_1\phi_2(y) = \phi_1(y) = x$. In a similar fashion, starting with $z \in G_2^*$, we can show that $\theta_2\theta_1(z) = z$. Thus both $\theta_1\theta_2$ and $\theta_2\theta_1$ are identity maps. Since $\theta_2\theta_1$ is the identity, θ_2 must be an onto map and θ_1 must be one to one. Since $\theta_1\theta_2$ is the identity, θ_1 is an onto map and θ_2 is one to one. Thus both θ_1 and θ_2 are one to one and onto, and hence are isomorphisms. Thus G_1^* and G_2^* are isomorphic groups.

b) Let C be the commutator subgroup of G and let G^* be G/C and let $\phi: G \longrightarrow G/C$ be the canonical homomorphism (which we have usually called γ). If $\psi: G \longrightarrow G'$ is any homomorphism of G into an abelian group G', then the kernel K of ψ contains C by Theorem 3.15. Let $\delta: G/K \longrightarrow \psi[G]$ be the isomorphism given by Theorem 4.1. Let $\gamma: (G/C) \longrightarrow (G/C)/(K/C)$ be the canonical homomorphism, and let $\mu: (G/C)/(K/C) \longrightarrow G/K$ be the isomorphism given by Theorem 4.3. Then $\theta = \delta\mu\gamma$ is a homomorphism of G/C into G' and for $x \in G$, we have $\theta\phi(x) = \delta\mu\gamma\phi(x) = \delta\mu\gamma(x + C) = \delta\mu((x + C) + K/C) = \delta(x + K) = \psi(x)$ so $\psi = \theta\phi$.

c) A blip group of G is isomorphic to the *abelianized version* of G, that is, to G modulo its commutator subgroup.

11. a) Consider the blop group G_1 on S and let G' be the free group $F[S]$, with $f: S \longrightarrow F[S]$ given by $f(s) = s$ for $s \in S$. Since f is one to one and $f = \phi_f g_1$, we see that g_1 must be one to one. By a similar argument, g_2 must be one to one.

To see that $g_1[S]$ generates G_1, we first let G' be the subgroup of G_1 generated by $g_1[S]$ and let $f(s) = g_1(s)$ for $s \in S$. By hypothesis, there exists a homomorphism $\phi_f: G_1 \longrightarrow G'$ such that $\phi_f g_1 = g_1$. As a second choice for

G', we take $G' = G_1$ and again, we let $f(s) = g_1(s)$ for $s \in S$. Clearly this time the identity map $\iota: G_1 \longrightarrow G_1$ is a homomorphism satisfying $\iota(g_1) = g_1$. But the homomorphism ϕ_f for our first choice of G' is also a homomorphism of G_1 into G_1 satisfying satisfying $\phi_f g_1 =$ $= g_1$. By the *uniqueness* hypothesis, we must have $\phi_f = \iota$ on G_1, and since the image of ϕ_f is the subgroup generated by $g_1[S]$, we see that $g_1[S]$ does indeed generate all of G_1. Similarly, $g_2[S]$ generates all of G_2.

Taking $G' = G_2$ and $f = g_2$, we obtain a homomorphism $\phi_{g_2}: G_1 \longrightarrow G_2$ such that $\phi_{g_2} g_1 = g_2$. Taking $G' = G_1$ and $f = g_1$, we obtain a homomorphism $\phi_{g_1}: G_2 \longrightarrow G_1$ such that $\phi_{g_1} g_2 = g_1$. Let $g_1(s)$ be a generator of G_1. Then $\phi_{g_1}\phi_{g_2} g_1(s) = \phi_{g_1} g_2(s) = g_1(s)$, so $\phi_{g_1}\phi_{g_2}$ maps G_1 into itself and acts as the identity on a generating set for G_1. Therefore $\phi_{g_1}\phi_{g_2}$ is the identity map of G_1 onto itself. Similarly, $\phi_{g_2}\phi_{g_1}$ is the identity map of G_2 onto itself. As in Exercise 10, we conclude that ϕ_{g_1} and ϕ_{g_2} are isomorphisms, so that $G_1 \simeq G_2$.

b) Let $G = F[S]$, the free group on S, and let $g: S \longrightarrow G$ be defined by $g(s) = s$ for $s \in S$. Let a group G' and a function $f: S \longrightarrow G'$ be given. Let $\phi_f: G \longrightarrow G'$ be the the unique homomorphism given by Theorem 4.24 such that $\phi_f(s) = f(s)$. Then $\phi_f g(s) = \phi_f(s) = f(s)$ for all $s \in S$, so $\phi_f g = f$.

c) A blop group on S is isomorphic to the *free group* [S] *on* S.

12. The characterization is just like that in Exercise 11 with the requirement that both G and G' be abelian groups.

Section 4.6

SECTION 4.6 - Group Presentations

1. (See the text answer.)
2. Thinking of $a = \rho_1$, $b = \mu_1$, and $c = \rho_2$, we obtain the presentation $(a,b,c\colon a^3 = 1,\ b^2 = 1,\ c = a^2,\ ba = cb)$. Starting with this presentation, the relations can be used to express every word in one of the forms 1, a, b, a^2, ab, or a^2b, so a group with this presentation has at most 6 elements. Since the relations are satisfied by S_3, we know it must be a presentation of a group isomorphic to S_3. (Many other answers are possible.)
3. See the text answer.
4. Let G be nonabelian of order 14. By Sylow theory, there exists a unique subgroup H of order 7. Let b be an element of G that is not in H. Since $G/H \simeq \mathbb{Z}_2$, we see that $b^2 \in H$. If $b^2 \neq 1$, then b^2 has order 14 and G would be cyclic and abelian. Thus $b^2 = 1$. Let a be a generator for the cyclic group H. Now $bHb^{-1} = H$ so $bab^{-1} \in H$, so $ba = a^rb$ for some value of r where $1 \leq r \leq 6$. If $r = 1$, then $ba = ab$ and G would be abelian. By Exercise 11, part (b), the presentation

$$(a,b\colon\ a^7 = 1,\ b^2 = 1,\ ba = a^rb)$$

gives a group of order $2\cdot 7 = 14$ if and only if $r^2 \equiv 1 \pmod 7$. Of the values $r = 2,3,4,5,6$ only $r = 6$ satisfies this condition. Thus every nonabelian group of order 14 is isomorphic to the group with presentation $(a,b\colon a^7 = 1,\ b^2 = 1,\ ba = a^6b)$. Of course, $\mathbb{Z}_{14}$ is the only abelian group of order 14.
5. Let G be nonabelian of order 21. By Sylow theory, there exists a unique subgroup $H = \langle a\rangle$ of order 7. Let b be an element of G that is not in H. Since $G/H \simeq \mathbb{Z}_3$, we see that $b^3 \in H$. If $b^3 \neq 1$, then b has order 21 and G would by cyclic and abelian. Thus $b^3 = 1$. Now $bHb^{-1} = H$ so $bab^{-1} \in H$, so $ba = a^rb$ for some value of r where $1 \leq r \leq 6$. If $r = 1$, then

$ba = ab$ and G would be abelian. By Exercise 11, part (b), the presentation

$$(a,b\colon a^7 = 1,\ b^3 = 1,\ ba = a^r b)$$

gives a group of order $3 \cdot 7 = 21$ if and only if $r^3 \equiv 1 \pmod 7$. Of the values $r = 2,3,4,5,6$ both $r = 2$ and $r = 4$ satisfy this condition. To see that the presentations with $r = 2$ and $r = 4$ yield isomorphic groups, consider the group having this presentation with $r = 2$, and let us form a new presentation of it, taking the same a but replacing b by $c = b^2$. We then have $a^7 = 1$ and $c^3 = 1$, but now $ca = b^2 a = a^4 b^2 = a^4 c$. Thus in terms of the elements a and c, this group has presentation $(a,c\colon a^7 = 1,\ c^3 = 1,\ ca = a^4 c)$. This shows the two values $r = 2$ and $r = 4$ lead to isomorphic presentations. Thus every group of order 21 is isomorphic to either $\mathbb{Z}_{21}$ or to the group with presentation

$$(a,b\colon a^7 = 1,\ b^3 = 1,\ ba = a^2 b).$$

6. T T F F F T T F T F

7. Let G be a group of order 15. By Sylow theory, G has a unique subgroup H of order 5, which must be normal. Let $H = \langle a \rangle$ and let $b \in G$ but not in H. Then $a^5 = 1$, and $b \in H$. If $b^3 \neq 1$, then b has order 15 and G would be cyclic and hence abelian. Thus we suppose $b^3 = 1$. Now $bab^{-1} \in H$ since H is normal. If $bab^{-1} = a$, then $ba = ab$ and G would be abelian. Thus the possibilities are $ba = a^r b$ for $r = 2$, 3, or 4. Exercise 11, part (b), shows that $(a,b\colon a^5 = 1,\ b^3 = 1,\ ba = a^r b)$ is a presentation of a group group of order 15 if and only if $r^3 \equiv 1 \pmod 5$. But none of 2^3, 3^3, or 4^3 are congruent to 1 modulo 5, so there are no nonabelian groups of order 15.

8. Exercise 11, part (b), shows that the given presentation is a group of order $2 \cdot 3 = 6$ if and only if $2^2 \equiv 1 \pmod 3$, which is the case. Thus we do have a group. Since the elements $1,\ a,\ b,\ a^2,\ ab,\ a^2 b$ are all distinct and since $ba = a^2 b$, the group is not abelian, for if $ba = ab$, then $a^2 b = ab$ from which we deduce that $a = 1$.

Section 4.6

9. Let G be nonabelian of order 6. By Sylow theory, G has a unique subgroup H of order 3, which must be normal. Let $H = \langle a \rangle$ and let $b \in G$ not in H. Then $a^3 = 1$ and $b^2 \in H$. If $b^2 \neq 1$, then b has order 6 and G is cyclic and abelian. Thus $b^2 = 1$. Now $bab^{-1} \in H$, and if $bab^{-1} = a$, then $ba = ab$ and G is abelian. Thus $ba = a^2$. The preceding exercise shows that the presentation $(a,b\colon a^3 = 1,\ b^2 = 1,\ ba = a^2b)$ gives a nonabelian group of order 6, and this exercise shows that a nonabelian group of order 6 is isomorphic to one with this presentation. Thus every nonabelian group of order 6 is isomorphic to S_3.

10. Every element of A_4 can be written as a product of disjoint cycles involving some of the numbers 1, 2, 3, 4 and each element is also an even permutation. Since no product of such disjoint cycles can give an element of order 6, we see that A_4 has no elements of order 6 and hence no subgroup isomorphic to $\mathbb{Z}_6$, the only possibility for an abelian subgroup of order 6. Therefore, any subgroup of order 6 of A_4 must be nonabelian, and hence isomorphic to S_3 by the preceding exercise. Now S_3 has two elements of order 3 and three elements of order 2. The only *even* permutations in S_4 of order 2 are products of two disjoint transpostions, and the only such permutations are $(1,2)(3,4)$ and $(1,3)(2,4)$ and $(1,4)(2,3)$. Thus a subgroup of A_4 isomorphic to S_3 must contain all three of these elements. It must also contain an element of order 3 ; we might as as well assume that it is the 3-cycle $(1,2,3)$. Then it must contain $(1,2,3)^2 = (1,3,2)$, and the identity would be the sixth element. But this set is not closed under multiplication, for $(1,2)(3,4)(1,2,3) = (2,4,3)$. Thus A_4 has no nonabelian subgroup of order 6 either.

11. a) We know that when computing integer sums modulo n, we may either reduce modulo n after each addition, or add in $\mathbb{Z}$ and reduce modulo n at the end. The same is true for products, as we now show. Suppose $c = nq_1 + r_1$ and $d = = nq_2 + r_2$, both in accord with the division algorithm. Then

$$cd = n(nq_1q_2) + n(q_1r_2 + r_1q_2) + r_1r_2,$$

showing that the remainder of cd modulo n is the remainder of r_1r_2 modulo n. That is, it does not matter whether we first reduce modulo n and then multiply and reduce, or whether we multiply in $\mathbf{Z}$ and then reduce.

Turning to our problem and delaying reduction modulo m and n of sums and products in exponents to the end, we have

$$a^sb^t[(a^ub^v)(a^wb^z)] = a^sb^t[a^{u+wr^v}b^{v+z}]$$
$$= a^{s+(u+wr^v)r^t}b^{t+v+z} \qquad (1)$$

and

$$[(a^sb^t)(a^ub^v)]a^wb^z = [a^{s+ur^t}b^{t+v}]a^wb^z. \qquad (2)$$

Before we can continue this last computation, we must reduce the exponent $t + v$ modulo n, for in the next step, $t + v$ would appear as an exponent of an exponent, rather than as a sum or product of first exponents. Let $t + v = nq_1 + r_1$. Note that since both t and v lie in the range from 0 to $n - 1$, either $q_1 = 1$ or $q_1 = 0$. Continuing, we see the expression (1) is equal to

$$a^{s+ur^t+wr^{r_1}}b^{t+v+z}. \qquad (3)$$

Comparing (1) and (3), we see the associative law holds if and only if

$$s + (u + wr^v)r^t \equiv s + ur^t + wr^{r_1} \pmod{m}$$

and

$$t + v + z \equiv t + v + z \pmod{n}.$$

Of course this second condition is true, and the first one reduces to

$$wr^{v+t} \equiv wr^{r_1} \pmod{m}.$$

Now this relation must hold for all w where $0 \le w < m$ and for all v and t from 0 to $n - 1$. Taking $w = 1$ and $v + t = n$ so that $r_1 = 0$, we see that we must have $r^n \equiv 1 \pmod{m}$. On the other hand, if this is true, then

$$wr^{v+t} \equiv wr^{nq_1+r_1} \equiv w(r^n)^{q_1} r^{r_1} \equiv wr^{r_1} \pmod{m}.$$

This completes the proof.

b) Part (a) proved the associative law, and given $a^u b^v$, we can determine t and s successively by finding $t \equiv -v \pmod{n}$ and $s \equiv -u(r^t) \pmod{m}$ to produce $a^s b^t$ which is a left inverse of $a^u b^v$. Thus the "left" group axioms hold, so we have a group of order mn.

12. Let G be a group of order pq for p and q primes and $q > p$ and $q \equiv 1 \pmod{p}$. By Sylow theory, G contains a unique subgroup H of order q which is normal in G and cyclic, being of prime order. Let a be a generator of H and let $b \in G$ not in H. Then G/H is of order p so $b^p \in H$. If $b^p \neq 1$, then b is of order pq and G is cyclic and thus abelian, so for nonabelian G we must have $b^p = 1$. Now $bab^{-1} \in H$. If $bab^{-1} = a$, then $ba = ab$ and G is abelian. Thus bab^{-1} must be either a^2, a^3, $\cdots$, a^{q-1}. By Exercise 11, part (b), the exponents x from 2 to $q - 1$ such that the presentation $(a,b\colon a^q = 1,\ b^p = 1,\ ba = a^x b)$ gives a group of order pq are those such that $x^p \equiv 1 \pmod{q}$. Assuming as stated that the integers $1, 2, 3, \cdots, q - 1$ form a cyclic group of order $q - 1$ under multiplication $\pmod{q}$, we see that we are hunting for elements x in this group of order p. Since $q \equiv 1 \pmod{p}$, we see that $q - 1$ is divisible by p, so such elements exist, and the theory of cyclic groups shows that they are of the form $1, r, r^2, \cdots, r^{p-1}$. Thus the presentations

$$(a,b\colon a^q = 1,\ b^p = 1,\ ba = a^{(r^j)} b)$$

give groups of order pq for $j = 1, 2, \cdots, p - 1$. The hint in the exercises proves that all these presentations are isomorphic, so there is only one nonabelian group of order pq, up to isomorphism.

CHAPTER 5

INTRODUCTION TO RINGS AND FIELDS

SECTION 5.1 - Rings and Fields

1. 0 2. 16 3. 1 4. 22 5. (1, 6) 6. (2, 2)

7. Yes, commutative, no unity unless $n = 1$

8. No, no identity for addition

9. Yes, commutative with unity, not a field

10. Yes, commutative, no unity

11. Yes, commutative with unity, not a field

12. Yes, commutative with unity, a field since

$$\frac{1}{a + b\sqrt{2}} = \frac{1}{a + b\sqrt{2}} \cdot \frac{a + b\sqrt{2}}{a - b\sqrt{2}} = \frac{a}{a^2 + 2b^2} + \frac{b}{a^2 + 2b^2}\sqrt{2}.$$

13. No, not closed under multiplication

14. 1, -1 15. (1,1), (-1,-1), (1,-1), (-1,1)

16. 1, 2, 3, 4 17. All nonzero elements

18. $(1,q,1)$, $(1,q,-1)$, $(-1,q,1)$, $(-1,q,-1)$ for any $q \neq 0$ in $\mathbb{Q}$

19. 1,3

20. a) Each of the four entries can be any one of two elements, so there are $2^4 = 16$ matrices in $M_2(\mathbb{Z}_2)$.

    b) $\begin{bmatrix} 1 & 1 \\ 1 & 0 \end{bmatrix}\begin{bmatrix} 0 & 1 \\ 1 & 1 \end{bmatrix} = I_2$ while $\begin{bmatrix} 1 & 0 \\ 1 & 1 \end{bmatrix}$, $\begin{bmatrix} 1 & 1 \\ 0 & 1 \end{bmatrix}$, $\begin{bmatrix} 1 & 0 \\ 0 & 1 \end{bmatrix}$, $\begin{bmatrix} 0 & 1 \\ 1 & 0 \end{bmatrix}$ are all their own inverse. These six matrices are the units.

21. Let $\phi: \mathbb{Z} \to \mathbb{Z} \times \mathbb{Z}$ be defined by $\phi(n) = (n, 0)$. It is easily checked that ϕ is a homomorphism, but $\phi(1) = (1, 0)$ while the unity of $\mathbb{Z} \times \mathbb{Z}$ is $(1, 1)$.

22. Since $\det(A + B)$ need not equal $\det(A) + \det(B)$, we see that det is not a ring homomorphism. For example, $\det(I_n + I_n) = 2^n$ while $\det(I_n) + \det(I_n) = 1 + 1 = 2$.

23. Let $\phi: \mathbb{Z} \to \mathbb{Z}$ be a ring homomorphism. Then $\phi(1)$ must be an integer whose square is itself, namely either 0 or 1. If

$\phi(1) = 0$, then $\phi(n) = \phi(n \cdot 1) = n$, so ϕ is the identity map of $\mathbb{Z}$ onto itself which is a homomorphism. If $\phi(1) = 0$ then $\phi(n) = \phi(n \cdot 1) = 0$, so ϕ maps everything onto zero, which also yields a homomorphism.

24. As in the preceding solution, we see that for a ring homomorphism $\phi: \mathbb{Z} \longrightarrow \mathbb{Z} \times \mathbb{Z}$, we must have $\phi(1)^2 = \phi(1^2) = \phi(1)$. The only elements of $\mathbb{Z} \times \mathbb{Z}$ that are their own squares are (0, 0), (1, 0), (0, 1), and (1, 1). Thus the possibilities are $\phi_1(n) = (0, 0)$, $\phi_2(n) = (n, 0)$, $\phi_3(n) = (0, n)$, and $\phi_4(n) = (n, n)$. It is easily checked that these four maps are ring homomorphisms.

25. Since both (1, 0) and (0, 1) are their own squares, their images under a ring homomorphism $\phi: \mathbb{Z} \times \mathbb{Z} \longrightarrow \mathbb{Z}$ must also have this property, and thus must be either 0 or 1. Since (1, 0) and (0, 1) generate $\mathbb{Z} \times \mathbb{Z}$ as an additive group, this determines the possible values of the homomorphism on $(n, m) \in \mathbb{Z} \times \mathbb{Z}$. Thus the possibilities are given by $\phi_1(n, m) = 0$, $\phi_2(n, m) = n$, $\phi_3(n, m) = m$, and $\phi_4(n, m) = n + m$. It is easily checked that ϕ_1, ϕ_2, and ϕ_3 are homomorphisms. However, ϕ_4 is not a homomorphism since $n + m = \phi_4(n, m) = \phi_4((1, 1)(n, m)) \neq \phi_4(1, 1)\phi_4(n, m) = (1 + 1)(n + m) = 2(n + m)$.

26. As in Exercise 25, we see that the images of the additive generators (1, 0, 0), (0, 1, 0), and (0, 0, 1) under a ring homomorphism $\phi: \mathbb{Z} \times \mathbb{Z} \times \mathbb{Z} \longrightarrow \mathbb{Z}$ can only be 0 or 1. As shown in the argument for ϕ_4 in that exercise, mapping more than one of these generators into 1 will not give a ring homomorphism. Thus either they are all mapped into 0 giving the trivial homomorphism, or we have a projection homomorphism where one of the generators is mapped into 1 and the other two are mapped into 0. Thus there are four such ring homomorphisms ϕ.

27. (See the text answer.)

28. We have $x^2 + x - 6 = (x + 3)(x - 2)$. Trying all integers x from -6 to 7 to see if $(x + 3)(x - 2)$ is a multiple of 14, we we find that this happens for $x = -5, -3, 2$, and 4. Thus the elements 2, 4, 9, and 11 in $\mathbb{Z}_{14}$ are solutions. As in Exercise 27, it is possible to have a product of two nonzero

elements be 0.

29. $a = 2$, $b = 3$ in $\mathbb{Z}_6$

30. $\mathbb{Z} \times \mathbb{Z}$ has unity (1,1) while the subring $\mathbb{Z} \times \{0\}$ has unity (1,0). Also, $\mathbb{Z}_6$ has unity 1 while the subring {0, 2, 4} has unity 4 and the subring {0, 3} has unity 3.

31. T F F F T F T T T T

32. Let $f, g, h \in F$. Now $[(fg)h](x) = [(fg)(x)]h(x) = [f(x)g(x)]h(x)$. Since multiplication in $\mathbb{R}$ is associative, we continue with $[f(x)g(x)]h(x) = f(x)[g(x)h(x)] = f(x)[(gh)(x)] = [f(gh)](x)$. Thus $(fg)h$ and $f(gh)$ have the same value on each $x \in \mathbb{R}$, so they are the same function and axiom 2 holds. For axiom 3, we use the distributive laws in $\mathbb{R}$ and we have $[f(g + h)](x) = f(x)[(g + h)(x)] = f(x)[g(x) + h(x)] = f(x)g(x) + f(x)h(x) = (fg)(x) + (fh)(x) = (fg + fh)(x)$ so $f(g + h)$ and $fg + fh$ are the same function and the left distributive law holds. The right distributive law is proved similarly.

33. For $f, g \in F$, we have $\phi_a(f + g) = (f + g)(a) = f(a) + g(a) = \phi_a(f) + \phi_a(g)$. Turning to the multiplication, we have $\phi_a(fg) = (fg)(a) = f(a)g(a) = \phi_a(f)\phi_a(g)$. Thus ϕ_a is a homomorphism.

34. We need check only the multiplicative property.
Reflexive: The identity map ι of a ring R into itself satisfies $\iota(ab) = ab = \iota(a)\iota(b)$, so the reflexive property is satisfied.
Symmetric: Let $\phi: R \longrightarrow R'$ be an isomorphism. We know from group theory that $\phi^{-1}: R' \longrightarrow R$ is an isomorphism of additive group of R' into the additive group of R. For $\phi(a), \phi(b) \in R'$, we have $\phi^{-1}(\phi(a)\phi(b)) = \phi^{-1}(\phi(ab)) = ab = \phi^{-1}(\phi(a))\phi^{-1}(\phi(b))$.
Transitive: Let $\phi: R \longrightarrow R'$ and $\psi: R' \longrightarrow R''$ be ring isomorphisms. We know from Theorem 3.3 that $\psi\phi$ is an isomorphism of the additive group structure. For $a, b \in R$, we have $\psi\phi(ab) = \psi(\phi(ab)) = \psi(\phi(a)\phi(b)) = (\psi\phi(a))(\psi\phi(b))$, so $\psi\phi$ is again a ring isomorphism.

35. Let $u, v \in U$. Then there exist $s, t \in R$ such that $us = su = 1$ and $vt = tv = 1$. These equations show that s and t are also units in U. Then $(ts)(uv) = t(su)v = t1v = tv = 1$ and $(uv)(ts) = u(vt)s = u1s = us = 1$, so uv is again a unit and U

is closed under multiplication. Of course multiplication in U is associative since multiplication in R is associative. The equation $(1)(1) = 1$ shows that 1 is a unit. We showed above that a unit u in U has a multiplicative inverse s in U. Thus U is a group under multiplication.

36. Now $(a + b)(a - b) = a^2 + ab - ba - b^2$ is equal to $a^2 - b^2$ if if and only if $ab - ba = 0$, that is, if and only if $ab = ba$. But $ab = ba$ in R if and only if R is commutative.

37. We need only check the second and third ring axioms. For axiom 2, we have $(ab)c = 0c = 0 = a0 = a(bc)$. For axiom 3, we have $a(b + c) = 0 = 0 + 0 = ab + ac$ and $(a + b)c = 0 = 0 + 0 = ac + bc$.

38. If $\phi: 2\mathbb{Z} \longrightarrow 3\mathbb{Z}$ is an isomorphism, then by group theory for the additive groups we know that $\phi(2) = 3$ or $\phi(3) = -2$, so that either $\phi(2n) = 3n$ or $\phi(2n) = -3n$. Suppose $\phi(2n) = 3n$. Then $\phi(4) = 6$ while $\phi(2)\phi(2) = (3)(3) = 9$. Thus $\phi(2n) = 3n$ does not give an isomorphism and a similar computation shows that $\phi(2n) = -3n$ does not give an isomorphism either.

39. In a *commutative* ring, we have $(a + b)^2 = a^2 + ab + ba + b^2 = a^2 + ab + ab + b^2 = a^2 + 2ab + b^2$. Now the binomial theorem simply counts the number of each type of product $a^i b^{n-i}$ appearing in $(a + b)^n$. As long as our ring is commutative, every summand of $(a + b)^n$ can be written as a product of factors a and b with all the factors a written first, so the usual binomial expansion is valid in a commutative ring.

In $\mathbb{Z}_p$, the coefficient $\binom{p}{i}$ of $a^i b^{p-i}$ in the expansion of $(a + b)^p$ is a multiple of p if $1 \le i \le p - 1$. Since $p \cdot a = 0$ for all $a \in \mathbb{Z}_p$, we see that the only nonzero terms in the expansion are those corresponding to $i = 0$ and $i = p$, namely b^p and a^p.

40. Let F be a field, and suppose that $u^2 = u$ for nonzero $u \in F$. Multiplying by u^{-1}, we find that $u = 1$. This shows that 0 and 1 are the only idempotents in a field. Now let K be a subfield of F. The unity of K must be an idempotent in K, and hence also an idempotent in F, and hence must be the unity 1 of F.

41. Let u be a unit in a ring R. Suppose that $su = us = 1$ and $tu = ut = 1$. Then $s = s1 = s(ut) = (su)t = 1t = t$. Thus the inverse of a unit is unique.

42. a) If $a^2 = a$ and $b^2 = b$ and if the ring is commutative, then $(ab)^2 = abab = aabb = a^2b^2 = ab$, showing that the idempotents are closed under multiplication.

 b) By trying all elements, we find that the idempotents in $\mathbb{Z}_6$ are 0, 1, 3, and 4 while the idempotents in $\mathbb{Z}_{12}$ are 0, 1, 4, and 9. Thus the idempotents in $\mathbb{Z}_6 \times \mathbb{Z}_{12}$ are (0, 0), (0, 1), (0, 4), (0, 9), (1, 0), (1, 1), (1, 4), (1, 9), (3, 0), (3, 1), (3, 4), (3, 9), (4, 0), (4, 1), (4, 4), and (4, 9).

43. We have $P^2 = [A(A^TA)^{-1}A^T][A(A^TA)^{-1}A^T] =$ $A[(A^TA)^{-1}(A^TA)](A^TA)^{-1}A = AI_n(A^TA)^{-1}A^T = A(A^TA)^{-1}A^T = P$.

44. As explained in the answer to Exercise 39, the binomial expansion is valid in a commutative ring. Suppose that $a^n = 0$ and $b^m = 0$ in R. Now $(a + b)^{m+n}$ is a sum of terms containing as a factor a^ib^{m+n-i} for $0 \le i \le m + n$. If $i \ge n$, then $a^i = 0$ so each term with factor a^ib^{m+n-i} is zero. On the other hand, if $i < n$ then $m + n - i > m$ so $b^{m+n-i} = 0$ and each term a^ib^{m+n-i} is zero. Thus $(a + b)^{m+n} = 0$ so $a + b$ is nilpotent.

45. If R has no nonzero nilpotent element, then the only solution of $x^2 = 0$ is 0, for any nonzero solution would be a nilpotent element. Conversely, suppose that the only solution of $x^2 = 0$ is 0, and suppose that $a \neq 0$ is nilpotent. Let n be the smallest positive integer such that $a^n = 0$. If n is even, then $a^{n/2} \neq 0$ but $(a^{n/2})^2 = a^n = 0$ so $a^{n/2}$ is a nonzero solution of $x^2 = 0$, contrary to assumption. If n is odd, then $(a^{(n+1)/2})^2 = a^{n+1} = a^na = 0a = 0$ so $a^{(n+1)/2}$ is a nonzero solution of $x^2 = 0$ contrary to assumption. Thus R has no nonzero nilpotent elements.

Section 5.1

46. It is clear that if S is a subring of R, then all three of the conditions must hold. Conversely, suppose the conditions hold. The first two conditions and Exercise 30 of Section 1.3 show that $\langle S, +\rangle$ is an additive group. The final condition shows that multiplication is closed on S. Of course the associative and distributive laws hold for elements from S, since they actually hold for all elements in R. Thus S is a subring of R.

47. a) Let R be a ring and let $H_i \leq R$ for $i \in I$. Theorem 1.10 shows that $H = \bigcap_{i\in I} H_i$ is an additive group. Let $a, b \in H$. Then $a, b \in H_i$ for $i \in I$, so $ab \in H_i$ for $i \in I$ and therefore $ab \in H$. Thus H is closed under multiplication. Clearly the associative and distributive laws hold for elements from H, since they actually hold for all elements in R. Thus H is a subring of R.

    b) Let F be a field and let $K_i \leq F$ for $i \in I$. Part (a) shows that $K = \bigcap_{i\in I} K_i$ is a ring. Let $a \in K$. Then $a \in K_i$ for $i \in I$ so $a^{-1} \in K_i$ for $i \in I$ because Exercises 40 and 41 show that the unity in K is the same as in F and that inverses are unique. Therefore $a^{-1} \in K$. Of course multiplication in K is commutative since multiplication in F is commutative. Therefore K is a subfield of F.

48. We show that I_a satisfies the conditions of Exercise 46. Since $a0 = 0$ we see that $0 \in I_a$. Let $c, d \in I_a$. Then $ac = ad = 0$ so $a(c - d) = ac - ad = 0 - 0 = 0$; thus $(c - d) \in I_a$. Also $(cd)x = c(dx) = c0 = 0$ so $cd \in I_a$. This completes the check of the properties in Exercise 46.

49. Clearly a^n is in every subring containing a, so R_a contains a^n for every positive integer n. Consequently $\langle R_a, +\rangle$ contains the additive group G generated by $S = \{a^n \mid n \in \mathbb{Z}^+\}$. We claim that $G = R_a$. We need only show that G is closed under multiplication. Now G consists of zero and all sums of positive powers of a. By the distributive laws, the product of two elements that are sums of positive powers of a can again be written as a sum of positive powers of a, and is

hence again in G. Thus G is actually a subring containing a and contained in R_a so we must have $G = R_a$.

50. Example 10 shows that the function $\phi: \mathbb{Z}_{rs} \to \mathbb{Z}_r \times \mathbb{Z}_s$ where $\phi(a) = a\cdot(1, 1)$ is an isomorphism. Let $b = \phi^{-1}(m, n)$. Computing $b\cdot(1, 1)$ by components, we see that $1 + \cdots + 1$ for b summands yields m in $\mathbb{Z}_r$ and yields n in $\mathbb{Z}_s$. Thus, viewing b as an integer in $\mathbb{Z}$, we see that $b \equiv m \pmod{r}$ and $b \equiv n \pmod{s}$.

51. a) *Statement:* Let $b_1, b_2, \cdots, b_k$ be integers such that $\gcd(b_i, b_j) = 1$ for $i \neq j$. Then $\mathbb{Z}_{b_1 b_2 \cdots b_3}$ is isomorphic to $\mathbb{Z}_{b_1} \times \mathbb{Z}_{b_2} \times \cdots \times \mathbb{Z}_{b_n}$ with an isomorphism ϕ where $\phi(1) = (1, 1, \cdots, 1)$.

   Proof: By the hypotheses that $\gcd(b_i, b_j) = 1$ for $i \neq j$, we know that the image group is cyclic and that $(1, 1, \cdots, 1)$ generates the group. Since the domain group is cyclic generated by 1, we know that ϕ is an additive group isomorphism. It remains to show that $\phi(ms) = \phi(m)\phi(s)$ for m and s in the domain group. This follows from the fact that the ith component of $\phi(ms)$ in the image group is $(ms)\cdot 1$ which is equal to the product of m summands of 1 times s summands of 1 by the distributive laws in a ring.

   b) Let ϕ be the isomorphism in part (a), and let $c = \phi^{-1}(a_1, a_2, \cdots, a_n)$. Computing $\phi(c) = \phi(c\cdot 1)$ in the ith component, we see that the $1 + 1 + \cdots + 1$ for c summands in the ring $\mathbb{Z}_{b_i}$ yields a_i. Viewing c as an integer, this means that $c \equiv a_i \pmod{b_i}$.

52. All of the axioms need to verify that S is a division ring follow at once from the two given group statements and the given distributive laws except for the commutativity of addition. The left and then the right distributive laws yield $(1 + 1)(a + b) = (1 + 1)a + (1 + 1)b = a + a + b + b$. The right and then the left distributive laws yield $(1 + 1)(a + b) = 1(a + b) + 1(a + b) = a + b + a + b$. Thus $a + a + b + b = a + b + a + b$ and by cancellation in the additive group, we obtain $a + b = b + a$.

Section 5.2

53. Let a, $b \in R$ where R is a Boolean ring. We have $a + b = (a + b)^2 = a^2 + ab + ba + b^2 = a + ab + ba + b$. Thus in a Boolean ring, $ab = -ba$. Taking $b = a$, we see that $aa = -aa$, so $a = -a$. Thus every element is its own additive inverse, so $-ba = ba$ and from $ab = -ba$ we obtain $ab = ba$, showing that R is commutative.

54. a)

$+$	$\emptyset$	$\{a\}$	$\{b\}$	S
$\emptyset$	$\emptyset$	$\{a\}$	$\{b\}$	S
$\{a\}$	$\{a\}$	$\emptyset$	S	$\{b\}$
$\{b\}$	$\{b\}$	S	$\emptyset$	$\{a\}$
S	S	$\{b\}$	$\{a\}$	$\emptyset$

$\cdot$	$\emptyset$	S	$\{a\}$	$\{b\}$
$\emptyset$	$\emptyset$	$\emptyset$	$\emptyset$	$\emptyset$
S	$\emptyset$	S	$\{a\}$	$\{b\}$
$\{a\}$	$\emptyset$	$\{a\}$	$\{a\}$	$\emptyset$
$\{b\}$	$\emptyset$	$\{b\}$	$\emptyset$	$\{b\}$

b) Let A, $B \in P(S)$. Then $A + B = (A \cup B) - (A \cap B) = (B \cup A) - (B \cap A) = B + A$, so addition is commutative.

We check associativity of addition; it is easiest to think in terms of the elements in $(A + B) + C$ and the elements in $A + (B + C)$. By definition, the sum of two sets contains the elements in precisely one of the sets. Thus $A + B$ consists of the elements that are in either one of the sets A or B, but not in the other. Therefore, $(A + B) + C$ consists of the elements that are in precisely one of the three sets A, B, C. Clearly $A + (B + C)$ yields this same set, so addition is associative.

The empty set $\emptyset$ acts as additive identity, for $A + \emptyset = (A \cup \emptyset) - (A \cap \emptyset) = A - \emptyset = A$ for all $A \in P(S)$.

For $A \in S$, we have $A + A = (A \cup A) - (A \cap A) = A - A = \emptyset$, so each element of $P(S)$ is its own additive inverse. This shows that $\langle P(S), +\rangle$ is an abelian group.

For associativity of multiplication, we see that $(A\cdot B)\cdot C = (A \cap B) \cap C = A \cap (B \cap C) = A\cdot(B\cdot C)$.

For the left distributive law, we again think in terms of the elements in the sets. The set $A\cdot(B + C) = A \cap (B + C)$ consists of all elements of A that are in precisely one of the two sets B, C. This set thus contains all the elements in $A \cap B$ or in $A \cap C$, but not in both sets. This is precisely the set $(A \cap B) + (A \cap C) = (A\cdot B) + (A\cdot C)$. The right distributive law can be demonstrated by a similar argument.

SECTION 5.2 - Integral Domains

1. 0, 3, 5, 8, 9, 11 2. 3 in $\mathbb{Z}_7$, 16 in $\mathbb{Z}_{23}$

3. There are no solutions. 4. 2

5. 0 6. 0 7. 0 8. 3 9. 12 10. 30

11. $(a + b)^4 = a^4 + 4a^3b + 6a^2b^2 + 4ab^3 + b^4 = a^4 + 2a^2b^2 + b^4$

12. Since all the binomial coefficients $\binom{9}{i}$ for $1 \leq i \leq 8$ have 3 as a factor, we see that $(a + b)^9 = a^9 + b^9$.

13. $(a + b)^6 = a^6 + 6a^5b + 15a^4b^2 + 20a^3b^3 + 15a^2b^4 + 6ab^5 + b^6$
$= a^6 + 2a^3b^3 + b^6$

14. We have $\begin{bmatrix} 2 & -1 \\ 2 & -1 \end{bmatrix}\begin{bmatrix} 1 & 2 \\ 2 & 4 \end{bmatrix} = \begin{bmatrix} 0 & 0 \\ 0 & 0 \end{bmatrix}$.

15. F(n=1) T F(n=1) F T T F T F F

16. *Noncommutative ring without unity:* Upper-triangular 3x3 matrices with integer entries and all zeros on the main diagonal.
Commutative ring without unity: $2\mathbb{Z}$
Noncommutative ring with unity: $M_2(\mathbb{R})$
Commutative ring with unity, not an integral domain: $\mathbb{Z}_4$
Integral domain, not a field: $\mathbb{Z}$
Field: $\mathbb{Q}$

17. (See the text answer.)

18.

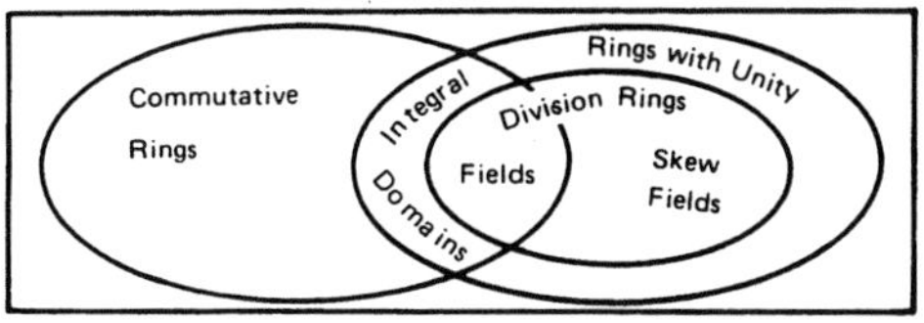

19. If $a^2 = a$, then $a^2 - a = a(a - 1) = 0$. If $a \neq 0$, then a^{-1} exists in R and we have $a^{-1}[a(a - 1)] = a^{-1}0 = 0$, from which we deduce that $a - 1 = 0$ so $a = 1$. Thus 0, 1 are the only two idempotent elements in a division ring.

20. Exercise 47(a) in Section 5.1 showed that an intersection of subrings of a ring R is again a subring of R. Thus an intersection of subdomains D_i for $i \in I$ of an integral domain D is at least a ring. The preceding exercise shows that unity in an integral domain can be characterized as the

nonzero idempotent. This shows that the unity in each D_i must be the unity 1 in D, so 1 is in the intersection of the D_i. Of course multiplication is commutative in the intersection since it is commutative in D and the operation is induced. Finally, if $ab = 0$ in the intersection, then $ab = 0$ in D so either $a = 0$ or $b = 0$, that is, the intersection has no divisors of zero, and is thus a subdomain of D.

21. Since R has no divisors of zero, multiplicative cancellation of nonzero elements is valid. The construction in the proof of Theorem 5.6 goes through and shows that each nonzero $a \in R$ has a right inverse, say a_i. A similar construction where the elements of R are all multiplied on the *right* by a shows that a has a left inverse, say a_j. By associativity of multiplication, we have $a_j = a_j(aa_i) = (a_ja)a_i = a_i$. Thus every nonzero $a \in R$ is a unit, so R is a division ring.

22. a) Let $a \neq 0$. We wish to show that a is not a divisor of zero. Let b be the unique element such that $aba = a$. Suppose $ac = 0$ or $ca = 0$. Then $a(b + c)a = aba + aca = a + 0 = a$. By uniqueness, $b + c = b$ so $c = 0$.
    b) From $aba = a$, we know that $b \neq 0$ also. Multiplying on the left by b, we obtain $baba = ba$. Since R has no divisors of zero by part (a), multiplicative cancellation is valid, and we see that $bab = b$.
    c) We claim that ab is unity for nonzero a and b given in the statement. Let $c \in R$. From $aba = a$, we see that $ca = caba$. Cancelling a, we obtain $c = c(ab)$. From part (b), we have $bc = babc$, and cancelling b yields $c = (ab)c$. Thus ab satisfies $(ab)c = c(ab) = c$ for all $c \in R$, so ab is unity.
    d) Let a be a nonzero element of the ring. By part (a), $aba = a$. By part (c), $ab = 1$ so b is a right inverse of a. Since the elements a and b behave in a symmetric fashion by part (b), an argument symmetric to that in part (c) shows that $ba = 1$ also. Thus b is also a left inverse of a, so a is a unit. This shows that R is a division ring.

23. By Exercise 19, we see that the unity in an integral domain can be characterized as the unique nonzero idempotent. This element in D must then also be the unity in every subdomain.

24. Let $R = \{n\cdot 1 \mid n \in \mathbb{Z}\}$. We have $n\cdot 1 + m\cdot 1 = (n+m)\cdot 1$ so R is

closed under addition. Taking $n = 0$, we see $0 \in R$. Since the inverse of $n \cdot 1$ is $(-n) \cdot 1$, we see that R contains all additive inverses of elements, so $\langle R, +\rangle$ is an abelian group. The distributive laws show that $(n \cdot 1)(m \cdot 1) = (nm) \cdot 1$, so R is closed under multiplication. Since $1 \cdot 1 = 1$, we see that $1 \in R$. Thus R is a commutative ring with unity. Since a product $ab = 0$ in R can also be viewed as a product in D, we see that R also has no divisors of zero. Thus R is a subdomain of D.

25. Suppose the characteristic is mn for $m > 1$ and $n > 1$. Following the hint, the distributive laws show that $(m \cdot 1)(n \cdot 1) = (mn) \cdot 1 = 0$. Since we are in an integral domain we must have either $m \cdot 1 = 0$ or $n \cdot 1 = 0$. But if $m \cdot 1 = 0$ then Theorem 5.7 shows that the characteristic of D is at most m. If $n \cdot 1 = 0$, the characteristic of D is at most n. Thus the characteristic can't be a composite positive integer, so it must either be 0 or a prime p.

26. a) From group theory, we know that S is an abelian group under addition. We check the associativity of multiplication, the fact that $n \cdot (m \cdot r) = (nm) \cdot r$, $n \cdot (r + s) = n \cdot r + n \cdot s$, $r(n \cdot s) = n \cdot (rs)$, and $(n \cdot r)s = n \cdot (rs)$ which follow from commutativity of addition and the distributive laws.

$$(r_1,n_1)[(r_2,n_2)(r_3,n_3)] = (r_1,n_1)(r_2r_3 + n_2 \cdot r_3 + n_3 \cdot r_2,\ n_2n_3)$$
$$= (r_1(r_2r_3 + n_2 \cdot r_3 + n_3 \cdot r_2) + n_1 \cdot (r_2r_3 + n_2 \cdot r_3 + n_3 \cdot r_2) + n_2n_3 \cdot r_1,\ n_1n_2n_3)$$
$$= (r_1r_2r_3 + n_1 \cdot r_2r_3 + n_2 \cdot r_1r_3 + n_3 \cdot r_1r_2 + n_1n_2 \cdot r_3 + n_1n_3 \cdot r_2 + n_2n_3 \cdot r_1,\ n_1n_2n_3).$$

and

$$[(r_1,n_1)(r_2,n_2)](r_3,n_3) = (r_1r_2 + n_1 \cdot r_2 + n_2 \cdot r_1,\ n_1n_2)(r_3,n_3)$$
$$= ((r_1r_2 + n_1 \cdot r_2 + n_2 \cdot r_1)r_3 + n_1n_2 \cdot r_3 + n_3(r_1r_2 + n_1 \cdot r_2 + n_2 \cdot r_1,\ n_1n_2n_3)$$
$$= (r_1r_2r_3 + n_1 \cdot r_2r_3 + n_2 \cdot r_1r_3 + n_3 \cdot r_1r_2 + n_1n_2 \cdot r_3 + n_1n_3 \cdot r_2 + n_2n_3 \cdot r_1,\ n_1n_2n_3).$$

Thus S is a ring.

b) We have $(0,1)(r,n) = (0r + 1 \cdot r + n \cdot 0,\ 1n) = (r,n) = (r0 + n \cdot 0 + 1 \cdot r,\ n1) = (r,n)(0,1)$, so $(0,1) \in S$ is unity.

c) Using Theorem 5.7 and (b), the ring S either has

characteristic 0 or the smallest positive integer n such that $(0,0) = n\cdot(0,1) = (0,n)$. Clearly n has this property if and only if $S = R \times \mathbb{Z}_n$. Since we chose $\mathbb{Z}$ or $\mathbb{Z}_n$ to form S according as R has characteristic 0 or n, we are done.

d) We have $\phi(r_1 + r_2) = \phi(r_1 + r_2,\ 0) = (r_1,\ 0) + (r_2,\ 0)$ and $\phi(r_1) + \phi(r_2)$ and $\phi(r_1r_2) = (r_1r_2,\ 0) = (r_1r_2 + 0\cdot r_2 + 0\cdot r_1, 00) = (r_1,\ 0)(r_2,\ 0) = \phi(r_1)\phi(r_2)$ so ϕ is a homomorphism. If $\phi(r_1) = \phi(r_2)$ then $(r_1,\ 0) = (r_2,\ 0)$ so $r_1 = r_2$; thus ϕ is one to one. Thus ϕ maps R isomorphically onto the subring $\phi[R]$ of S.

27. $3^2 = 9$ code words, $3^4 = 81$ received words

28. $16^2 = 256$ code words, $16^4 = 39936$ received words

SECTION 5.3 - Fermat's and Euler's Theorems

1. 3 (also 5) 2. 2 (also 6, 7, 8)

3. 3 (also 5,6,7,10,11,12,14)

4. $3^{47} \equiv (3^{22})^2\cdot 3^3 \equiv 1^2\cdot 27 \equiv 27 \equiv 4 \pmod{23}$

5. $37^{49} \equiv 2^{49} \equiv (2^6)^8\cdot 2 \equiv 1^8\cdot 2 \equiv 2 \pmod{7}$

6. $2^{17} \equiv (2^4)^4\cdot 2 \equiv (-2)^4\cdot 2 \equiv 16\cdot 2 \equiv 14 \pmod{18}$. Thus $2^{17} = 18q + 14$. Then $2^{(2^{17})} \equiv 2^{18q+14} \equiv (2^{18})^q\cdot 2^{14} \equiv 1^q\cdot 2^{14}$ $(2^4)^3\cdot 2^2 \equiv (-3)^3\cdot 2^2 \equiv -27\cdot 4 \equiv -8\cdot 4 \equiv 6 \pmod{19}$ so the answer is $6 + 1 = 7$.

7. (See the text answer.)

8. All positive integers less than p^2 that are not divisible by p are relatively prime to p. Thus we delete from the $p^2 - 1$ integers less than p^2 the integers $p, 2p, 3p, \cdots, (p - 1)p$. There are $p - 1$ integers deleted, so $\phi(p^2) = (p^2 - 1) - (p - 1) = p^2 - p$.

9. We delete from the $pq - 1$ integers less than pq those that

are multiples of p or of q to obtain those relatively prime to pq. The multiples of p are p, $2p$, $3p$, $\cdots$, $(q - 1)p$ and the multiples of q are q, $2q$, $3q$, $\cdots$, $(p - 1)q$. Thus we delete a total of $(q - 1) + (p - 1) = p + q - 2$ elements, so $\phi(pq) = (pq - 1) - (p + q + 2) = pq - p - q + 1 = (p - 1)(q - 1)$.

10. From Exercise 7, we find that $\phi(24) = 8$, so $7^8 \equiv 2 \pmod{24}$. Then $7^{1000} \equiv (7^8)^{125} \equiv 1^{127} \equiv 1 \pmod{24}$.

11. We can reduce the congruence to $2x \equiv 2 \pmod 4$. The gcd of 4 and 2 is $d = 2$ which divides $b = 2$. We divide by 2 and solve instead the congruence $x \equiv 1 \pmod 2$. Of course $x = 1$ is a solution. Another incongruent (mod 4) solution is $x = 1 + 2 = 3$. Thus the solutions are the numbers in $1 + 4\mathbf{Z}$ or $2 + 4\mathbf{Z}$.

12. We can reduce the congruence to $7x \equiv 5 \pmod{15}$. The gcd of 15 and 7 is $d = 1$ which divides $b = 5$. By inspection, $x = 5$ is a solution, and all solutions must be congruent to 5. Thus the solutions are the numbers in $5 + 15\mathbf{Z}$.

13. The congruence can be reduced to $12x \equiv 15 \pmod{24}$. The gcd of 24 and 12 is $d = 12$ which does not divide $b = 15$, so there are no solutions.

14. The congruence can be reduced to $21x \equiv 15 \pmod{24}$. The gcd of 24 and 21 is $d = 3$ which divides $b = 15$. We divide by 3 and solve instead the congruence $7x \equiv 5 \pmod 8$. By inspection, $x = 3$ is a solution. Other incongruent (mod 24) solutions are given by $x = 3 + 8 = 11$ and $x = 3 + 2 \cdot 8 = 19$. Thus the solutions are the number in $3 + 24\mathbf{Z}$, $11 + 24\mathbf{Z}$, or $19 + 24\mathbf{Z}$.

15. The congruence can be reduced to $3x \equiv 8 \pmod 9$. The gcd of 9 and 3 is $d = 3$ which does not divide $b = 8$, so there are no solutions.

16. The congruence can be reduced to $5x \equiv 8 \pmod 9$. The gcd of 9 and 5 is 1 which divides 8. By inspection $x = 7$ is a solution, and there are no other incongruent (mod 9) solutions, so solutions are the numbers in $7 + 9\mathbf{Z}$.

17. The congruence can be reduced to $25x \equiv 10 \pmod{65}$. The gcd of 65 and 25 is $d = 5$ which divides $b = 10$. We divide by 5 and solve instead the congruence $5x \equiv 2 \pmod{13}$. By inspection, $x = 3$ is a solution. Other solutions that are incongruent (mod 65) are $3 + 13 = 16$, $3 + 2 \cdot 13 = 29$, $3 + 3 \cdot 13 = 42$, and $3 + 4 \cdot 13 = 55$. Thus the solutions are the numbers in $3 + 65\mathbf{Z}$, $16 + 65\mathbf{Z}$, $29 + 65\mathbf{Z}$, $42 + 65\mathbf{Z}$, $55 + 65\mathbf{Z}$.

Section 5.3

18. The gcd of 130 and 39 is $d = 13$ which divides $b = 10$. We divide by 13 and solve instead $3x \equiv 4 \pmod{10}$. By inspection, $x = 8$ is a solution. Repeatedly adding 10 eleven times, we see that the solutions are the numbers in $8 + 130\mathbf{Z}$, $18 + 130\mathbf{Z}$, $28 + 130\mathbf{Z}$, $38 + 130\mathbf{Z}$, $48 + 130\mathbf{Z}$, $58 + 130\mathbf{Z}$, $68 + 130\mathbf{Z}$, $78 + 130\mathbf{Z}$, $88 + 130\mathbf{Z}$, $98 + 130\mathbf{Z}$, $108 + 130\mathbf{Z}$, $118 + 130\mathbf{Z}$, or $128 + 130\mathbf{Z}$.

19. Since $(p - 1)! = (p - 1)\cdot(p - 2)!$, Exercise 26 shows that $-1 \equiv (p - 1)\cdot(p - 2)! \pmod{p}$. Reducing $\pmod{p}$, we have the congruence $-1 \equiv (-1)(p - 2)! \pmod{p}$, so we must have $(p - 2)! \equiv 1 \pmod{p}$.

20. Taking $p = 37$ and using Exercise 26, we have

$$36! \equiv (36)(35)(34!) \equiv -1 \pmod{37}$$

so $(-1)(-2)(34!) \equiv -1 \pmod{37}$ and $2(34!) \equiv 36 \pmod{37}$. Thus $34! \equiv 18 \pmod{37}$.

21. Taking $p = 53$ and using Exercise 26, we have

$$52! \equiv (52)(51)(50)(49!) \equiv -1 \pmod{53}$$

so

$$(-1)(-2)(-3)(49!) \equiv -1 \pmod{53}$$

and $6(49!) \equiv 1 \pmod{53}$. By inspection, we see that $49! \equiv 9 \pmod{37}$.

22. Taking $p = 29$ and using Exercise 26, we have

$$28! \equiv (28)(27)(26)(25)(24!) \equiv -1 \pmod{29}$$

so

$$(-1)(-2)(-3)(-4)(24!) \equiv -1 \pmod{29}$$

and $(-5)(24!) \equiv -1 \pmod{29}$. By inspection, we see that $24! \equiv 6 \pmod{29}$.

23. F T T F T T F T F T

24.

	1	5	7	11
1	1	5	7	11
5	5	1	11	7
7	7	11	1	5
11	11	7	5	1

This group is isomorphic to $\langle \mathbf{Z}_2 \times \mathbf{Z}_2, +\rangle$.

25. If $a^2 = 1$, then $a^2 - 1 = (a - 1)(a + 1) = 0$. Since a field has no divisors of 0, either $a - 1 = 0$ or $a + 1 = 0$. Thus either $a = 1$ or $a = p - 1$.

26. Since $\mathbf{Z}_p$ is a field, for each factor in $(p - 1)!$, its inverse in $\mathbf{Z}_p$ is also a factor. In two cases, namely for the

factors 1 and $p - 1$, the inverse is the same factor while in the other cases, the inverse is a different factor. For $p \geq 3$ we see that $(p - 1)! \equiv (p - 1)\cdot\underbrace{(1)(1)\cdots(1)}_{\frac{p-2}{2}\ (1)\text{'s}} \pmod{p}$

so $(p - 1)! \equiv p - 1 \equiv -1 \pmod{p}$. For the case $p = 2$, we have $p - 1 \equiv 1 \equiv -1 \pmod{2}$.

27. We show that $n^{37} - n$ is divisible by each of the primes 37, 19, 13, 7, 3, and 2 for every positive integer n. By the corollary of Theorem 5.8, $a^p \equiv a \pmod{p}$ so of course $n^{37} \equiv n \pmod{37}$ for all n, so $n^{37} - n \equiv 0 \pmod{37}$ for all n, that is, $n^{37} - n$ is divisible by 37 for all n.

Working modulo 19, we have $n^{37} - n = n[(n^{18})^2 - 1]$. If n is divisible by 19, then so is $n^{37} - n$. If n is not divisible by 19, then by Fermat's theorem, $(n^{18})^2 - 1 \equiv 1^2 - 1 \equiv 0 \pmod{19}$, so again, $n^{37} - n$ is divisible by 19. Notice that the reason this argument works is that $36 = 37 - 1$ is a multiple of $18 = 19 - 1$.

Divisibility by 13, 7, 3, and 2 are handled in the same way, and the computatations are successful since

36 is a multiple of $13 - 1 = 12$,
36 is a multiple of $7 - 1 = 6$,
36 is a multiple of $3 - 1 = 2$,
36 is a multiple of $2 - 1 = 1$.

28. Looking at the argument in Exercise 27, we try to find still another prime p less than 37 such that 36 is divisible by $p - 1$. We see that $p = 5$ fills the bill, so $n^{37} - n$ is actually divisible by $5(383838) = 1919190$ for all integers n.

SECTION 5.4 - The Field of Quotients of an Integral Domain

1. $\{q_1 + q_2 i \mid q_1, q_2 \in \mathbb{Q}\}$

2. Since $\dfrac{1}{a + b\sqrt{2}} = \dfrac{1}{a + b\sqrt{2}} \cdot \dfrac{a + b\sqrt{2}}{a - b\sqrt{2}} = \dfrac{a}{a^2 + 2b^2} + \dfrac{b}{a^2 + 2b^2}\sqrt{2}$, we

see that $\{q_1 + q_2\sqrt{2} \mid q_1, q_2 \in \mathbb{Q}\}$ is a field, and must be the field of quotients.

3. T F T F T T F T T T

4. Let $D = \{q \in \mathbb{Q} \mid q = m/2^n$ for $m, n \in \mathbb{Z}\}$, that is, the set of all rational numbers that can be written as a quotient of integers with denominator a power of 2. It is easy to see that D is an integral domain. Let $D' = \mathbb{Z}$. Then $\mathbb{Q}$ is a field of quotients of both D and D'.

5. We have $[(a,b)] + ([c,d] + [(e,f)]) = [(a,b)] + [(cf+de,df)]$ $= [(adf+bcf+bde,bdf)] = [(ad+bc,bd)] + [(e,f)] =$ $([(a,b)] + [c,d]) + [(e,f)]$. Thus addition is associative.

6. We have $[(0,1)] + [(a,b)] = [(0b+1a,1b)] = [(a,b)]$. By Part 1 of Step 3, we also have $[(a,b)] + [(0,1)] = [(a,b)]$.

7. We have $[(-a,b)] + [(a,b)] = [(-ab+ba,b^2)] = [(0,b^2)]$. But $[(0,b^2)] \sim [(0,1)]$ since $(0)(1) = (b^2)(0) = 0$. Thus $[(-a,b))] + [(a,b)] = [(0,1)]$. By Part 1 of Step 3, $[(a,b)] + [(-a,b)] = [(0,1)]$ also.

8. We have $[(a,b)]([(c,d)][(e,f)]) = [(a,b)][(ce,df)] =$ $[(ace,bdf)] = [(ac,bd)][(e,f)] = ([(a,b)][(c,d)])[(e,f)]$. Thus multiplication is associative.

9. We have $[(a,b)][(c,d)] = [(ac,bd)] = [(ca,db)] =$ $[(c,d)][(a,b)]$ so multiplication is commutative.

10. For the left distributive law, we have $[(a,b)]([(c,d)] + [(e,f)]) = [(a,b)][(cf+de,df)] =$ $[(acf+ade,bdf)]$. Also, $[(a,b)][(c,d)] + [(a,b)][(e,f)]$ $= [(ac,bd] + [(ae,bf)] = [(acbf+bdae,bdbf)] \sim$ $[(acf+ade,bdf)]$ since $(acbf+bdae)bdf = acbfbdf+bdaebdf =$ $bdbf(acf+ade)$, for multiplication in D is commutative. The right distributive law then follows from Part 6.

11. a) Since T is nonempty, there exists $a \in T$. Then $[(a,a)]$ is unity in $Q(R,T)$, for $[(a,a)][(b,c)] = [(ab,ac)] \sim [(b,c)]$ since $abc = acb$ in the commutative ring R.
    b) A nonzero element $a \in T$ is identified with $[(aa,a)]$ in $Q(R,T)$. Since T has no divisors of zero, $[(a,aa)] \in Q(R,T)$, and we see that $[(aa,a)][(a,aa)] = [(aaa,aaa)] \sim$ $[(a,a)]$ since $aaaa = aaaa$. We saw in (a) that $[(a,a)]$ is unity in $Q(R,T)$. Commutativity of $Q(R,T)$ shows that $[(a,aa)][(aa,a)]$ is unity also, so $a \in T$ has an inverse in $Q(R,T)$ if $a \neq 0$.

12. We need only take $T = \{a\}$ in Exercise 11. This construction is entirely different from the one in Exercise 26 of Section 5.2.

13. There are four elements, for 1 and 3 are already units in $\mathbb{Z}_4$.

14. It is isomorphic to the ring D of all rational numbers that can be expressed as a quotient of integers with denominator a power of 2, as described in the answer to Exercise 4.

15. It is isomorphic to the ring of all rational numbers that can be expressed as a quotient of integers with denominator a power of 6. The 3 in the $3\mathbb{Z}$ does not restrict the numerator, since 1 can be recovered as $[(6,6)]$, 2 as $[(12,6)]$, etc. We also see that the denominator need not be restricted to a positive power of 6; for example, $3/6^{-2}$ can be recovered as $[(3 \cdot 6^3, 6)]$.

16. It runs into trouble when we try to prove the transitive property in the proof of Lemma 5.1, for multiplicative cancellation may not hold. For $R = \mathbb{Z}_6$ and $T = \{1,2,4\}$ we have $(1,2) \sim (2,4)$ since $(1)(4) = (2)(2) = 4$ and $(2,4) \sim (2,1)$ since $(2)(1) = (4)(2)$ in $\mathbb{Z}_6$. However, $(1,2)$ is not equivalent to $(2,1)$ since $(1)(1) \neq (2)(2)$ in $\mathbb{Z}_6$.

SECTION 5.5 - Rings of Polynomials

1. $f(x) + g(x) = 2x^2 - 3$, $f(x)g(x) = 6x^2 + 4x + 6$

2. $f(x) + g(x) = 0$, $f(x)g(x) = x^2 + 1$

3. $f(x) + g(x) = 5x^2 + 5x + 1$, $f(x)g(x) = x^3 + 5x$

4. $f(x) + g(x) = 3x^4 + 2x^3 + 4x^2 + 1$,
   $f(x)g(x) = x^7 + 2x^6 + 4x^5 + x^3 + 2x^2 + x + 3$

5. Such a polynomial is of the form $ax^3 + bx^2 + cx + d$ where each of a, b, c, d may be either 0 or 1. Thus there are $2 \cdot 2 \cdot 2 \cdot 2 = 16$ such polynomials in all.

6. Such a polynomial is of the form $ax^2 + bx + c$ where each of a, b, c may be either 0, 1, 2, 3 or 4. Thus there are

$5 \cdot 5 \cdot 5 = 125$ such polynomials in all.

7. $\phi_2(x^2 + 3) = 2^2 + 3 = 4 + 3 = 0$

8. $\phi_0(2x^3 - x^2 + 3x + 2) = 0 - 0 + 0 + 2 = 2$

9. $\phi_3[(x^4+2x)(x^3-3x^2+3)] = \phi_3(x^4+2x)\phi_3(x^3-3x^2+3) =$
$(3^4+6)(3^3-3^3+3) = (4+6)(3) = 2$

10. $\phi_5[(x^3+2)(4x^2+3)(x^7+3x^2+1)] =$
$\phi_5(x^3+2)\phi_5(4x^2+3)\phi_5(x^7+3x^2+1) = (6+2)(2+3)(5+5+1)$
$= (1)(5)(4) = 6$

11. $\phi_4(3x^{106} + 5x^{99} + 2x^{53}) = 3(4)^{106} + 5(4)^{99} + 2(4)^{53} =$
$3(4^6)^{17}4^4 + 5(4^6)^{16}4^3 + 2(4^6)^8 4^5 = 3(1)4 + 5(1)1 + 2(1)2$
$= 5 + 5 + 4 = 0$

12. $1^2 + 1 = 0$ but $0^2 + 1 \neq 0$, so 1 is the only zero.

13. Let $f(x) = x^3 + 2x + 2$. Then $f(0) = 2$, $f(1) = 5$, $f(2) = 0$, $f(3) = 0$, $f(-3) = 4$, $f(-2) = 4$, and $f(-1) = 6$ so 2 and 3 are the only zeros.

14. Let $f(x) = x^5 + 3x^3 + x^2 + 2x$. Then $f(0) = 0$, $f(1) = 2$, $f(2) = 4$, $f(-2) = 4$, and $f(-1) = 0$ so 0 and 5 are the only zeros.

15. Since $\mathbb{Z}_7$ is a field, $f(a)g(a) = 0$ if and only if either $f(a) = 0$ or $g(a) = 0$. Let $f(x) = x^3 + 2x^2 + 5$ and $g(x) = 3x^2 + 2x$. Then $f(0) = 5$, $f(1) = 1$, $f(2) = 0$, $f(3) = 1$, $f(-3) = 3$, $f(-2) = 5$, and $f(-1) = 6$ while $g(0) = 0$, $g(1) = 5$, $g(2) = 2$, $g(3) = 5$, $g(-3) = 0$, $g(-2) = 1$, and $g(-1) = 1$. Thus $f(x)g(x)$ has 0, 2, and 4 as its only zeros.

16. $\phi_3(x^{231} + 3x^{117} - 2x^{53} + 1) = 3^{231} + 3^{118} - 2(3^{53}) + 1$
$= (3^4)^{57}3^3 + (3^4)^{29}3^2 - 2(3^4)^{13}3 + 1 = 3^3 + 3^2 - 2(3) + 1$
$= 2 + 4 - 1 + 1 = 1$

17. Let $f(x) = 2x^{219} + 3x^{74} + 2x^{57} + 3x^{44}$
$= 2(x^4)^{54}x^3 + 3(x^4)^{18}x^2 + 2(x^4)^{14}x + 3(x^4)^{11}$. Then $f(0) = 0$,

$f(1) = 2 + 3 + 2 + 3 = 0$, $f(2) = 1 + 2 + 4 + 3 = 0$, $f(-2) = -1 + 2 - 4 + 3 = 0$, and $f(-1) = -2 + 3 - 2 + 3 = 2$. Thus 0, 1, 2, and 3 are zeros of $f(x)$.

18. $f(x,y) = (3x^3+2x)y^3 + (x^2-6x+1)y^2 + (x^4-2x)y + (x^4-3x^2+2) = (y+1)x^4 + 3y^3x^3 + (y^2-3)x^2 + (2y^3-6y^2-2y)x + (y^2+2)$

19. (See the text answer.)

20. $2x + 1$ is a unit since $(2x + 1)^2 = 1$.

21. T T T T F F T T T F

22. Let $f(x) = a_nx^n + a_{n-1}x^{n-1} + \cdots + a_1x + a_0$ and $g(x) = b_mx^m + b_{m-1}x^{m-1} + \cdots + b_1x + b_0$ be polynomials in $D[x]$ with a_n and b_m both nonzero. Since D is an integral domain, we know that $a_nb_m \neq 0$, so $f(x)g(x)$ is nonzero since its term of highest degree has coefficient a_nb_m. As stated in the text, $D[x]$ is a commutative ring with unity, and we have shown it has no divisors of zero, so it is an integral domain.

23. a) The units in $D[x]$ are the units in D since a polynomial of degree n times a polynomial of degree m is a polynomial of degree nm, as proved for the preceding exercise. Thus a polynomial of degree 1 cannot be multiplied by anything in $D[x]$ to give 1, which is a polynomial of degree 0.
b) They are the units in $\mathbb{Z}$, namely 1 and -1.
c) They are the units in $\mathbb{Z}_7$, namely 1, 2, 3, 4, 5, and 6.

24. Let

$$f(x) = \sum_{i=0}^{\infty} a_ix^i, \quad g(x) = \sum_{i=0}^{\infty} b_ix^i, \quad \text{and } h(x) = \sum_{j=0}^{\infty} c_jx^j.$$

Then $h(x)(f(x) + g(x)) = \left[\sum_{j=0}^{\infty} c_jx^j\right]\left[\sum_{i=0}^{\infty} (a_i + b_i)x^i\right]$

$$= \sum_{n=0}^{\infty}\left[\sum_{i=0}^{n} c_i(a_{n-i} + b_{n-i})x^n\right]$$

$$= \sum_{n=0}^{\infty}\left[\sum_{i=0}^{n} c_ia_{n-i}x^i\right] + \sum_{n=0}^{\infty}\left[\sum_{i=0}^{n} c_ib_{n-i}x^i\right] = h(x)f(x) + h(x)g(x)$$

so the left distributive law holds.

Section 5.5

25. a) Let $f(x)$ and $g(x)$ be polynomials in $F[x]$ and let n be the maximum of the degrees of the two polynomials. Allowing zero coefficients, we can write

$$f(x) = a_n x^n + a_{n-1}x^{n-1} + \cdots + a_1 x + a_0$$

and

$$g(x) = b_n x^n + b_{n-1}x^{n-1} + \cdots + b_1 x + b_0.$$

Since F is of characteristic zero, we can regard $n\cdot 1$ in F as the positive integer n, that is, we can consider $\mathbf{Q}$ as being contained in $F[x]$. Then

$$D[f(x) + g(x)] = D[(a_n+b_n)x^n+(a_{n-1}+b_{n-1})x^{n-1}+\cdots+(a_1+b_1)x + (a_0+b_0)]$$

$$= n(a_n+b_n)x^{n-1} + (n-1)(a_{n-1}+b_{n-1})x^{n-2}+ \cdots + (a_1+b_1)$$

$$= [na_n x^{n-1}+(n-1)a_{n-1}x^{n-2}+\cdots+a_1] + [nb_n x^{n-1}+(n-1)b_{n-1}x^{n-2}+ \cdots + b_1]$$

$$= D[f(x)] + D[g(x)]$$

so D is a homomorphism of $\langle F[x], +\rangle$.

b) The kernel of D is F. [This is not true if F has characteristic p, for then $D(x^p) = 0$.]

c) The image of D is $F[x]$ since D is additively a homomorphism with $D(1) = 0$ and $D(\frac{1}{i+1}a_i x^{i+1}) = a_i x^i$.

26. a) $\phi_{\alpha_1,\cdots,\alpha_n}(f(x_1, \cdots, x_n))$ is the element of F obtained by replacing each x_i by α_i in the polynomial and computing in E the resulting sum of products. That is, $\phi_{\alpha_1,\cdots,\alpha_n}(f(x_1, \cdots, x_n)) = f(\alpha_1, \cdots, \alpha_n)$. This gives a map $\phi_{\alpha_1,\cdots,\alpha_n} : F[x_1, \cdots, x_n] \rightarrow E$ which is a homomorphism and maps F isomorphically by the identity map, that is, $\phi_{\alpha_1,\cdots,\alpha_n}(a) = a$ for $a \in F$.

b) $\phi_{-3,2}(x_1^2 x_2^3 + 3x_1^4 x_2) = (9)(8) + 3(81)(2) = 72 + 486 = 558$.

c) Let F be a subfield of E. Then $(\alpha_1, \cdots, \alpha_n)$ in

$\underbrace{E \times E \times \cdots \times E}_{n \text{ factors}}$ is a zero of $f(x_1, \cdots, x_n) \in$
$F[x_1, \cdots, x_n]$ if $\phi_{\alpha_1, \cdots, \alpha_n}(f(x_1, \cdots, x_n)) = 0$.

27. *(Addition associative)* Let $\phi, \psi, \mu \in R^R$. Then $[(\phi + \psi) + \mu](r) = (\phi + \psi)(r) + \mu(r) = \phi(r) + \psi(r) + \mu(r) = \phi(r) + (\psi + \mu)(r) = [\phi + (\psi + \mu)](r)$ since addition in R is associative. Since $(\phi + \psi) + \mu$ and $\phi + (\psi + \mu)$ have the same value on each $r \in R$, they are the same function.
(Identity for +) The function ϕ_0 such that $\phi_0(r) = 0$ for all $r \in R$ acts as additive identity, for $(\phi_0 + \psi)(r) = \phi_0(r) + \psi(r) = 0 + \psi(r) = \psi(r)$. Since ϕ_0 and $\phi_0 + \psi$ have the same value on each $r \in R$, we see that they are the same function. A similar argument shows that $\psi + \phi = \psi_0$.
(Additive inverse) Given $\phi \in R^R$, the function $-\phi$ defined by $(-\phi)(r) = -(\phi(r))$ for $r \in R$ is the additive inverse of ϕ, for $(\phi + (-\phi))(r) = \phi(r) + (-\phi)(r) = \phi(r) + -(\phi(r)) = 0 = \phi_0(r)$, so $\phi + (-\phi) = \phi_0$. A similar argument shows that $(-\phi) + \phi = \phi_0$.
(Addition commutative) We have $(\phi + \psi)(r) = \phi(r) + \psi(r) = \psi(r) + \phi(r) = (\psi + \phi)(r)$ since addition in R is commutative. Thus $\phi + \psi$ and $\psi + \phi$ are the same function, so $\phi + \psi = \psi + \phi$.
(Multiplication associative) We have $[(\phi\cdot\psi)\cdot\mu](r) = [(\phi\cdot\psi)(r)]\mu(r) = [\phi(r)\psi(r)]\mu(r) = \phi(r)[\psi(r)\mu(r)] = \phi(r)[(\psi\cdot\mu)(r)] = [\phi\cdot(\psi\cdot\mu)](r)$ since multiplication in R is associative. Thus $\phi\cdot(\psi\cdot\mu) = (\phi\cdot\psi)\cdot\mu$ since the functions have the same value on each $r \in R$.
(Left distributive law) We have $[\phi\cdot(\psi + \mu)](r) = \phi(r)[(\psi + \mu)(r)] = \phi(r)[\psi(r) + \mu(r)] = \phi(r)\psi(r) + \phi(r)\mu(r) = (\phi\cdot\psi)(r) + (\phi\cdot\mu)(r) = [(\phi\cdot\psi) + (\phi\cdot\mu)](r)$ since the left distributive law holds in R. Thus $\phi\cdot(\psi + \mu) = \phi\cdot\psi + \phi\cdot\mu$ since these functions have the same value at each $r \in R$.
(Right distributive law) The proof is analogous to that for the left distributive law.

28. a) The map $\mu\colon F[x] \to P_F$ where $\mu(f(x))$ is the function ϕ such that $\phi(a) = f(a)$ for all $a \in F$ is easily seen to be a homomorphism, and by definition, $P_F = \phi[F[x]]$. Thus P_F is is the homomorphic image of a ring, which we know is at least an abelian group under addition, and is easily seen to be a ring. (See Theorem 6.1.)

b) Let F be the finite field $\mathbb{Z}_2$. A function in $\mathbb{Z}_2^{\mathbb{Z}_2}$ has just has just two elements in both its domain and range. Thus there are only $2^2 = 4$ such functions in all. However, $\mathbb{Z}_2[x]$ is an infinite set, so it isn't isomorphic to $P_{\mathbb{Z}_2}$.

29\. a) $2^2 = 4$ and $3^3 = 27$ respectively.

b) Since $(\phi + \phi)(a) = \phi(a) + \phi(a) = 2\cdot\phi(a) = 0$ in $\mathbb{Z}_2$, we see that every element of $\mathbb{Z}_2^{\mathbb{Z}_2}$ is its own additive inverse, so this additive group of order 4 must be isomorphic to $\mathbb{Z}_2 \times \mathbb{Z}_2$. Similarly, if $\phi \in \mathbb{Z}_3^{\mathbb{Z}_3}$ then $3\cdot\phi(a) = 0$ for all $a \in \mathbb{Z}_3$. Since this group is abelian, we see that it must be isomorphic to $\mathbb{Z}_3 \times \mathbb{Z}_3 \times \mathbb{Z}_3$.

c) Note that the polynomial $f_i(x)$ defined in the hint satisfies $f_i(a_j) = 0$ if $j \neq i$ while $f_i(a_i) = c$. Let $\phi \in F^F$ and suppose that $\phi(a_i) = c_i$. Let $f(x) \in F$ be defined by $f(x) = \sum_{i=1}^{n} c_i f_i(x)$. Then $f(a_k) =$ $f_1(a_k) + f_2(a_k) + \cdots + f_k(a_k) + \cdots + f_n(a_k) =$ $0 + 0 + \cdots + c_k + \cdots + 0 = c_k$. Thus $f(a_k) = \phi(a_k)$ for $k = 1, 2, \cdots, n$ so every function ϕ in F^F is a polynomial function. That is, $F^F = P_F$.

SECTION 5.6 - Factorization of Polynomials Over a Field

1.

$$
\begin{array}{r}
x^4 + x^3 + x^2 + x - 2 = q(x) \\
x^2 + 2x - 3 \,\big)\, \overline{x^6 + 3x^5 + 4x^2 - 3x + 2} \\
\underline{x^6 + 2x^5 - 3x^4} \\
x^5 + 3x^4 \\
\underline{x^5 + 2x^4 - 3x^3} \\
x^4 + 3x^3 + 4x^2 \\
\underline{x^4 + 2x^3 - 3x^2} \\
x^3 - 3x \\
\underline{x^3 + 2x^2 - 3x} \\
-2x^2 + 2 \\
\underline{-2x^2 - 4x + 6} \\
4x + 3 = r(x)
\end{array}
$$

2.

$$
\begin{array}{r}
5x^4 + 5x^2 - x = q(x) \\
3x^2 + 2x - 3 \,\big)\, \overline{x^6 + 3x^5 + 4x^2 - 3x + 2} \\
\underline{x^6 + 3x^5 + 6x^4} \\
x^4 + 4x^2 \\
\underline{x^4 + 3x^3 + 6x^2} \\
-3x^3 - 2x^2 - 3x \\
\underline{-3x^3 - 2x^2 + 3x} \\
x + 2 = r(x)
\end{array}
$$

3.

$$
\begin{array}{r}
6x^4 + 7x^3 + 2x^2 - x + 2 = q(x) \\
2x + 1 \,\big)\, \overline{x^5 - 2x^4 + 3x - 5} \\
\underline{x^5 + 6x^4} \\
3x^4 \\
\underline{3x^4 + 7x^3} \\
4x^3 \\
\underline{4x^3 + 2x^2} \\
-2x^2 + 3x \\
\underline{-2x^2 - x} \\
4x - 5 \\
\underline{4x + 2} \\
4 = r(x)
\end{array}
$$

Section 5.6

4.

$$
\begin{array}{r|l}
 & 9x^2 + 5x + 10 = q(x) \\
\hline
5x^2 - x + 2 & x^4 + 5x^3 - 3x^2 \\
 & \underline{x^4 + 2x^3 + 7x^2} \\
 & \qquad 3x^3 + x^2 \\
 & \qquad \underline{3x^3 - 5x^2 + 10x} \\
 & \qquad\qquad 6x^2 + x \\
 & \qquad\qquad \underline{6x^2 - 10x + 9} \\
 & \qquad\qquad\qquad\quad 2 = r(x)
\end{array}
$$

5. Trying $2 \in \mathbb{Z}_5$, we find that $2^2 = 4$, $2^3 = 3$, $2^4 = 1$, so 2 generates the multiplicative subgroup $\langle 2 \rangle = \{1, 2, 3, 4\}$ of all units in $\mathbb{Z}_5$. By the corollary of Theorem 1.9, the only generators are $2^1 = 2$ and $2^3 = 3$.

6. Trying $2 \in \mathbb{Z}_7$, we find that $2^3 = 1$, so 2 does not generate. Trying 3, we find that $3^2 = 2$, $3^3 = 6$, $3^4 = 4$, $3^5 = 5$, and $3^6 = 1$, so 3 generates the six units 1, 2, 3, 4, 5, 6 in $\mathbb{Z}_7$. By the corollary of Theorem 1.9, the only generators are $3^1 = 3$ and $3^5 = 5$.

7. Trying $2 \in \mathbb{Z}_{17}$, we find that $2^4 = -1$ so $2^8 = 1$ and 2 does not generate. Trying 3, we find that $3^2 = 9$, $3^3 = 10$, $3^4 = 13$, $3^5 = 5$, $3^6 = 15$, $3^7 = 11$, $3^8 = 15 = -1$. Since the order of 3 must divide 16, we see that 3 must be of order 16, so 3 generates the units in $\mathbb{Z}_{17}$. By the corollary of Theorem 1.9, the only generators are $3^1 = 3$, $3^3 = 10$, $3^5 = 5$, $3^7 = 11$, $3^9 = 14$, $3^{11} = 7$, $3^{13} = 12$, and $3^{15} = 6$.

8. Trying $2 \in \mathbb{Z}_{23}$, we find that $2^2 = 4$, $2^3 = 8$, $2^4 = 16$, $2^5 = 9$, $2^6 = 18$, $2^7 = 13$, $2^8 = 3$, $2^9 = 6$, $2^{10} = 12$, and $2^{11} = 1$, so 2 does not generate. However, this computation shows that $(-2)^{11} = -1$. Since the order of -2 must divide 22, we see that $21 = -2$ must be of order 22, so 21 generates the

units of $\mathbb{Z}_{23}$. By the corollary of Theorem 1.9, the only generators are $21^1 = 21$, $21^3 = 15$, $21^5 = 14$, $21^7 = 10$, $12^9 = 17$, $21^{13} = 19$, $21^{15} = 7$, $21^{17} = 5$, $21^{19} = 20$, and $21^{21} = 11$.

9. In $\mathbb{Z}_5[x]$, we have $x^4 + 4 = x^4 - 1 = (x^2 + 1)(x^2 - 1) = (x^2 + 1)(x - 1)(x + 1)$. Since 2 is a zero of $x^2 + 1$, we divide by $x - 2$ and discover that $x^2 + 1 = (x - 2)(x + 2)$. The complete factorization is therefore
$x^4 + 4 = (x - 2)(x + 2)(x - 1)(x + 1)$.

10. By inspection, -1 is a zero of $x^3 + 2x^2 + 2x + 1$ in $\mathbb{Z}_7[x]$. We find that

$$\begin{array}{r|l} & x^2 + x + 1 \\ \hline x + 1 & x^3 + 2x^2 + 2x + 1 \\ & \underline{x^3 + x^2} \\ & x^2 + 2x \\ & \underline{x^2 + x} \\ & x + 1 \\ & \underline{x + 1} \\ & 0 \end{array}$$

and by inspection, 2 and 4 are zeros of $x^2 + x + 1$. Thus the factorization is
$x^3 + 2x^2 + 2x + 1 = (x + 1)(x - 4)(x - 2)$.

11. By inspection, 3 is a zero of $2x^3 + 3x^2 - 7x - 5$ in $\mathbb{Z}_{11}[x]$. Dividing by $x - 3$, we obtain

$$\begin{array}{r|l} & 2x^2 - 2x - 2 \\ \hline x - 3 & 2x^3 + 3x^2 - 7x - 5 \\ & \underline{2x^3 + 5x^2} \\ & -2x^2 - 7x \\ & \underline{-2x^2 + 6x} \\ & -2x - 5 \\ & \underline{-2x + 6} \\ & 0 \end{array}$$

so the polynomial factors into $(x - 3)2(x^2 - x - 1)$. By inspection, -3 and 4 are zeros of $x^2 - x - 1$, so the factorization is
$(x - 3)(x + 3)(2x - 8)$

12. By inspection, -1 is a zero of $x^3 + 2x + 3$ in $\mathbb{Z}_5[x]$, so the polynomial is not irreducible. We divide by $x + 1$. By

Section 5.6

$$
\begin{array}{r|l}
 & x^2 - x + 3 \\
\hline
x + 1 & x^3 \qquad\quad + 2x + 3 \\
 & \underline{x^3 + x^2} \\
 & \quad -x^2 + 2x \\
 & \quad \underline{-x^2 - \ x} \\
 & \qquad\qquad 3x + 3 \\
 & \qquad\qquad \underline{3x + 3} \\
 & \qquad\qquad\qquad 0
\end{array}
$$

inspection, -1 and 2 are solutions of $x^2 - x + 3$, so the factorization is $(x + 1)(x + 1)(x - 2)$.

13. Let $f(x) = 2x^3 + x^2 + 2x + 2$ in $\mathbb{Z}_5[x]$. Then $f(0) = 2$, $f(1) = 2$, $f(-1) = -1$, $f(2) = 1$, and $f(-2) = 1$ so $f(x)$ has no zeros in $\mathbb{Z}_5$. Since $f(x)$ is of degree 3, Theorem 5.19 shows that $f(x)$ is irreducible over $\mathbb{Z}_5$.

14. $f(x) = x^2 + 8x - 2$ satisfies the Eisenstein criterion for irreducibility over $\mathbb{Q}$ with $p = 2$. It is not irreducible over $\mathbb{R}$ since the quadratic formula shows that it has the real zeros $(-8 \pm \sqrt{72})/2$.

15. $g(x) = x^2 + 6x + 12$ is irreducible over $\mathbb{Q}$ since it satisfies the Eisenstein criterion with $p = 3$. It is also irreducible over $\mathbb{R}$ since the quadratic formula shows that is zeros are $(-6 \pm \sqrt{-12})/2$ which are not in $\mathbb{R}$.

16. If $x^3 + 3x^2 - 8$ is reducible over $\mathbb{Q}$, then by Theorem 5.20, it factors in $\mathbb{Z}[x]$, and must therefore have a linear factor of the form $x - a$ in $\mathbb{Z}[x]$. Then a must be a root of the polynomial and must divide -8, so the possibilities are $a = \pm 1, \pm 2, \pm 4, \pm 8$. Computing the polynomial at these eight values, we find none of them is a zero of the polynomial, which is therefore irreducible over $\mathbb{Q}$.

17. If $x^4 - 22x^2 + 1$ is reducible over $\mathbb{Z}$, then by Theorem 5.20, it factors in $\mathbb{Z}[x]$, and must therefore either have a linear factor in $\mathbb{Z}[x]$ or factor into two quadratics in $\mathbb{Z}[x]$. The only possibilities for a linear factor are $x \pm 1$, and clearly neither 1 nor -1 is a zero of the polynomial, so a linear factor is impossible. Suppose $x^2 - 22x^2 + 1 = (x^2 + ax + b)(x^2 + cx + d)$. Equating coefficients, we see that

x^3 *coefficient:* $0 = a + c$

x^2 *coefficient:* $-22 = ac + b + d$

x *coefficient:* $0 = bc + ad$

constant: $1 = bd$, so either $b = d = 1$ or $b = d = -1$.

Suppose $b = d = 1$. Then $-22 = ac + 1 + 1$ so $ac = -24$.

Since $a + c = 0$, we have $a = -c$ so $-c^2 = -24$ which is impossible for an integer c. Similarly if $b = d = -1$, we deduce that $-c^2 = -20$, which is also impossible. Thus the polynomial is irreducible.

18. Yes. $p = 3$ **19.** Yes. $p = 3$

20. No, for 2 divides the coefficient 4 of x^{10} and 3^2 divides the constant term -18.

21. Yes. $p = 5$

22. Let this polynomial be $f(x)$. If $f(x)$ has a rational zero, then this zero can be expressed as a fraction with numerator dividing 10 and denominator dividing 6. The possibilities are ±10, ±5, ±10/3, ±5/2, ±2, ±5/3, ±1, ±5/6, ±2/3, ±1/2, ±1/3, and ±1/6.

Experimentation with a calculator shows that there is a negative real zero between -2 and -3 since $f(-2) < 0$ and $f(-3) > 0$. (Recall the intermediate value theorem.) The only possible rational candidate is -5/2. We reach for our calculator and find that $f(-5/2) = 0$, so -5/2 is a zero and $(2x + 5)$ is a linear factor.

Since $f(0) < 0$ and $f(1) > 0$, the intermediate value theorem shows there is a real zero a satisfying $0 < a < 1$. The possibilities are 5/6, 2/3, 1/2, 1/3, and 1/6. Since $2x + 5$ is a factor, accounting for the factor 2 of 6 and the factor 5 of 10, we can discard 5/6, 1/2, and 1/6, leaving 2/3 and 1/3 to try. We reach for our calculator and compute $f(2/3) = 0$, so $3x - 2$ is also a factor. Since we have accounted for the 6 and the 10 with these linear factors, the only other possible rational zeros would have to be 1 or -1, and we easily find that these are not zeros. Thus the zeros are 2/3 and -5/2.

23. T T T F T F T T T T

24. Considering $f(x) = x^4 + x^3 + x^2 - x + 1$ in $\mathbb{Z}[x]$, we find that $f(-2) = 16 - 8 + 2 + 2 + 1 = 15$. Thus $p = 3$ and $p = 5$ are primes such that -2 is a zero of $f(x)$ in $\mathbf{Z}_p$, that is, such that $x + 2$ is a factor of $f(x)$ in $\mathbf{Z}_p[x]$.

25. The polynomials of degree 2 in $\mathbf{Z}_2[x]$ are

x^2: not irreducible since 0 is a zero,

$x^2 + 1$: not irreducible since 1 is a zero,

$x^2 + x$: not irreducible since 0 is a zero,

$x^2 + x + 1$: irreducible since neither 0 nor 1 are zeros.

Thus our answer is $x^2 + x + 1$.

26. The polynomials of degree 3 in $\mathbf{Z}_2[x]$ are

x^3: not irreducible since 0 is a zero,

$x^3 + 1$: not irreducible since 1 is a zero,

$x^3 + x$: not irreducible since 0 is a zero,

$x^3 + x^2$: not irreducible since 0 is a zero,

$x^3 + x + 1$: irreducible, neither 0 nor 1 is a zero,

$x^3 + x^2 + 1$: irreducible, neither 0 nor 1 is a zero,

$x^3 + x^2 + x$: not irreducible since 0 is a zero,

$x^3 + x^2 + x + 1$: not irreducible, 1 is a zero.

The irreducible cubics are $x^3 + x + 1$ and $x^3 + x^2 + 1$.

27. The polynomials of degree 2 in $\mathbf{Z}_3[x]$ include

x^2, $x^2 + x$, $x^2 + 2x$, $2x^2$, $2x^2 + x$, $2x^2 + 2x$ all reducible since 0 is a zero,

$x^2 + 2$, $x^2 + x + 1$, $2x^2 + 1$, $2x^2 + 2x + 2$ all reducible since 1 is a zero,

$x^2 + 2x + 1$, $2x^2 + x + 2$ both reducible since 2 is a zero.

The remaining six polynomials, $x^2 + 1$, $x^2 + x + 2$, $x^2 + 2x + 2$, $2x^2 + 2$, $2x^2 + x + 1$, and $2x^2 + 2x + 1$ are irreducible since none of 0, 1, or 2 is a zero.

28. An irreducible polynomial must have a nonzero constant term or 0 is a zero, this eliminates 18 of the 54 cubic polynomials in $\mathbf{Z}_3[x]$. Now a is a zero of $f(x)$ if and only if a is a zero of $2f(x)$, so we can consider just the 18 cubics

with leading coefficient 1 and constant term nonzero.
$x^3 + 2$, $x^3 + x^2 + 1$, $x^3 + x + 1$, $x^3 + 2x^2 + 2x + 1$, $x^3 + x^2 + 2x + 2$, $x^3 + 2x^2 + x + 2$ have 1 as a zero, so they are reducible.
$x^3 + 1$, $x^3 + 2x^2 + 1$, $x^3 + x + 2$, $x^3 + x^2 + x + 1$ have -1 as a zero so they are reducible.

The remaining eight cubics with leading coefficient 1 and nonzero constant, namely:
$x^3 + 2x + 1$, $x^3 + 2x + 2$, $x^3 + x^2 + 2$, $x^3 + 2x^2 + 1$,
$x^3 + 2x^2 + 2$, $x^3 + x^2 + x + 2$, $x^3 + x^2 + 2x + 1$,
$x^3 + 2x^2 + x + 1$, $x^3 + 2x^2 + 2x + 2$

and their doubles
$2x^3 + x + 2$, $2x^3 + x + 1$, $2x^3 + 2x^2 + 1$, $2x^3 + x^2 + 2$,
$2x^3 + x^2 + 1$, $2x^3 + 2x^2 + 2x + 1$, $2x^3 + 2x^2 + x + 2$,
$2x^3 + x^2 + 2x + 2$, $2x^3 + x^2 + x + 1$
are irreducible.

29. Following the hint, each reducible quadratic of the form $x^2 + ax + b$ is a product $(x + c)(x + d)$ for $c, d \in \mathbb{Z}_p$. There are $\binom{p}{2} = p(p - 1)/2$ such products (neglecting order of factors) where $c \neq d$. There are p such products where $c = d$. Thus there are $p(p - 1)/2 + p = p^2/2 + p/2 = p(p + 1)/2$ reducible quadratics with leading coefficient 1. Since the leading coefficient can be any one of $p - 1$ nonzero elements, there are $(p - 1)p(p + 1)/2$ reducible quadratics altogether. The total number of quadratic polynomials in $\mathbb{Z}_p[x]$ is $(p - 1)p^2$. Thus the number of reducible quadratics is $(p - 1)p^2 - (p - 1)p(p + 1)/2 = p(p - 1)[p - (p + 1)/2] = p(p - 1)^2/2$.

30. Note that $x^2 = xx$ and $x^2 + 1 = (x + 1)^2$ are reducible in $\mathbb{Z}_2[x]$. For an odd prime p and $a \in \mathbb{Z}_p$, we know that $(-a)^p + a = -a^p + a = -a^p + a = 0$ by the corollary of Theorem 5.8. Thus $x^p + a$ is reducible over $\mathbb{Z}_p$ for every prime p.

Section 5.6

31. We are given that $f(a) = a_0 + a_1a + \cdots + a_na^n = 0$ and $a \neq 0$.

Dividing by a^n, we find that

$$a_0(\frac{1}{a})^n + a_1(\frac{1}{a})^{n-1} + \cdots + a_n = 0$$

which is just what we wanted to show.

32. By Theorem 5.18, we know that $f(x) = q(x)(x - a) + c$ for some constant $c \in F$. Applying the evaluation homomorphism ϕ_a to both sides of this equation, we find that

$$f(a) = q(a)(a - a) + c = q(a)0 + c = c$$

so the remainder $r(x) = c$ is actually $f(a)$.

33. a) Let $f(x) = \sum_{i=0}^{\infty} a_ix^i$ and $g(x) = \sum_{i=0}^{\infty} b_ix^i$. Then

$$\overline{\sigma_m}(f(x) + g(x)) = \overline{\sigma_m}\left[\sum_{i=0}^{\infty} (a_i + b_i)x^i\right] = \sum_{i=0}^{\infty} \overline{\sigma_m}(a_i + b_i)x^i$$

$$= \sum_{i=0}^{\infty} [\overline{\sigma_m}(a_i) + \overline{\sigma_m}(b_i)]x^i = \overline{\sigma_m}(f(x)) + \overline{\sigma_m}(g(x)) \quad \text{and}$$

$$\overline{\sigma_m}(f(x)g(x)) = \overline{\sigma_m}\left[\sum_{n=0}^{\infty} (\sum_{i=0}^{n} a_ib_{n-i}x^i)\right] =$$

$$= \sum_{n=0}^{\infty} \overline{\sigma_m}\left[\sum_{i=0}^{n} a_ib_{n-i}x^i\right] = \sum_{n=0}^{\infty}\left[\sum_{i=0}^{n} \overline{\sigma_m}(a_ib_{n-i})\right]$$

$$= \sum_{n=0}^{\infty}\left[\sum_{i=0}^{n} \overline{\sigma_m}(a_i)\overline{\sigma_m}(b_{n-i})\right] = \overline{\sigma_m}(f(x))\overline{\sigma_m}(g(x)) \text{ , so } \overline{\sigma_m}$$

is a homomorphism.

b) Let $f(x) = g(x)h(x)$ for $g(x)$, $h(x) \in \mathbb{Z}[x]$ with the degrees of both $g(x)$ and $h(x)$ less than the degree n of $f(x)$.

Applying the homomorphism $\overline{\sigma_m}$, we see that $\overline{\sigma_m}(f(x)) = \overline{\sigma_m}(f(x))\overline{\sigma_m}(g(x))$ is a factorization of $\overline{\sigma_m}(f(x))$ into two polynomials of degree less than the degree n of $\overline{\sigma_m}(f(x))$, contrary to hypothesis. Thus $f(x)$ is irreducible in $\mathbb{Z}[x]$, and therefore in $\mathbb{Q}[x]$ by Theorem 5.20.

c) Taking $m = 5$, we see that $\overline{\sigma_5}(x^3 + 17x + 36) = x^3 + 2x + 1$ which does not have any of the five elements 0, 1, -1, 2, -2 of $\mathbb{Z}_5$ as a zero, and is thus irreducible over $\mathbb{Z}_5$ by Theorem 5.19. By Part (a), we conclude that $x^3 + 17x + 36$ is irreducible over $\mathbb{Q}$.

SECTION 5.7 - Noncommutative Examples

1. $(2e + 3a + 0b) + (4e + 2a + 3b) = e + 0a + 3b.$

2. $(2e + 3a + 0b)(4e + 2a + 3b) =$
$2e(4e + 2a + 3b) + 3a(4e + 2a + 3b) + 0b(4e + 2a + 3b) =$
$(3e + 4a + 1b) + (4e + 2a + 1b) + (-e + 0a + 0b) =$
$2e + a + 2b.$

3. $(3e + 3a + 3b)^2 =$
$3e(3e + 3a + 3b) + 3a(3e + 3a + 3b) + 3b(3e + 3a + 3b) =$
$(4e + 4a + 4b) + (4e + 4a + 4b) + (4e + 4a + 4b) =$
$2e + 2a + 2b$, so $(3e + 3a + 3b)^4 = [(3e + 3a + 3b)^2]^2 =$
$(2e + 2a + 2b)^2 = 4\cdot(1e + 1a + 1b)^2 = 4\cdot(3e + 3a + 3b) =$
$2e + 2a + 2b.$

4. $(i + 3j)(4 + 2j - k) = 4i + 2ij - ik + 12j + 6jj - 3jk =$
$4i + 2k + j + 12j - 6 - 3i = -6 + i + 13j + 2k.$

5. $i^2j^3kji^5 = (-1)(-j)kji = (jk)(ji) = i(-k) = j.$

6. $(i + j)^{-1} = \dfrac{1}{i + j}\cdot\dfrac{-i - j}{-i - j} = \dfrac{-i - j}{2} = -\frac{1}{2}i - \frac{1}{2}j.$

7. $[(1 + 3i)(4j + 3k)]^{-1} = (4j + 3k + 12k - 9j)^{-1} =$
$(-5j + 15k)^{-1} = \dfrac{1}{-5j + 15k}\cdot\dfrac{5j - 15k}{5j - 15k} = \dfrac{5j - 15k}{25 + 225} = \dfrac{j - 3k}{50}$
$= \frac{1}{50}j - \frac{3}{50}k.$

8. $(0\rho_0 + 1\rho_1 + 0\rho_2 + 0\mu_1 + 1\mu_2 + 1\mu_3)(1\rho_0 + 1\rho_1 + 0\rho_2 + 1\mu_1 + 0\mu_2 + 1\mu_3)$
$= (1\rho_1 + 1\rho_2 + 1\mu_3 + 1\mu_2) + (1\mu_2 + 1\mu_3 + 1\rho_2 + 1\rho_1) +$

$(1\mu_3 + 1\mu_1 + 1\rho_1 + 1\rho_0) = 1\rho_0 + 1\rho_1 + 0\rho_2 + 1\mu_1 + 0\mu_2 + 1\mu_3.$

9. The center is $\{r + 0i + 0j + 0k \mid r \in \mathbb{R},\ r \neq 0\}$ since nonzero coefficients of i, j, or k lead to an element that does not not commute with j, k, or k respectively.

10. Clearly $\{a + bi \mid a, b \in \mathbb{R}\}$ and $\{a + bj \mid a, b \in \mathbb{R}\}$ are subrings of the quaternions that are actually fields isomorphic to $\mathbb{C}$,

11. F F F F F[0 $\in$ Hom(A) is not an isomorphism] F T F T F

12. a) The polynomial $x^2 + 1$ has i, $-i$, j, $-j$, k, and $-k$ as solutions in the quaternions.
    b) The subset $\{1, -1, i, -i, j, -j, k, -k\}$ of the quaternions is a group under quaternion multiplication, and is not cyclic since every element is of order 2.

13. Let $\psi(1,0) = (1,-1)$ and $\psi(0,1) = (0,0)$. Then $\phi\psi(m,n) = \phi(\psi(m,n)) = \phi(m,-m) = (0,0)$, so ϕ is a left divisor of zero.

14. $\begin{bmatrix}1 & 0\\0 & 1\end{bmatrix}, \begin{bmatrix}0 & 1\\1 & 0\end{bmatrix}, \begin{bmatrix}1 & 1\\0 & 1\end{bmatrix}, \begin{bmatrix}1 & 0\\1 & 1\end{bmatrix}, \begin{bmatrix}0 & 1\\1 & 1\end{bmatrix}, \begin{bmatrix}1 & 1\\1 & 0\end{bmatrix}$
    have inverses

    $\begin{bmatrix}1 & 0\\0 & 1\end{bmatrix}, \begin{bmatrix}0 & 1\\1 & 0\end{bmatrix}, \begin{bmatrix}1 & -1\\0 & 1\end{bmatrix}, \begin{bmatrix}1 & 0\\-1 & 1\end{bmatrix}, \begin{bmatrix}-1 & 1\\1 & 0\end{bmatrix}, \begin{bmatrix}0 & 1\\1 & -1\end{bmatrix}$

    respectively in every field. (Note that if the field has characteristic 2, we have $-1 = 1$ so the last four matrices in the second row may be the same as four in the top row.)

15. Let $m \in \mathbb{Z}$ [or $m \in \mathbb{Z}_n$ as the case may be]. Let ϕ_m be the endomorphism of the abelian group of the ring such that $\phi_m(1) = m$. Then $\{\phi_m \mid m \in \mathbb{Z}$ [or $m \in \mathbb{Z}_n$]$\}$ is the entire homomorphism ring, for a homomorphism of each of these cyclic groups is entirely determined by its value on the generator 1 of the group. Define ψ: Hom($\mathbb{Z}$) $\longrightarrow \mathbb{Z}$ [or Hom($\mathbb{Z}_n$) $\longrightarrow \mathbb{Z}_n$] by $\psi(\phi_m) = m$. Now $(\phi_i + \phi_j)(1) = \phi_i(1) + \phi_j(1) = i + j = \phi_{i+j}(1)$, so $\phi_i + \phi_j = \phi_{i+j}$ since these homomorphism agree on the generator 1. Consequently $\psi(\phi_i + \phi_j) = \psi(\phi_{i+j}) = i + j = \psi(\phi_i) + \psi(\phi_j)$, so ψ is an additive homomorphism. Also, $(\phi_i\phi_j)(1) = \phi_i(\phi_j(1)) = \phi_i(j) = ij = \phi_{ij}(1)$, so $\phi_i\phi_j = \phi_{ij}$.

Therefore $\psi(\phi_i\phi_j) = \psi(\phi_{ij}) = ij = \psi(\phi_i)\psi(\phi_j)$. Hence ψ is a ring homomorphism. By definition, the image under ψ is the entire ring $\mathbb{Z}$ [or $\mathbb{Z}_n$]. If $\psi(\phi_i) = \psi(\phi_j)$ then $i = j$ in $\mathbb{Z}$ [or $\mathbb{Z}_n$] so ϕ_i and ϕ_j map the generator 1 into the same element and thus are the same homomorphism. Thus ψ is an isomorphism.

16. A homomorphism of $\mathbb{Z}_2 \times \mathbb{Z}_2$ can map the generators (1,0) and (0,1) onto any elements of the ring. Thus there are a total of (4)(4) = 16 homomorphisms of $\mathbb{Z}_2 \times \mathbb{Z}_2$ into itself, while the ring itself has only 4 elements. Thus the ring of all homomorphisms cannot be isomorphic to the ring itself, for they have different numbers of elements.

17. $(YX - XY)(a_0 + a_1x + \cdots + a_nx^n) =$

$(YX)(a_0 + a_1x + \cdots + a_nx^n) - (XY)(a_0 + a_1x + \cdots + a_nx^n) =$

$Y[X(a_0 + a_1x + \cdots + a_nx^n)] - X[Y(a_0 + a_1x + \cdots + a_nx^n)] =$

$Y(a_0x + a_1x^2 + \cdots + a_nx^{n+1}) - X(a_1 + 2a_2x^2 + \cdots + na_nx^{n-1}) =$

$(a_0 + 2a_1x + \cdots + (n+1)a_nx^n) - (a_1x + 2a_2x^2 + \cdots + na_nx^n) =$

$1(a_0 + a_1x + \cdots + a_nx^n)$ so $YX - XY = 1$.

18. Let $\phi\colon R(G) \longrightarrow R$ be defined by $\phi(re) = r$. Then $\phi(re + se) = \phi((r+s)e) = r \quad s = \phi(re) + \phi(se)$ and $\phi(r(se)) = \phi((rs)e) = rs = \phi(r)\phi(s)$ so ϕ is a homomorphism. Clearly, the image of $R(G)$ under ϕ is all of R. If $\phi(re) = \phi(se)$ then $r = s$, so ϕ is one to one. Thus ϕ is an isomorphism.

19. $(a+bi+cj+dk)[(e+fi+gj+hk)(r+si+tj+uk)] =$
$(a+bi+cj+dk)[(er-fs-gt-hu)+(es+fr+gu-ht)i+(et-fu+gr+hs)j$
$+(eu+ft-gs+hr)k] =$
$(aer-afs-agt-ahu-bes-bfr-bgu+bht-cet+cfu-cgr-chs-deu-dft+$
$dgs-dhr) +$
$(aes+afr+agu-aht+ber-bfs-bgt-bhu+ceu+cft-cgs+chr-det+dfu-$
$dgr-dhs)i +$
$(aet-afu+agr+ahs-beu-bft+bgs-bhr+cer-cfs-cgt-chu+des+dfr+$
$dgu-dht)j +$
$(aeu+aft-ags+ahr+bet-bfu+bgr+bhs-ces-cfr-cgu+cht+der-dfs-$
$dgt-dhu)k$.

Also,
$[(a+bi+cj+dk)(e+fi+gj+hk)](r+si+tj+uk) =$

Section 5.7

$$[(ae-bf-cg-dh)+(af+be+ch-dg)i+(ag-bh+ce+df)j+(ah+bg-cf+de)k](r+si+tj+uk) =$$

$$(aer-bfr-cgr-dhr-afs-bes-chs+dgs-agt+bht-cet-dft-ahu-bgu+cfu-deu) +$$

$$(aes-bfs-cgs-dhs+afr+ber+chr-dgr+agu-bhu+ceu+dfu-aht-bgt+cft-det)i +$$

$$(aet-bft-cgt-dht-afu-beu-chu+dgu+agr-bhr+cer+dfr+ahs+bgs-cfs+des)j +$$

$$(aeu-bfu-cgu-dhr+aft+bet+cht-dgt-ags+bhs-ces-dfs+ahr+bgr-cfr+der)k.$$

We obtained the same quaternions, proving associativity.

CHAPTER 6

FACTOR RINGS AND IDEALS

SECTION 6.1 - Homomorphisms and Factor Rings

1. Let ϕ: $\mathbb{Z} \longrightarrow \mathbb{Z}$ be a homorphism such that $\phi(1) = n$. By group theory for $\langle \mathbb{Z}, +\rangle$, we know that we must then have $\phi(m) = mn$ for all $n \in \mathbb{Z}$. To satisfy the requirement $\phi(rs) = \phi(r)\phi(s)$, we see that we need to have $rsn = (rn)(sn)$ for all $r, s \in \mathbb{Z}$. Consequently, we must have $n = n^2$, so the only possible choices for n are 0 and 1. Thus the only homomorphisms of $\mathbb{Z}$ into $\mathbb{Z}$ are the identity map and the trivial homomorphism mapping all elements into 0.

2. a) Let ϕ be a homomorphism of $\mathbb{Z} \times \mathbb{Z}$ into $\mathbb{Z}$. Then ϕ induces homomorphisms of the subrings $\mathbb{Z} \times \{0\}$ and $\{0\} \times \mathbb{Z}$ into $\mathbb{Z}$. By the preceding exercise, we see that the only possible values for both $\phi(1,0)$ and $\phi(0,1)$ are 0 and 1. Thus there are a total of four possibilities: $\phi_1(m,n) = (0,0)$, $\phi_2(m, n) = (m,0)$, $\phi_3(m,n) = (0,n)$, and $\phi_4(m,n) = (m,n)$.

   b) Let ϕ be a homomorphism of $\mathbb{Z} \times \mathbb{Z}$ into $\mathbb{Z} \times \mathbb{Z}$. Suppose that $\phi(1,0) = (m,n)$. From $\phi(1,0) = \phi[(1,0)(1,0)]$ we see that $m^2 = m$ and $n^2 = n$, so $\phi(1,0)$ must be one of the elements $(0,0)$, $(1,0)$, $(0,1)$, or $(1,1)$. By a similar argument, $\phi(0,1)$ must be one of these same four elements. We also must have $\phi(1,0)\phi(0,1) = \phi(0,0) = (0,0)$. This gives just nine possibilities:

   $\phi(1,0) = (1,0)$ while $\phi(0,1) = (0,0)$ or $(0,1)$,
   $\phi(1,0) = (0,1)$ while $\phi(0,1) = (0,0)$ or $(1,0)$,
   $\phi(1,0) = (1,1)$ while $\phi(0,1) = (0,0)$, and
   $\phi(1,0) = (0,0)$ while $\phi(0,1) = (1,0)$, $(0,1)$, or $(1,1)$.

   It is easily checked that each of these does give rise to a homomorphism.

3. In order for $\mathbb{Z}_n$ to contain a subring isomorphic to $\mathbb{Z}_2$, there must be a nonzero element of $\mathbb{Z}_n$ which when added to itself is zero, to play the role of $1 \in \mathbb{Z}_2$. Thus n must be even. Let $n = 2m$ so that the subset $\langle \{0, m\}, +_n\rangle \simeq \langle \mathbb{Z}_2, +_2\rangle$. In order to have $\langle \{0, m\}, \cdot_n\rangle \simeq \langle \mathbb{Z}, \cdot_2\rangle$, we must have $mm = m$

in $\mathbb{Z}_n$. Since in $\mathbb{Z}_n$ we have $2 \cdot m = 0$, $3 \cdot m = (2 \cdot m) + m = m$, $4 \cdot m = 0$, $5 \cdot m = m$, etc., we see that we have $mm = m$ in $\mathbb{Z}_n$ if and only if m is an odd integer. Hence $\mathbb{Z}_n$ contains a subring isomorphic to $\mathbb{Z}_2$ if and only if $n = 2m$ for an odd integer m.

4. Since the ideals must be additive subgroups, by group theory we see that possibilities are restricted to the cyclic additive subgroups

$\langle 0 \rangle = \{0\}$,
$\langle 1 \rangle = \{0, 1, 2, 3, 4, 5, 6, 7, 8, 9, 10, 11\}$,
$\langle 2 \rangle = \{0, 2, 4, 6, 8, 10\}$,
$\langle 3 \rangle = \{0, 3, 6, 9\}$,
$\langle 4 \rangle = \{0, 4, 8\}$, and
$\langle 6 \rangle = \{0, 6\}$.

It is easily checked that each of these is closed under multipication by any element of $\mathbb{Z}_{12}$. For example, multiplication in $\mathbb{Z}$ of any element in $\langle 3 \rangle$ by any number in $\mathbb{Z}_{12}$ gives a multiple of 3, which we may write as $3n = 12q + r$ by the division algorithm. Since 3 is a divisor of 12, we see that 3 is a divisor of r, that is, $r \in \langle 3 \rangle$.

5. (See the text answer.)

6. The differentiation map δ is not a homomorphism, for $\delta[f(x)g(x)] = f(x)g'(x) + g'(x)f(x) \neq f'(x)g'(x) = \delta[f(x)]\delta[g(x)]$. To connect it with Example 6, we note that the kernel of δ as an additive group homomorphism is the set of all constant functions. Example 6 shows that this is not an ideal, so δ cannot be a ring homomorphism.

7. Let $\phi: \mathbb{Z} \longrightarrow \mathbb{Z} \times \mathbb{Z}$ be defined by $\phi(n) = (n,0)$. Then $\mathbb{Z}$ has unity 1 but $\phi(1) = (1,0)$ is not the unit of $\mathbb{Z} \times \mathbb{Z}$; the unity of $\mathbb{Z} \times \mathbb{Z}$ is $(1,1)$.

8. T F T F T F T T T T

9. (See the text answer.)

10. $\mathbb{Z}/2\mathbb{Z} \simeq \mathbb{Z}_2$ which is a field.

11. (See the text answer.)

12. $\mathbb{Z} \times \mathbb{Z}$ has divisors of zero, but $(\mathbb{Z} \times \mathbb{Z})/(\mathbb{Z} \times \{0\}) \simeq \mathbb{Z}$ which has no divisors of zero.

13. (See the text answer.)

14. a) r and s are elements of R, not of R/N. The student probably does not understand the structure of a factor ring.

b) Assume R/N is commutative. Then

$$(r + N)(s + N) = (s + N)(r + N) \text{ for all } r,\ s \in R.$$

c) Let $r,\ s \in N$. Then $(r + N)(s + N) = (s + N)(r + N)$ for all $r,\ s \in R$ if and only if $rs + N = sr + N$ for all $r,\ s \in R$ if and only if $(rs + N) - (sr + N) = N$ for all $r,\ s \in R$ if and only if $(rs - sr) + N = N$ for all $r,\ s \in R$ if and only if $(rs - sr) \in N$.

15. Since $(a + b\sqrt{2}) + (c + d\sqrt{2}) = (a + c) + (b + c)\sqrt{2}$ and $0 = 0 + 0\sqrt{2}$ and $-(a + b\sqrt{2}) = (-a) + (-b)\sqrt{2}$, we see that R is closed under addition, has an additive identity, and contains additive inverses. Thus $\langle R, +\rangle$ is a group. Since $(a + b\sqrt{2})(c + d\sqrt{2}) = (ac + 2bd) + (ad + bc)\sqrt{2}$, we see that R is closed under multiplication and is thus a ring. We will show that R' is a ring by showing that it is the image of R under a homomorphism $\phi: R \longrightarrow M_2$. Let

$$\phi(a + b\sqrt{2}) = \begin{bmatrix} a & 2b \\ b & a \end{bmatrix}.$$

Then $\phi((a + b\sqrt{2}) + (c + d\sqrt{2})) = \phi((a + c) + (b + d)\sqrt{2}) =$

$$\begin{bmatrix} a+c & 2(b+d) \\ b+d & a+c \end{bmatrix} = \begin{bmatrix} a & 2b \\ b & a \end{bmatrix} + \begin{bmatrix} c & 2d \\ d & c \end{bmatrix} = \phi(a + b\sqrt{2}) + \phi(c + d\sqrt{2})$$

and $\phi((a + b\sqrt{2})(c + d\sqrt{2})) = \phi((ac + 2bd) + (ad + bc)\sqrt{2}) =$

$$\begin{bmatrix} ac+2bd & 2(ad+bc) \\ ad+bc & ac+2bd \end{bmatrix} = \begin{bmatrix} a & 2b \\ b & a \end{bmatrix}\begin{bmatrix} c & 2d \\ d & c \end{bmatrix} = \phi(a + b\sqrt{2})\phi(c + d\sqrt{2}).$$

This shows that ϕ is a homomorphism. Since $R' = \phi[R]$, we see that R' is a ring. If $\phi(a + b\sqrt{2})$ is the matrix with all entries zero, then we must have $a = b = 0$, so $\text{Ker}(\phi) = 0$ and ϕ is one to one. Hence ϕ is an isomorphism of R onto R'.

16. For any ring R in the collection, the identity map $\iota: R \longrightarrow R$ is an isomorphism, so $R \simeq R$.

17. Suppose $R \simeq R'$. Then by definition, there exists an isomorphism $\phi: R \longrightarrow R'$. Since ϕ is one-to-one and onto R', it has an inverse ϕ^{-1} that maps R' onto R in a one-to-one

fashion. We know from our work in groups that $\phi^{-1}(r' + s') = \phi^{-1}(r') + \phi^{-1}(s')$ for all $r', s' \in R'$. Now $\phi(\phi^{-1}(r's')) = r's'$ and $\phi(\phi^{-1}(r')\phi^{-1}(s')) = \phi(\phi^{-1}(r'))\phi(\phi^{-1}(s')) = r's'$. Since ϕ is one to one, this shows that $\phi^{-1}(r's') = \phi^{-1}(r')\phi^{-1}(s')$, so $\phi^{-1}: R' \longrightarrow R$ is an isomorphism. Thus $R' \simeq R$.

18. Suppose that $R \simeq R'$ and $R' \simeq R''$. Then there exists an isomorphism $\phi: R \longrightarrow R'$ and an isomorphism $\psi: R' \longrightarrow R''$. From our work in group theory, we know that the composite map $\psi\phi: R \longrightarrow R''$ is one to one, onto R'', and satisfies $\psi\phi(r + s) = \psi\phi(r) + \psi\phi(s)$ for all $r, s \in R$. Now $\psi\phi(rs) = \psi(\phi(rs)) = \psi(\phi(r)\phi(s)) = [\psi\phi(r)][\psi\phi(s)]$ for all $r, s \in R$. Thus $\psi\phi$ is an isomorphism, so $R \simeq R''$.

19. Let $\phi: F \longrightarrow F'$ be a homomorphism of a field F into a field F', and let $N = \text{Ker}(\phi)$. If $N \neq \{0\}$, then N contains a nonzero element u of F which is a unit. Since N is an ideal, we see that $u^{-1}u = 1$ is in N, and then N contains $a1 = a$ for all $a \in F$. Thus N is either $\{0\}$, in which case ϕ is one to one by group theory, or $N = F$, in which case ϕ maps every element of F onto 0.

20. Exercise 40 of Section 3.1 shows that $\psi\phi(r + s) = \psi\phi(r) + \psi\phi(s)$ for all $r, s \in R$. For multiplication, we note that $\psi\phi(rs) = \psi(\phi(rs)) = \psi(\phi(r)\phi(s)) = [\psi\phi(r)][\psi\phi(s)]$ since both ϕ and ψ are homomorphisms. Thus $\psi\phi$ is also a homomorphism.

21. In a *commutative* ring, the binomial expansion

$$(a+b)^n = a^n + \binom{n}{1}\cdot a^{n-1}b + \binom{n}{2}\cdot a^{n-2}n^2 + \cdots + \binom{n}{n-1}\cdot ab^{n-1} + b^n$$

is valid. If p is a prime and $n = p$, then all the binomial coefficients $\binom{p}{i}$ for $1 \leq i \leq n-1$ are divisible by p, and hence the term $\binom{p}{i}\cdot a^i b^{p-i} = 0$ for a and b in a commutative ring of characteristic p. This shows at once that $\phi_p(a + b) = (a + b)^p = a^p + b^p = \phi_p(a) + \phi_p(b)$. Also $\phi_p(ab) = (ab)^p = a^p b^p$ since ρ is commutative. But $a^p b^p = \phi_p(a)\phi_p(b)$, so ϕ is a homomorphism.

22. By Theorem 6.1, we know that $\phi(1)$ is unity for $\phi[R]$. Suppose

that R' has unity $1'$. Then $\phi(1) = \phi(1)1' = \phi(1)\phi(1)$ so $\phi(1)1' - \phi(1)\phi(1) = 0'$. Consequently, $\phi(1)(1' - \phi(1)) = 0'$. Now if $\phi(1) = 0'$, then $\phi(a) = \phi(1a) = \phi(1)\phi(a) = 0'\phi(a) = 0'$ so $\phi[R] = \{0'\}$ contrary to hypothesis. Thus $\phi(1) \neq 0'$. Since R' has no divisors of zero, we conclude from $\phi(1)(1' - \phi(1)) = 0'$ that $1' - \phi(1) = 0'$, so $\phi(1)$ is the unity $1'$ of R'.

23. Since the ideal N is also a subring of R, Theorem 6.1 shows that $\phi[N]$ is a subring of R'. To show that it is an ideal of $\phi[R]$, we need only show that $\phi(r)\phi[N] \subseteq \phi[N]$ and $\phi[N]\phi(r) \subseteq \phi[N]$ for all $r \in R$. Let $r \in R$ and let $s \in N$. Then $rs \in N$ and $sr \in N$ since N is an ideal. Applying ϕ, we see that $\phi(r)\phi(s) = \phi(rs) \in \phi[N]$ and $\phi(s)\phi(r) = \phi(sr) \in \phi[N]$.

    To see that $\phi[N]$ need not be an ideal of R', we let $\phi: \mathbb{Z} \to \mathbb{Q}$ be the injection map given by $\phi(n) = n$ for all $n \in \mathbb{Z}$. Now $2\mathbb{Z}$ is an ideal of $\mathbb{Z}$, but $2\mathbb{Z}$ is not an ideal of $\mathbb{Q}$ since $(1/2)2 = 1$ and 1 is not in $2\mathbb{Z}$.

    Now let N' be an ideal of R'. We know that $\phi^{-1}[N']$ is at least a subring of R by Theorem 6.1. We must show that $r\phi^{-1}[N'] \subseteq \phi^{-1}[N']$ and $\phi^{-1}(N']r \subseteq \phi^{-1}[N']$ for all $r \in R$. Let $s \in \phi^{-1}[N']$, so that $\phi(s) \in N'$. Then $\phi(rs) = \phi(r)\phi(s)$ and $\phi(r)\phi(s) \in N'$ since N' is an ideal. This shows that $rs \in \phi^{-1}[N']$, so $r\phi^{-1}[N'] \subseteq \phi^{-1}[N']$. Also $\phi(sr) = \phi(s)\phi(r)$ and $\phi(s)\phi(r) \in N'$ since N' is an ideal. This shows that $sr \in \phi^{-1}[N']$, so $\phi^{-1}[N']r \subseteq \phi^{-1}[N']$.

24. If $f(x_1, \cdots, x_n)$ and $g(x_1, \cdots, x_n)$ both have every element of S as a zero, then so do their sum, product, and any multiple of one of them by any element $h(x_1, \cdots, x_n)$ in $F[x_1, \cdots, x_n]$. Since the possible multipliers from $F[x_1, \cdots, x_n]$ include 0 and -1, we see that the set N_S is indeed a subring closed under multiplication by elements of $F[x_1, \cdots, x_n]$, and thus is an ideal of this polynomial ring.

25. Let N be an ideal of a field F. If N contains a nonzero element a, then N contains $(1/a)a = 1$, since N is an ideal. But then N contains $s1 = s$ for every $s \in F$ so $N = F$. Thus N is either $\{0\}$ or F. If $N = F$, then $F/N = F/F$ is the trivial ring of one element. If $N = \{0\}$, then $F/N = F/\{0\}$ is isomorphic to F, since each element $s + \{0\}$ of $F/\{0\}$ can simply be renamed s.

26. If $N \neq R$, then the unity 1 of R is not an element of N, for if $1 \in N$, then so is $r1 = 1$ for all $r \in R$. Thus $1 + N \neq N$, that is, $1 + N$ is not the zero element of R/N. Clearly $(1 + N)(r + N) = (r + N) = (r + N)(1 + N)$ in R/N, which shows that $1 + N$ is unity for R/N.

27. Let $x, y \in I_a$ so $ax = ay = 0$. Then $a(x + y) = ax + ay = 0 + 0 = 0$ so $(x + y) \in I_a$. Also, $a(xy) = (ax)y = 0y = 0$ so $xy \in I_a$. Since $a0 = 0$ and $a(-x) = -(ax) = -0 = 0$, we see that I_a contains 0 and additive inverses of each of its elements, so I_a is a subring of R. (Note that thus far, we have not used commutativity in R.) Let $r \in R$. Then $a(xr) = (ax)r = 0r = 0$, and since R is commutative, we see that $a(rx) = r(ax) = r0 = 0$. Thus I_a is an ideal of R.

28. Let $\{N_i \mid i \in I\}$ be a collection of ideals in R. Each of these ideals is a subring of R, and Exercise 47 of Section 5.1 shows that $N = \bigcap_{i \in I} N_i$ is also a subring of R. We need only show that N is closed under multiplication by elements of R. Let $r \in R$ and let $s \in N$. Then $s \in N_i$ for all $i \in I$. Since each N_i is an ideal of R, we see that $rs \in N_i$ and $sr \in N_i$ for all $i \in I$. Thus $rs \in N$.

29. Exercise 34 of Section 3.3 shows that the map $\phi_*: R/N \longrightarrow R'/N'$ defined by $\phi_*(r + N) = \phi(r) + N'$ is well defined and satisfies the additive requirements for a homomorphism. Now $\phi_*((r + N)(s + N)) = \phi_*(rs + N) = \phi(rs) + N' = [\phi(r)\phi(s)] + N' = [\phi(r) + N'][\phi(s) + N'] = [\phi_*(r + N)][\phi_*(s + N)]$ so ϕ_* also satisfies the multiplicative condition, and is a ring homomorphism.

30. If any unit of R is in the kernel N of ϕ, then $1 \in N$ and consequently $N = R$. In this case, $\phi(u) = 0'$ in R', which is not a unit. On the other hand, suppose u is a unit not in $\text{Ker}(\phi)$. Since ϕ is *onto* R', we know that $\phi(1)$ is unity $1'$ in R'. From $uu^{-1} = u^{-1}u = 1$, we obtain $\phi(uu^{-1}) = \phi(u)\phi(u^{-1}) = 1'$ and $\phi(u^{-1}u) = \phi(u^{-1})\phi(u) = 1'$. Thus $\phi(u)$ is a unit of R', and its inverse is $\phi(u^{-1})$.

31. Let $\sqrt{\{0\}}$ be the collection of all nilpotent elements of R. Let $a, b \in \sqrt{\{0\}}$. Then there exist positive integers m and n such that $a^m = b^n = 0$. In a *commutative* ring, the binomial expansion is valid. Consider $(a + b)^{m+n}$. In the binomial expansion, each summand contains a term $a^i b^{m+n-i}$. Now either $i \geq m$ so that $a^i = 0$, or $m + n - i \geq n$ so that $b^{m+n-i} = 0$. Thus each summand of $(a + b)^{m+n}$ is zero, so $(a + b)^{m+n} = 0$ and $\sqrt{\{0\}}$ is closed under addition. For multiplication, we note that since R is *commutative*, $(ab)^{mn} = (a^m)^n(b^n)^m = (0)(0) = 0$, so $ab \in \sqrt{\{0\}}$. If $s \in R$ then $(sa)^m = s^m a^m = s^m 0 = 0$ and $(as)^m = a^m s^m = 0a^m = 0$ so $\sqrt{\{0\}}$ is also closed under left and right multiplication by elements of R. Taking $s = 0$, we see that $0 \in \sqrt{\{0\}}$. Also $(-a)^m$ is either a^m or $-a^m$, so $(-a)^m = 0$ and $-a \in \sqrt{\{0\}}$. Thus $\sqrt{\{0\}}$ is an ideal of R.

32. The radical of $\mathbb{Z}_{12}$ is $\{0, 6\}$. The radical of $\mathbb{Z}$ is $\{0\}$ and the radical of $\mathbb{Z}_{32}$ is $\{0,2,4,6,8,\cdots,30\}$.

33. Suppose $(a + N)^m = N$ in R/N. Then $a^m \in N$. Since N is the radical of R, there exists $n \in \mathbb{Z}^+$ such that $(a^m)^n = 0$. But then $a^{mn} = 0$ so $a \in N$. Thus $a + N = N$ so $\{N\}$ is the radical of R/N.

34. Let $a \in R$. Since the radical of R/N is R/N, there is some positive integer m such that $(a + N)^m = N$. Then $a^m \in N$. Since every element of N is nilpotent, there exists a positive integer n such that $(a^m)^n = 0$ in R. But then $a^{mn} = 0$, so a is an element of the radical of R. Thus the radical of R is R.

35. Let $a, b \in \sqrt{N}$. Then $a^m \in N$ and $b^n \in N$ for some positive integers m and n. Precisely as in the answer to Exercise 31, we argue that $(a + b) \in N$, $ab \in N$, and also that $sa \in N$ and $as \in N$ for any $s \in R$. Since $0^1 \in N$ we see that $0 \in \sqrt{N}$. Also $(-a)^m$ is either a^m or $-(a^m)$, and both a^m and $-(a^m)$ are in N.

Thus $-a \in \sqrt{N}$. This shows that $\sqrt{N}$ is an ideal of R. This terminology is not consistent with that in Exercise 31, for in the notation of this exercise, taking R as an ideal in R, we have $\sqrt{R} = R$ for all commutative rings R. In the sense of Exercise 31, the radical of $\mathbb{Z}$ is $\{0\}$ and not $\mathbb{Z}$. The radical of R in Exercise 31 is $\sqrt{\{0\}}$ in the sense of this exercise which is why we used that notation in the solution to Exercise 31.

36. a) Let $R = \mathbb{Z}$ and let $N = 4\mathbb{Z}$. Then $\sqrt{N} = 2\mathbb{Z} \neq 4\mathbb{Z}$.

b) Let $R = \mathbb{Z}$ and let $N = 2\mathbb{Z}$. Then $\sqrt{N} = 2\mathbb{Z}$.

37. (See the text answer.)

38. Changing the statement of Theorem 4.2 to additive notation, $\langle M + N, +\rangle$ is an additive subgroup of $\langle R,+\rangle$ and we see that $\phi\colon M + N \longrightarrow M/(M \cap N)$ given by $\phi(m + n) = m + (M \cap N)$ is well defined, satisfies the additive homomorphism property, is onto $M/(M \cap N)$, and has kernel N. We need to show that $M + N$ and $M \cap N$ are ideals in R and that ϕ satisfies the multiplicative homomorphism property, and then apply Theorem 6.7.

Showing $M + N$ is an ideal: Let $m_1, m_2 \in M$ and $n_1, n_2 \in N$. Then $(m_1+n_1) + (m_2+n_2) = (m_1+m_2) + (n_1+n_2)$ since addition is commutative. This shows that $M + N$ is closed under addition. For $r \in R$, we know that $rm_1 \in M$ and $rn_1 \in N$, so $r(m_1 + n_1) = rm_1 + rn_1$ is in $M + N$. A similar argument with multiplication on the right shows that $(m_1 + n_1)r = m_1r + n_1r$ is in $M + N$. Thus $M + N$ is closed under multiplication on the left or right by elements of R, in particular, multiplication is closed on $M + N$. Since $0 = 0 + 0 \in M + N$ and $-(m_1 + n_1) = (-m_1) + (-n_1)$ is in $M + N$, we see that $M + N$ is an ideal.

Showing $M \cap N$ is an ideal: Exercise 28 shows that $M \cap N$ is an ideal.

Showing the multiplicative property for ϕ: Note that $(m_1 + n_1)(m_2 + n_2) = m_1m_2 + (m_1n_2 + n_1m_2 + n_1n_2)$ and that the sum in parentheses is in N since N is closed under addition and under left and right multiplication by elements in R. Of course $m_1m_2 \in M$. Thus we have $\phi[(m_1 + n_1)(m_2 + n_2)] =$

$\phi[m_1m_2 + (m_1n_2 + n_1m_2 + n_1n_2)] = m_1m_2 + M \cap N =$
$(m_1 + M \cap N)(m_2 + M \cap N) = \phi(m_1 + n_1)\phi(m_2 + n_2)$.

Since ϕ is a homomorphism of the ring $M + N$ onto the ring $M/(M \cap N)$ with kernel N, Theorem 6.7 shows that $(M + N)/N$ is naturally isomorphic to $M/(M \cap N)$.

39. Theorem 4.3 applied to the additive groups $\langle R,+\rangle$, $\langle M, +\rangle$, and $\langle N,+\rangle$ shows that the map $\phi\colon R \to (R/M)/(N/M)$ given by $\phi(a) = (a + M) + N/M$ is well defined, onto $(R/M)/(N/M)$, satisfies the additive homomorphism property, and has kernel N. We need only demonstrate the multiplicative homomorphism property and apply Theorem 6.7.

Showing the multiplicative property for ϕ: We have

$$\begin{aligned}\phi(ab) &= (ab + M) + N/M = (a + M)(b + M) + N/M\\ &= [(a + M) + N/M][(b + M) + N/M] = \phi(a)\phi(b).\end{aligned}$$

Theorem 6.7 then shows that R/N is naturally isomorphic to $(R/M)/(R/N)$.

40. We have

$$\phi[(a+bi) + (c+di)] = \phi[(a+c) + (b+d)i] = \begin{bmatrix} a+c & b+d \\ -b-d & a+c \end{bmatrix}$$

$$= \begin{bmatrix} a & b \\ -b & a \end{bmatrix} + \begin{bmatrix} c & d \\ -d & c \end{bmatrix} = \phi(a+bi) + \phi(c+di).$$

Also,

$$\begin{aligned}\phi[(a + bi)(c + di)] &= \phi[(ac - bd) + (ad + bc)i]\\ &= \begin{bmatrix} ac-bd & ad+bc \\ -ad-bc & ac-bd \end{bmatrix} = \begin{bmatrix} a & b \\ -b & a \end{bmatrix}\begin{bmatrix} c & d \\ -d & c \end{bmatrix}\\ &= \phi(a + bi)\phi(c + di).\end{aligned}$$

Thus ϕ is a homomorphism. It is obvious that ϕ is one to one. Thus ϕ exhibits an isomorphism of $\mathbb{C}$ with the subring $\phi[\mathbb{C}]$, which must therefore be a field.

41. a) *Homomorphism property*: For $x, y \in R$, we have

$$\lambda_a(x + y) = a(x + y) = ax + ay = \lambda_a(x) + \lambda_a(y).$$

Thus λ_a is a homomorphism of $\langle R, +\rangle$ into itself, that is, an endomorphism in Hom($\langle R, +\rangle$).

b) Note that for $a, b \in R$, we have $(\lambda_a\lambda_b)(x) = \lambda_a(\lambda_b(x)) = \lambda_a(bx) = a(bx) = (ab)x = \lambda_{ab}(x)$. Thus $\lambda_a\lambda_b = \lambda_{ab}$ and R' is closed under multiplication. We also have $(\lambda_a + \lambda_b)(x) =$

$\lambda_a(x) + \lambda_b(x) = ax + bx = (a + b)x = \lambda_{a+b}(x)$, so $\lambda_a + \lambda_b = \lambda_{a+b}$. Thus R' is closed under addition. From what we have shown, it follows that $\lambda_0 + \lambda_a = \lambda_{0+a} = \lambda_a$ and $\lambda_a + \lambda_0 = \lambda_{a+0} = \lambda_a$ so λ_0 acts as addditive identity. Finally, $\lambda_{-a} + \lambda_a = \lambda_{-a+a} = \lambda_0$ and $\lambda_a + \lambda_{-a} = \lambda_{a-a} = \lambda_0$ so R' contains an additive inverse of each element. Thus R' is a ring.

c) Let $\phi: R \longrightarrow R'$ be defined by $\phi(a) = \lambda_a$. By our work in part (b), we see that $\phi(a + b) = \lambda_{a+b} = \lambda_a + \lambda_b = \phi(a) + \phi(b)$ and $\phi(ab) = \lambda_{ab} = \lambda_a\lambda_b = \phi(a)\phi(b)$. Thus ϕ is a homomorphism, and is clearly onto R'. Suppose that $\phi(a) = \phi(b)$. Then $ax = bx$ for all $x \in R$. Since R has unity (and this is the only place where that hypothesis is needed), we have in particular $a1 = b1$ so $a = b$. Thus ϕ is one to one and onto R', so it is an isomorphism.

SECTION 6.2 - Prime and Maximal Ideals

1. Since a finite integral domain is a field, the prime and the maximal ideals coincide. The ideals $\{0,2,4\}$ and $\{0,3\}$ are both prime and maximal since the factor rings are isomorphic to the fields $\mathbb{Z}_2$ and $\mathbb{Z}_3$ respectively.

2. Since a finite integral domain is a field, the prime and the maximal ideals coincide. The prime and maximal ideals are $\{0,2,4,6,8,10\}$ and $\{0,3,6,9\}$ leading to factor rings isomorphic to $\mathbb{Z}_2$ and $\mathbb{Z}_3$ respectively.

3. Since a finite integral domain is a field, the prime and the maximal ideals coincide. The prime and maximal ideals are $\{(0,0), (1,0)\}$ and $\{(0,0), (0,1)\}$ leading to factor rings isomorphic to $\mathbb{Z}_2$.

4. Since a finite integral domain is a field, the prime and the maximal ideals coincide. The prime and maximal ideals are $\{(0,0), (0,1), (0,2), (0,3)\}$ and $\{(0,0), (1,0), (0,2), (1,2)\}$ leading to factor rings isomorphic to $\mathbb{Z}_2$.

5. By Theorem 6.14, we need only find all values c such that

$x^2 + c$ is irreducible over $\mathbb{Z}_3$. Let $f(x) = x^2$. Then $f(0) = 0$, $f(1) = 1$, and $f(2) = 1$. We must find $c \in \mathbb{Z}_3$ such that $0 + c$ and $1 + c$ are both nonzero. Clearly $c = 1$ is the only choice.

6. By Theorem 6.14, we need only find all values c such that $x^3 + x^2 + c$ is irreducible over $\mathbb{Z}_3$. Let $f(x) = x^3 + x^2$. Then $f(0) = 0$, $f(1) = 2$, and $f(3) = 0$. We must find $c \in \mathbb{Z}_3$ such that $0 + c$ and $2 + c$ are both nonzero. Clearly $c = 2$ is the only choice.

7. By Theorem 6.14, we need only find all values c such that $g(x) = x^3 + cx^2 + 1$ is irreducible over $\mathbb{Z}_3$. When $c = 0$, $g(2) = 0$, when $c = 1$, $g(1) = 0$, but when $c = 2$, $g(x)$ has no zeros. Thus $c = 2$ is the only choice.

8. By Theorem 6.14, we need only find all values c such that $x^2 + x + c$ is irreducible over $\mathbb{Z}_5$. If $f(x) = x^2 + x$, then $f(0) = 0$, $f(1) = 2$, $f(2) = 1$, $f(3) = 2$, and $f(4) = 0$. We must find $c \in \mathbb{Z}_5$ such that $f(x) + c$ has no zeros, that is, $0 + c$, $1 + c$, and $2 + c$ must all be nonzero. Clearly $c = 1$ and $c = 2$ both work.

9. By Theorem 6.14, we need only find all values c such that $g(x) = x^2 + cx + 1$ is irreducible over $\mathbb{Z}_5$. We compute that when $c = 0$, $g(2) = 0$, when $c = 1$, $g(x)$ has no zeros, when $c = 2$, $g(-1) = 0$, when $c = 3$, $g(1) = 0$, and when $c = 4$, $g(x)$ has no zeros. Thus c can be either 1 or 4.

10. F T T F T T T F T F

11. $\mathbb{Z} \times 2\mathbb{Z}$ is a maximal ideal of $\mathbb{Z} \times \mathbb{Z}$, for the factor ring is isomorphic to $\mathbb{Z}_2$, which is a field.

12. $\mathbb{Z} \times \{0\}$ is a prime ideal of $\mathbb{Z} \times \mathbb{Z}$ that is not maximal, for the factor ring is isomorphic to $\mathbb{Z}$ which is an integral domain, but not a field.

13. $\mathbb{Z} \times 4\mathbb{Z}$ is a proper ideal that is not prime, for the factor ring is isomorphic to $\mathbb{Z}_4$ which has divisors of zero.

14. $\mathbb{Q}[x]/\langle x^2 - 5x + 6\rangle$ is not a field, for $x^2 - 5x + 6 = (x - 2)(x - 3)$ is not an irreducible polynomial, so the ideal

$\langle x^2 - 5x + 6\rangle$ is not maximal.

15. $\mathbb{Q}[x]/\langle x^2 - 6x + 6\rangle$ is a field, for the polynomial $x^2 - 6x + 6$ is irreducible over $\mathbb{Q}$ by the Eisenstein criterion with $p = 3$, so $\langle x^2 - 6x + 6\rangle$ is a maximal ideal.

16. Theorem 5.6 shows that every finite integral domain is a field. Let N be a prime ideal in a finite commutative ring R with unity. Then R/N is a finite integral domain, and therefore a field, and therefore N is a maximal ideal.

17. Yes, it is possible. $\mathbb{Z}_2 \times \mathbb{Z}_3$ is an example.

18. Yes, it is possible. $\mathbb{Z}_2 \times \mathbb{Z}_3$ is an example.

19. No. Enlarging the integral domain to a field of quotients, we would have to have a field containing (up to isomorphism) two different prime fields $\mathbb{Z}_p$ and $\mathbb{Z}_q$. The unity of each of these fields is a zero of $x^2 - x$, but this polynomial has only one nonzero solution in a field, namely the unity of the field.

20. Let M be a maximal ideal of R and suppose that $ab \in M$ but a is not in M. Let $N = \{ra + m \mid r \in R,\ m \in M\}$. From $(r_1a + m_1) + (r_2a + m_2) = (r_1 + r_2)a + (m_1 + m_2)$, we see that N is closed under addition. From $r(r_1a + m_1) = (rr_1)a + (rm_1)$ and the fact that M is an ideal, we see that that N is closed under multiplication by elements of R, and is of course closed itself under multiplication. Also $0 = 0a + 0$ is in N and $(-r)a + (-m) = -(ra) - m = -(ra + m)$ is in N. Thus N is an ideal. Clearly N contains M, but $N \neq M$ since $1a + 0 = a$ is in N but a is not in M. Since M is maximal, we must have $N = R$. Therefore $1 \in N$, so $1 = ra25 + m$ for some $r \in R$ and $m \in M$. Multiplying by b, we find that $b = rab + mb$. But ab and mb are both in M, so $b \in M$. We have shown that if $ab \in M$ and a is not in M, then $b \in M$. This is the definition of a prime ideal.

21. We use the addendum to Theorem 6.1 stated in the final paragraph of Section 6.1 and proved in Exercise 23 of that section. Suppose that N is any ideal of R. By the addendum mentioned using the canonical homomorphism $\gamma\colon R \longrightarrow R/N$, if M is a proper ideal of R properly containing N then $\gamma[M]$ is a proper nontrivial ideal of R/N. This shows that if M is not

maximal, then R/N is not a simple ring. On the other hand, suppose that R/N is not a simple ring, and let N' be a proper nontrivial ideal of R/N. By the addendum mentioned, $\gamma^{-1}[N']$ is an ideal of R, and of course $\gamma^{-1}[N'] \neq R$ since N' is a proper ideal of R/N and also $\gamma^{-1}[N']$ properly contains N since N' is nontrivial in R/N. Thus $\gamma^{-1}[N']$ is a proper ideal of R that properly contains N, so N is not maximal. We have proved *p if and only if q* by proving *not p if and only if not q*.

This exercise is the straightforward analogue of Theorem 3.14 for groups, that is, a maximal ideal of a ring is analogous to a maximal normal subgroup of a group.

22. Every ideal of $F[x]$ is principal by Theorem 6.13. Suppose $\langle f(x)\rangle \neq \{0\}$ is a proper prime ideal of $F[x]$. Then every polynomial in $\langle f(x)\rangle$ has degree greater than or equal to $f(x)$. Thus if $f(x) = g(x)h(x)$ in $F[x]$ where the degree of both $g(x)$ and $h(x)$ is less than the degree of $f(x)$, neither $g(x)$ nor $h(x)$ can be in $\langle f(x)\rangle$. This would contradict the fact that $\langle f(x)\rangle$ is a prime ideal, so no such factorization of $f(x)$ in $F[x]$ can exist, that is, $f(x)$ is irreducible in $F[x]$. By Theorem 6.14, $\langle f(x)\rangle$ is therefore a maximal ideal of $F[x]$.

23. If $f(x)$ divides $g(x)$, then $g(x) = f(x)q(x)$ for some $q(x) \in F[x]$, so $g(x) \in \langle f(x)\rangle$ since this ideal consists of all multiples of $f(x)$. Conversely, if $g(x) \in \langle f(x)\rangle$ then $g(x)$ is some multiple $h(x)f(x)$ of $f(x)$ for $h(x) \in F[x]$. The equation $g(x) = h(x)f(x)$ is the definition of $f(x)$ dividing $g(x)$.

24. The equation

$$[r_1(x)f(x) + s_1(x)g(x)] + [r_2(x)f(x) + s_2(x)g(x)] = [r_1(x) + r_2(x)]f(x) + [s_1(x) + s_2(x)]g(x)$$

shows that N is closed under addition. The equation

$$[r(x)f(x) + s(x)g(x)]h(x) = h(x)[r(x)f(x) + s(x)g(x)] = [h(x)r(x)]f(x) + [h(x)s(x)]g(x)$$

shows that N is closed under multiplication by any $h(x) \in F[x]$; in particular, N is closed under multiplication. Since $0 = 0f(x) + 0g(x)$ and $-[r(x)f(x) + s(x)g(x)] = [-r(x)]f(x) + [-s(x)]g(x)$ are in N, we see that N is an ideal.

Suppose now that $f(x)$ and $g(x)$ have different degrees and that $N \neq F[x]$. Suppose that $f(x)$ is irreducible. By Theorem 6.14, we know that then $\langle f(x) \rangle$ is a maximal ideal of $F[x]$. But clearly $\langle f(x) \rangle \subseteq N$. Since $N \neq F[x]$, we must have $\langle f(x) \rangle = N$. In partcular $g(x) \in N$ so $g(x) = f(x)q(x)$. Since $f(x)$ and $g(x)$ have different degrees, we see that $g(x) = f(x)q(x)$ must be a factorization of $g(x)$ into polynomials of smaller degree than the degree of $g(x)$. Hence $g(x)$ is not irreducible.

25. Given that the Fundamental Theorem of Algebra holds, let N be the smallest ideal of $\mathbb{C}[x]$ containing r polynomials $f_1(x)$, $f_2(x)$, $\cdots$, $f_r(x)$. Since every ideal in $\mathbb{C}[x]$ is a principal ideal, we have $N = \langle h(x) \rangle$ for some polynomial $h(x) \in \mathbb{C}[x]$. Let α_1, α_2, $\cdots$, α_s be all the zeros in $\mathbb{C}$ of $h(x)$, and let α_i be a zero of multiplicity m_i. By the Fundamental Theorem of of Algebra, $h(x)$ must factor into linear factors in $\mathbb{C}[x]$, so that

$$h(x) = c(x - \alpha_1)^{m_1}(x - \alpha_2)^{m_2}\cdots(x - \alpha_s)^{m_s}.$$

Since each α_i is also a zero of $g(x)$ by hypothesis, the Fundamental Theorem of Algebra shows that $g(x) = k(x)(x - \alpha_1)(x - \alpha_2)\cdots(x - \alpha_s)$ for some polynomial $k(x) \in \mathbb{C}[x]$. Let m be the maximum of m_1, m_2, $\cdots$, m_s. Then $g(x)^m =$ has each $(x - \alpha_i)^{m_i}$ as a factor, and thus has $h(x)$ as a factor, so $g(x)^m \in \langle h(x) \rangle = N$.

Conversely, let the Nullstellensatz for $\mathbb{C}[x]$ hold. Suppose that the Fundamental Theorem of Algebra does not hold, so that there exists a nonconstant polynomial $f_1(x)$ in $\mathbb{C}[x]$ having no zero in $\mathbb{C}$. Then every zero of $f_1(x)$ is also a zero of every polynomial in $\mathbb{C}[x]$, since there are no zeros of $f_1(x)$. By the Nullstellensatz for $\mathbb{C}[x]$, every element of $\mathbb{C}[x]$ has the property that some power of it is in $\langle f_1(x) \rangle$, so that some power of every polynomial in $\mathbb{C}[x]$ has $f_1(x)$ as a factor. This is certainly impossible, since $1 \in \mathbb{C}[x]$ and $f_1(x)$ is a nonconstant polynomial and thus is not a factor of

$1^n = 1$ for any positive integer n. Thus there can be no such polynomial $f_1(x)$ in $\mathbb{C}[x]$, and the Fundamental Theorem of Algebra holds.

26. a) The proof that $A + B$ is an ideal is identical with the proof that $M + N$ is an ideal in Exercise 38 of Section 6.1.
b) Since $a + 0 = a$ is in $A + B$ and $0 + b = b$ is in $A + B$ for all $a \in A$ and $b \in B$, we see that $A \subseteq (A+B)$ and $B \subseteq (A+B)$.

27. a) It is clear that AB is closed under multiplication, for the sum of [a sum of m products of the form a_ib_i] and [a sum of n products of the form a_jb_j] is a sum of $m + n$ products of this form, and hence is in AB. Since A and B are ideals, we see that $r(a_ib_i) = (ra_i)b_i$ and $(a_ib_i)r = a_i(b_ir)$ are again of the form a_jb_j. The distributive laws then show that each sum of products a_ib_i when multiplied on the left or right by $r \in R$ produces again a sum of such products. Thus AB is closed under multiplication by elements of R, and hence is closed itself under multiplication. Since $0 = 00$ and $-(a_ib_i) = (-a_i)b_i$ are in AB, we see that AB is indeed an ideal.
b) Exercise 28 of Section 6.1 shows that $A \cap B$ is an ideal. Regarding a_ib_i as a_i multiplied by an element of R on the right, we see that $a_ib_i \in A$. Regarding a_ib_i as b_i multiplied by an element of R on the left, we see that $a_ib_i \in B$. Hence $a_ib_i \in A \cap B$. Since $A \cap B$ is an ideal, and thus closed under addition, we see that $AB \subseteq A \cap B$.

28. Let $x, y \in A:B$, and let $b \in B$. Then $xb \in A$ and $yb \in A$ so $(x + y)b = xb + yb$ is in A, because A is closed under addition. Thus $A:B$ is closed under addition.

Turning to multiplication, let $r \in R$. We want to show that xr and rx are in $A:B$, that is, that $(xr)b$ and $(rx)b$ are in A for all $b \in B$. Since multiplication is commutative by hypothesis, it suffices to show that xbr is in A for all $b \in B$. But since $x \in A:B$, we know that $xb \in A$, and A is an ideal so $(xb)r \in A$. Thus $A:B$ is closed under multiplication by elements in R; in particular, it is itself closed under multiplication.

Since $0b = 0$ and $0 \in A$, we see that $0 \in A:B$. Since $(-x)b = -(xb)$ and $xb \in A$ implies $(-xb) \in A$, we see that $A:B$

contains the additive identity and additive inverses, so it is an ideal.

29. Clearly S is closed under addition, contains the zero matrix, and contains the additive inverse of each of its elements. The computation

$$\begin{bmatrix} a & b \\ 0 & 0 \end{bmatrix}\begin{bmatrix} c & d \\ 0 & 0 \end{bmatrix} = \begin{bmatrix} ac & ad \\ 0 & 0 \end{bmatrix}$$

shows that S is closed under multiplication, so it is a subring of $M_2(F)$. The computations

$$\begin{bmatrix} 0 & 1 \\ 1 & 0 \end{bmatrix}\begin{bmatrix} a & b \\ 0 & 0 \end{bmatrix} = \begin{bmatrix} 0 & 0 \\ a & b \end{bmatrix} \quad \text{and} \quad \begin{bmatrix} a & b \\ 0 & 0 \end{bmatrix}\begin{bmatrix} c & d \\ e & f \end{bmatrix} = \begin{bmatrix} ac+be & ad+bf \\ 0 & 0 \end{bmatrix}$$

show that S is not closed under left multiplication by elements of $M_2(F)$, but is closed under right multiplication by those elements. Thus S is a right ideal, but not a left ideal, of $M_2(F)$.

30. Let $R = M_2(\mathbb{Z}_2)$. The computations

$$\begin{bmatrix} 0 & 1 \\ 1 & 0 \end{bmatrix}\begin{bmatrix} a & b \\ c & d \end{bmatrix} = \begin{bmatrix} c & d \\ a & b \end{bmatrix} \quad \text{and} \quad \begin{bmatrix} a & b \\ c & d \end{bmatrix}\begin{bmatrix} 0 & 1 \\ 1 & 0 \end{bmatrix} = \begin{bmatrix} c & d \\ a & b \end{bmatrix}$$

show that for every matrix in an ideal N of R, the matrix obtained by interchanging its rows and the matrix obtained by interchanging it columns are again in N. Thus if N contains any one of the four matrices having 1 for one entry and 0 for all the others, then N contains all four such matrices, and hence all nonzero matrices since any matrix in R is a sum of such matrices. By interchanging rows and columns every nonzero matrix with at least two nonzero entries can be brought to one of the following forms:

two zero entries: $\begin{bmatrix} 1 & 1 \\ 0 & 0 \end{bmatrix}, \begin{bmatrix} 1 & 0 \\ 1 & 0 \end{bmatrix}, \begin{bmatrix} 1 & 0 \\ 0 & 1 \end{bmatrix}$

one zero entry: $\begin{bmatrix} 0 & 1 \\ 1 & 1 \end{bmatrix}$ no zero entry: $\begin{bmatrix} 1 & 1 \\ 1 & 1 \end{bmatrix}$.

The following computations then show that every nontrivial ideal of R must contain one of the four matrices with only one nonzero entry, and hence must be all of R:

$$\begin{bmatrix}1 & 1\\0 & 0\end{bmatrix}\begin{bmatrix}1 & 0\\0 & 0\end{bmatrix} = \begin{bmatrix}1 & 0\\0 & 0\end{bmatrix}, \qquad \begin{bmatrix}1 & 0\\0 & 0\end{bmatrix}\begin{bmatrix}1 & 0\\1 & 0\end{bmatrix} = \begin{bmatrix}1 & 0\\0 & 0\end{bmatrix}$$

$$\begin{bmatrix}1 & 0\\0 & 0\end{bmatrix}\begin{bmatrix}1 & 0\\0 & 1\end{bmatrix} = \begin{bmatrix}1 & 0\\0 & 0\end{bmatrix}, \qquad \begin{bmatrix}1 & 0\\0 & 0\end{bmatrix}\begin{bmatrix}0 & 1\\1 & 1\end{bmatrix} = \begin{bmatrix}0 & 1\\0 & 0\end{bmatrix}.$$

$$\begin{bmatrix}1 & 0\\0 & 0\end{bmatrix}\begin{bmatrix}1 & 1\\1 & 1\end{bmatrix}\begin{bmatrix}1 & 0\\0 & 0\end{bmatrix} = \begin{bmatrix}1 & 0\\0 & 0\end{bmatrix}.$$

CHAPTER 7

FACTORIZATION

SECTION 7.1 - Unique Factorization Domains

1. Yes. 5 is a prime in $\mathbb{Z}$.
2. Yes. -17 is an associate of 17 which is a prime in $\mathbb{Z}$.
3. No. $14 = (2)(7)$ and neither 2 nor 7 is a unit in $\mathbb{Z}$.
4. Yes. $2x - 3$ is a primitive irreducible polynomial in $\mathbb{Q}[x]$.
5. No. $2x - 10 = 2(x - 5)$ and neither 2 nor $x - 5$ is a unit in $\mathbb{Z}[x]$.
6. Yes. $2x - 3$ does not factor into a product of nonconstant polynomials of lower degree in $\mathbb{Q}[x]$.
7. Yes. $2x - 10$ does not factor into a product of nonconstant polynomials of lower degree in $\mathbb{Q}[x]$.
8. Yes. $2x - 10$ does not factor into a product of nonconstant polynomials of lower degree in $\mathbb{Z}_{11}[x]$.
9. In $\mathbb{Z}[x]$, the only associates of $2x - 7$ are $2x - 7$ and $-2x + 7$, for 1 and -1 are the only units.

   In $\mathbb{Q}[x]$, every multiple $q(2x - 7)$ for $q \in \mathbb{Q}$, $q \neq 0$, is an associate of $2x - 7$ since all nonzero elements of $\mathbb{Q}$ are units in $\mathbb{Q}[x]$. In particular, $2x - 7$, $-10x + 35$, $x - 7/2$, and $8x - 28$ are associates of $2x - 7$.

   In $\mathbb{Z}_{11}[x]$, every multiple $a(2x - 7)$ for $a \in \mathbb{Z}_{11}$, $a \neq 0$, is an associate of $2x - 7$ since all nonzero elements of $\mathbb{Z}_{11}$ are units in $\mathbb{Z}_{11}[x]$. In particular, $2x - 7$, $2(2x - 7) = 4x + 8$, $5(2x - 7) = 10x + 9$, and $7(2x - 7) = 3x + 6$ are associates of $2x - 7$.
10. In $\mathbb{Z}[x]$: $4x^2 - 4x + 8 = (2)(2)(x^2 - x + 2)$. The quadratic is irreducible since its roots are complex.

    In $\mathbb{Q}[x]$: $4x^2 - 4x + 8$ is already irreducible since its roots are complex.

    In $\mathbb{Z}_{11}[x]$: $4x^2 - 4x + 8 = (4x + 2)(x + 4)$. We found this factorization by discovering that -4 and 5 are zeros of the polynomial.

11. $6(3x^2 - 2x + 8)$

12. Since every nonzero $q \in \mathbf{Q}$ is a unit in $\mathbf{Q}[x]$, we can "factor out" any nonzero rational constant as the (unit) content of this polynomial. For example,

$$(1)(18x^2 - 12x + 48) \quad \text{and} \quad (1/2)(36x^2 - 24x + 96)$$

are two of an infinite number of possible answers.

13. $(1)(2x^2 - 3x + 6)$, since the polynomial is primitive.

14. Since every nonzero $a \in \mathbf{Z}_7$ is a unit in $\mathbf{Z}_7[x]$, we can "factor out" any nonzero constant as the (unit) content of this polynomial. For example,

$$(1)(2x^2 - 3x + 6) \quad \text{and} \quad (5)(6x^2 + 5x + 4)$$

are two of six possible answers.

15. T T T F T F F T F T

16. They are the irreducibles of D, together with the irreducibles of $F[x]$ which are in $D[x]$ and are furthermore *primitive* polynomials in $D[x]$. (See the paragraph following following Lemma 7.6)

17. $2x + 4$ is irreducible in $\mathbf{Q}[x]$ but not in $\mathbf{Z}[x]$.

18. Not every nonzero nonunit of $\mathbf{Z} \times \mathbf{Z}$ has a factorization into irreducibles. For example, $(1, 0)$ is not a unit, and every factorization of $(1,0)$ has a factor of the form $(\pm 1,0)$, which is not irreducible since $(\pm 1, 0) = (\pm 1, 0)(1, 9)$. The only irreducibles of $\mathbf{Z} \times \mathbf{Z}$ are $(\pm 1, p)$ and $(q, \pm 1)$, where p and q are irreducibles in $\mathbf{Z}$.

19. Let p be a prime of D, and suppose that $p = ab$ for some $a, b \in D$. Then $ab = (1)p$, so p divides ab and thus divides either a or b, since p is a prime. Suppose that $a = pc$. Then $p = (1)p = pcb$ and cancellation in the integral domain D yields $1 = cb$, so b is a unit of D. Similarly, if p divides b, we conclude that a is a unit in D. Thus either a or b is a unit, so p is an irreducible.

20. Let q be an irreducible in a UFD, and suppose that q divides ab. We must show that either q divides a or q divides b. Let $ab = qc$, and factor ab into irreducibles by first factoring a into irreducibles, then factoring b into irreducibles, and finally taking the product of these two factorizations. Now ab could also be factored into irreducibles by taking q times a factorization of c into

irreducibles. Since factorization into irreducibles in a UFD is unique up to order and associates, it must be that an associate of q appears in the first factorization, formed by taking factors of a times factors of b. Thus an associate of q, say uq, appears in the factorization of a or in the factorization of b. It follows at once that q divides either a or b.

21. *Reflexive:* $a = a(1)$, so $a \sim a$.
Symmetric: Suppose $a \sim b$, so that $a = bu$ for a unit u. Then u^{-1} is a unit and $b = au^{-1}$ so $b \sim a$.
Transitive: Suppose that $a \sim b$ and $b \sim c$. Then there are units u_1 and u_2 such that $a = bu_1$ and $b = cu_2$. Substituting, we have $a = cu_2u_1 = c(u_2u_1)$. Since the product u_2u_1 of two units is again a unit, we find that $a \sim c$.

22. Let a and b be nonunits in $D^* - U$. Suppose that ab is a unit, so that $(ab)c = 1$ for some $c \in D$. Then $a(bc) = 1$ and a is a unit, contrary to our choice for a. Thus ab is again a nonunit, and $ab \neq 0$ since D has no divisors of zero. Hence $ab \in (D^* - U)$ also.

$D^* - U$ is not a group, for the multiplicative identity 1 is a unit, and hence is not in $D^* - U$.

23. Let $g(x)$ be a nonconstant divisor of the primitive polynomial $f(x)$ in $D[x]$. Suppose that $f(x) = g(x)q(x)$. Since D is a UFD, we know that $D[x]$ is a UFD also. Factor $f(x)$ into irreducibles by factoring each of $g(x)$ and $q(x)$ into irreducibles, and then taking the product of these factorizations. Each nonconstant factor appearing is an irreducible in $D[x]$, and hence is a primitive polynomial. Since the product of primitive polynomials is primitive by the corollary of Lemma 7.5, we see that the content of $g(x)q(x)$ is the product of the content of $g(x)$ and the content of $q(x)$, and must be the same (up to a unit factor) as the content of $f(x)$. But $f(x)$ has content 1 since it is primitive. Thus $g(x)$ and $q(x)$ both have content 1. In particular, $g(x)$ is a product of primitive polynomials, so it is primitive by the corollary of Lemma 7.5.

24. Let N be an ideal in a PID D. If N is not maximal, then there is a proper ideal N_1 of D such that $N \subset N_1$. If N_1 is

not maximal, we find a proper ideal N_2 such that $N_1 \subset N_2$. Continuing this process, we construct a chain $N \subset N_1 \subset N_2 \subset \cdots \subset N_i$ of ideals, each properly contained in the next except for the last ideal. Since a PID satisfies the ascending chain condition, we cannot have an infinite such chain, so after some finite number of steps we must encounter an ideal N_r that contains N and that is not properly contained in any proper ideal of D. That is, we must attain a maximal ideal N_r.

25. $x^3 - y^3 = (x - y)(x^2 + xy + y^2)$. We claim that $x^2 + xy + y^2$ is irreducible in $\mathbf{Q}[x,y]$. Suppose that $x^2 + xy + y^2$ factors into a product of two polynomials that are not units in $\mathbf{Q}[x,y]$. Such a factorization would have to be of the form $x^2 + xy + y^2 = (ax + by)(cx + dy)$ with a, b, c, and d all nonzero elements of $\mathbf{Q}$. Consider the evaluation homomorphism ϕ_1: $(\mathbf{Q}[x])[y] \rightarrow \mathbf{Q}[x]$ such that $\phi(y) = 1$. Applying ϕ_1 to both sides of such a factorization would yield $x^2 + x + 1 = (ax + b)(cx + d)$. But $x^2 + x + 1$ is irreducible in $\mathbf{Q}[x]$ since its zeros are complex, so no such factorization exists. This shows that $x^2 + xy + y^2$ is irreducible in $(\mathbf{Q}[x])[y]$ which is isomorphic to $\mathbf{Q}[x,y]$ under an isomorphism that identifies $y^2 + yx + x^2$ and $x^2 + xy + y^2$.

26. We show that ACC implies MC, that MC implies FBC, and that FBC implies ACC.

ACC implies MC: Suppose that MC does not hold for some set S of ideals of R, that is, suppose it is not true that S contains an ideal not properly contained in any other ideal of S. Then every ideal of S *is* properly contained in another ideal of S. We can then start with any ideal N_1 of S and find an ideal N_2 of S properly containing it, then $N_3 \in S$ properly containing N_2, etc. Thus we could construct an infinite chain of ideals $N_1 \subset N_2 \subset N_2 \subset \cdots$ which contradicts the ACC. Hence the ACC implies the MC.

MC implies FBC: Suppose the FBC does not hold, and let N be an ideal of R having no finite basis. Let $b_1 \in N$ and let

$N_1 = \langle b_1 \rangle$ be the smallest ideal of N containing b_1. Now $N_1 \neq N$ or $\{b_1\}$ would be a basis for N, so find $b_2 \in N$ such that b_2 is not in N_1. Let N_2 be the intersection of all ideals containing b_1 and b_2. Since N contains b_1 and b_2, we see that $N_2 \subseteq N$, but $N_2 \neq N$ since $\{b_1, b_2\}$ cannot be a basis for N. We then choose $n_3 \in N$ but not in N_2, and let N_3 be the intersection of all ideals containing b_1, b_2, and b_3. Continuing this process, using the fact that N has no finite basis, we can construct an infinite chain of ideals $N_1 \subset N_2 \subset N_3 \subset \cdots$ of R. But then the set $S = \{N_i \mid i \in \mathbb{Z}^+\}$ is a set of ideals, each of which is properly contained in another ideal of the set, for $N_i \subset N_{i+1}$. This would contradict the MC, so the FBC must be true.

FBC implies ACC: Let $N_1 \subseteq N_2 \subseteq N_3 \subseteq \cdots$ be a chain of ideals in R, and let $N = \bigcup_{i=0}^{\infty} N_i$. It is easy to see that N is an ideal of R. Let $B_N = \{b_1, b_2, \cdots, b_n\}$ be a finite basis for N. Let $b_j \in N_{i_j}$. If r is the maximum of the subscripts i_j, then $B_N \subseteq N_r$. Since $N_r \subseteq N$ and N is the intersection of all ideals containing B_N, we must have $N_r = N$. Hence $N_r = N_{r+1} = N_{r+2} = \cdots$ so the ACC is satisfied.

27. *DCC implies mC:* Suppose that mC does not hold in R, and let S be a set of ideals in R where the mC fails, so that every ideal in S does properly contain another ideal of S. Then starting with any ideal $N_1 \in S$, we can find an ideal $N_2 \in S$ properly contained in N_1, and then an ideal N_3 of S properly contained in N_2, etc. This leads to an unending chain of ideals, each properly containing the next, which would contradict the DCC. Thus the DCC implies the mC.

mC implies DCC: Let N_1, N_2, N_3, $\cdots$ be a sequence of ideals in R such that N_i contains N_{i+1} for $i \in \mathbb{Z}^+$. Let $S = \{N_i \mid i \in \mathbb{Z}^+\}$. By the mC, there is some ideal N_r of S that does not properly contain any other ideal in the set. Thus $N_r = N_{r+1} = N_{r+2} = \cdots$ so the ACC holds.

28. $\mathbb{Z}$ is a ring in which the ACC holds, since $\mathbb{Z}$ is a PID. However, $\mathbb{Z}$, $2\mathbb{Z}$, $4\mathbb{Z}$, $\cdots$, $2^{i-1}\mathbb{Z}$ is a sequence of ideals, each properly containing the next, so the DCC does not hold in $\mathbb{Z}$.

SECTION 7.2 - Euclidean Domains

1. Yes, it is a valuation. To see this, remember that we know $|\ |$ is a valuation on $\mathbb{Z}$. For Condition 1, find q and r such that $a = bq + r$ where either $r = 0$ or $|r| < |b|$. Then surely we have either $\nu(r) = 0$ or $\nu(r) = r^2 < b^2 = \nu(b)$, since r and b are integers. For Condition 2, note that $\nu(a) = a^2 \leq a^2b^2 = \nu(b)$ for nonzero a and b, since a and b are integers.

2. No, it is not a valuation. Let $a = x$ and $b = 2x$ in $\mathbb{Z}[x]$. There are no $q(x)$, $r(x) \in \mathbb{Z}[x]$ satisfying $x = (2x)q(x) + r(x)$ where the degree of $r(x)$ is less than 1.

3. No, it is not a valuation. Let $a = x$ and $b = x + 2$ in $\mathbb{Z}[x]$. There are no $q(x)$, $r(x) \in \mathbb{Z}[x]$ satisfying $x = (x + 2)q(x) + r(x)$ where $\nu(r(x)) < 1$.

4. No, it is not a valuation. Let $a = 1/2$ and $b = 1/3$. Then $\nu(a) = (1/2)^2 = 1/4 > 1/36 = \nu(1/6) = \nu(ab)$, so Condition 2 is violated.

5. Yes, it is a valuation, but not a useful one. Let $a, b \in \mathbb{Q}$. If $b \neq 0$, let $q = a/b$. Then $a = bq + 0$, which satisfies Condition 1. For Condition 2, if both a and b are nonzero, then $\nu(a) = 50 \leq 50 = \nu(ab)$.

6. $23 = 3(138) - 1(391)$ but $138 = 3{,}266 - 8(391)$,
$23 = 3[3{,}266 - 8(391)] - 1(391) = 3(3{,}266) - 25(391)$
but $391 = 7(3{,}266) - 22{,}471$,
$23 = 3(3{,}266) - 25[7(3{,}266) - 22{,}471]$
$= 25(22{,}471) - 172(3{,}266)$.

Section 7.2

7. 49,349 = (15,555)3 + 2,684,
   15,555 = (2,684)6 - 549,
   2,684 = (549)5 - 61,
   549 = (61)9 + 0, so the gcd is 61.

8. 61 = 5(549) - 2,684 but 549 = 6(2,684) - 15,555,
   61 = 5[6(2,684) - 15,555] - 2,684 = 29(2,684) - 5(15,555)
   but 2,684 = 49,349 - 3(15,555),
   61 = 29[49,349 - 3(15,555)] - 5(15,555)
   = 29(49,349) - 92(15,555).

9.

$$
\begin{array}{r|l}
 & x^4 - 2x \\
\hline
x^6-3x^5+3x^4-9x^3+5x^2-5x+2 & x^{10}-3x^9+3x^8-11x^7+11x^6-11x^5+19x^4-13x^3 \\
 & \qquad +8x^2-9x+3 \\
 & \underline{x^{10}-3x^9+3x^8-9x^7+5x^6-5x^5+2x^4} \\
 & -2x^7+6x^6-6x^5+17x^4-13x^3+8x^2-9x+3 \\
 & \underline{-2x^7+6x^6-6x^5+18x^4-10x^3+10x^2-4x} \\
 & -x^4-3x^3-2x^2-5x+3
\end{array}
$$

$$
\begin{array}{r|l}
 & -x^2 + 6x - 19 \\
\hline
-x^4 - 3x^3 - 2x^2 - 5x + 3 & x^6 - 3x^5 + 3x^4 - 9x^3 + 5x^2 - 5x + 2 \\
 & \underline{x^6 + 3x^5 + 2x^4 + 5x^3 - 3x^2} \\
 & -6x^5 + x^4 - 14x^3 + 8x^2 - 5x \\
 & \underline{-6x^5 - 18x^4 - 12x^3 - 30x^2 + 18x} \\
 & 19x^4 - 2x^3 + 38x^2 - 23x + 2 \\
 & \underline{19x^4 + 57x^3 + 38x^2 + 95x - 57} \\
 & -59x^3 - 118x + 59
\end{array}
$$

$-59x^3 - 118x + 59 = -59(x^3 + 2x - 1)$ and we alter the remainder by the unit factor -1/59 before the next step.

$$
\begin{array}{r|l}
 & -x - 3 \\
\hline
x^3 + 2x - 1 & -x^4 - 3x^3 - 2x^2 - 5x + 3 \\
 & \underline{-x^4 \qquad\quad - 2x^2 + x} \\
 & - 3x^3 \qquad - 6x + 3 \\
 & \underline{- 3x^3 \qquad - 6x + 3} \\
 & 0
\end{array}
$$

A gcd is $x^3 + 2x - 1$.

10. a) Yes, it is a UFD since $\mathbb{Z}$ is a UFD and Theorem 7.3 tells us that if D is a UFD, then $D[x]$ is a UFD.

b) The set described consists of all polynomials in $\mathbb{Z}[x]$ that have a constant term in $2\mathbb{Z}$, that is, an even number as constant term. This property of a polynomial is obviously satisfied by $0 \in 2\mathbb{Z}$ and is preserved under addition, subtraction, and multiplication by every element of $\mathbb{Z}[x]$, so the set is an ideal of $\mathbb{Z}[x]$.

c) $\mathbb{Z}[x]$ is not a PID since the ideal described in part (b) is not a principal ideal; a generating polynomial would have to have constant term 2, but could not yield, under multiplication by elements of $\mathbb{Z}[x]$, all the polynomials of the form $2 + nx$ in the ideal since n can be odd as well as even.

d) $\mathbb{Z}[x]$ is not a Euclidean domain, since every Euclidean domain is a PID by Theorem 7.4 but part (c) shows that $\mathbb{Z}[x]$ is not a PID.

11. T F T F T T T F T T

12. No, it does not. The arithmetic structure is completely determined by the binary operations of addition and of multiplication on any integral domain. A Euclidean valuation, if one exists, can be used to *study* the arithmetic structure, but it in no way changes it.

13. Let a and b be associates of a Euclidean domain D with Euclidean valuation ν. Since a and b are associates, there exists a unit u in D such that $a = bu$. Then u^{-1} is also a unit of D and $b = au^{-1}$. Condition 2 of a Euclidean valuation yields $\nu(b) \le \nu(bu) = \nu(a)$ and $\nu(a) \le \nu(au^{-1}) = \nu(b)$. Thus $\nu(a) = \nu(b)$.

14. Suppose that $\nu(a) < \nu(ab)$. If b were a unit, then a and ab would be associates, and by Exercise 13, we would have $\nu(a) = \nu(ab)$. Thus b is not a unit.

For the converse, suppose that $\nu(a) = \nu(ab)$. We claim that then $\langle a \rangle = \langle ab \rangle$, for the proof of Theorem 7.4 shows that a nonzero ideal in a Euclidean domain is generated by any element having minimum valuation in the ideal. Thus ab also generates $\langle a \rangle$, so $a = (ab)c$ for some c in D. Then, cancelling the a in the integral domain, we find that $1 = bc$, so b is a unit.

15. The statement is false. For example, let $D = \mathbb{Z}$ and let the valuation be $|\ |$. Then $|2| > 1$ and $|-3| > 1$, but $|2 + (-3)| = |-1| = 1$, so the given set is not closed under addition.

16. Let F be a field and let $\nu(a) = 1$ for all $a \in F$, $a \neq 0$. Since every nonzero element of F is a unit, we see that for nonzero $b \in F$ we have $a = b(a/b) + 0$, which satisfies Condition 1 for a Euclidean valuation. Also, $\nu(a) = \nu(ab) = 1$, so Condition 2 is satisfied.

17. a) For Condition 1, let $a, b \in D^*$ where $b \neq 0$. There exist $q, r \in D$ such that $a = bq + r$ where either $r = 0$ or $\nu(r) < \nu(b)$. But then either $r = 0$ or $\eta(r) = \nu(r) + s < \nu(b) + s = \eta(b)$, so Condition 1 holds. For $a, b \neq 0$, we have $\eta(a) = \nu(a) + s \leq \nu(ab) + s = \eta(ab)$, so Condition 2 holds. The hypothesis that $\nu(1) + s > 0$ guarantees that $\eta(a) > 0$ for all $a \in D^*$, since $\nu(1)$ is minimal among all $\nu(a)$ for $a \in D^*$ by Theorem 7.5.

    b) For Condition 1, let $a, b \in D^*$ where $b \neq 0$. There exist $q, r \in D$ such that $a = bq + r$ where either $r = 0$ or $\nu(r) < \nu(b)$. But then either $r = 0$ or $\lambda(r) = r\cdot\nu(a) < r\cdot\nu(b) = \lambda(b)$, so Condition 1 holds. For $a, b \neq 0$, we have $\lambda(a) = r\cdot\nu(a) \leq r\cdot\nu(ab) = \lambda(ab)$, so Condition 2 holds. The hypothesis that $r \in \mathbb{Z}^+$ guarantees that $\lambda(a) > 0$ and $\lambda(a)$ is an integer for all $a \in D^*$.

    c) Let ν be a Euclidean valuation on D. Then $\lambda(a) = 100\cdot\nu(a)$ for $a \in D^*$ is a Euclidean valuation on D by part (b). Let $s = 1 - \lambda(1)$. Then $\mu(a) = \lambda(a) + s$ for $a \in D^*$ is a valuation on D by part (a), and $\mu(1) = \lambda(1) + s = \lambda(1) + [1 - \lambda(1)] = 1$. If $a \neq 0$ is a nonunit in D, then $\nu(a) \geq \nu(1) + 1$ so $\lambda(a) = 100\cdot\nu(a) \geq 100\cdot[\nu(1) + 1]$ and $\mu(a) = \lambda(a) + s \geq 100\cdot[\nu(1) + 1] + 1 - \lambda(1) = 100\cdot[\nu(1) + 1] + 1 - 100\cdot\nu(1) = 100 + 1 = 101$.

18. We know that all multiples of $a \in D$ form the principal ideal $\langle a\rangle$ and all multiples of b form the principal ideal $\langle b\rangle$. By Exercise 28 of Section 6.1, the intersection of ideals is an ideal, so $\langle a\rangle \cap \langle b\rangle$ is an ideal, and consists of all common multiples of a and b. Since a Euclidean domain is a PID, this ideal has a generator c. Now $a|c$ and $b|c$ since c is a common multiple of a and of b. Since every common multiple of a and of b is in $\langle a\rangle \cap \langle b\rangle = \langle c\rangle$, we see that every common multiple is of the form dc, that is, every common multiple is a multiple of c. Thus c is a lcm of a and b.

19. The subgroup of $\langle \mathbb{Z}, +\rangle$ generated by $r, s \in \mathbb{Z}$ is $H = \{mr + ns \mid n, m \in \mathbb{Z}\}$. Now $H = G$ if and only if $1 \in H$, so

$H = G$ if and only if $1 = mr + ns$ for some $m, n \in \mathbb{Z}$. If r and s are relatively prime, then 1 is a gcd of r and of s, and the Euclidean algorithm together with the technique outlined in Exercise 6 show that 1 can be expressed in the form $1 = mr + ns$ for some $m, n \in \mathbb{Z}$. Conversely, if $1 = mr + ns$, then every integer dividing both m and n divides the righthand side of this equation, and hence divides 1, so 1 is a gcd of r and s.

20. If a and n are relatively prime, then a gcd of a and n is 1. By Theorem 7.6, we can express 1 in the form $1 = m_1a + m_2n$ for some $m_1, m_2 \in \mathbb{Z}$. Multiplying by b, we obtain $b = a(m_1b) + (bm_2)n$. Thus $x = m_1b$ is a solution of the congruence $ax \equiv b \pmod{n}$.

21. Suppose that the positive gcd d of a and n in $\mathbb{Z}$ divides b. By Theorem 7.6, we can express d in the form $d = m_1a + m_2n$ for some $m_1, m_2 \in \mathbb{Z}$. Multiplying by b/d we obtain $b = a(m_1b/d) + (bm_2/d)n$. Thus $x = m_1b/d$ is a solution of the congruence $ax \equiv b \pmod{n}$. Conversely, suppose that $ac \equiv b \pmod{n}$ so that n divides $ac - b$, say $ac - b = nq$. Then $b = ac - nq$. Since the positive gcd d of a and n divides the righthand side of this equation, it must be that d divides b also.

    In $\mathbb{Z}_n$, this result has the following interpretation: $ax = b$ has a solution in $\mathbb{Z}_n$ for nonzero $a, b \in \mathbb{Z}_n$ if and only if the positive gcd of a and n in $\mathbb{Z}$ divides b.

22. *Step 1.* Use the Euclidean algorithm to find the positive gcd d of a and n.
    Step 2. Use the technique of Exercise 6 to express d in the form $d = m_1a + m_2n$.
    Step 3. $x = m_1b/d$ is a solution of $ax \equiv b \pmod{n}$.

    We now illustrate with $22x \equiv 18 \pmod{42}$.
    Step 1. Find the gcd of 22 and 42:

$$\begin{aligned} 42 &= 1(22) + 20, \\ 22 &= 1(20) + 2, \\ 20 &= 10(2), \end{aligned}$$

    so 2 is a gcd of 22 and 42.
    Step 2. Express 2 in the form $m_1(22) + m_2(42)$:

$$2 = 22 - 1(20) \quad \text{but} \quad 20 = 42 - 1(22),$$
$$2 = 22 - 1[42 - 1(22)] = 2(22) + (-1)(42),$$

so $m_1 = 2$.

Step 3. A solution of $22x \equiv 18 \pmod{42}$ is $x = m_1 b/d = (2)(18)/2 = 18$.

SECTION 7.3 - Gaussian Integers and Norms

1. If α divides 5 in $\mathbb{Z}[i]$, then $N(\alpha)$ must divide $N(5) = 25$, so $N(\alpha)$ must be 1, 5, or 25. If $N(\alpha) = 1$, then α is a unit, and if $N(\alpha) = 25$, then $N(5/\alpha)$ is a unit, and we are not interested in such factorizations. Thus we must have $N(\alpha) = 5$, so the possibilities are $\alpha = a + bi$ where $a = \pm 1$ and $b = \pm 2$ or where $a = \pm 2$ and $b = \pm 1$. We note that $(1 + 2i)(1 - 2i) = 5$, and $1 + 2i$ and $1 - 2i$ are irreducibles since they have a prime, 5, as norm (Theorem 7.9). Since $\mathbb{Z}[i]$ is a UFD, this factorization of 5 is unique up to unit factors. For example, $[i(1 + 2i)][-i(1 - 2i)] = (-2 + i)(-2 - i)$ is another factorization of 5 into irreducibles.

2. Proceeding as in the answer to Exercise 1, we find that $N(\alpha)$ must be 7 if α is an irreducible dividing 7. Since the equation $a^2 + b^2$ has no solutions in integers, it must be that 7 is already irreducible in $\mathbb{Z}[i]$.

3. Proceeding as in the answer to Exercise 1, we find that $N(\alpha)$ must be a divisor of 25, and 5 is the only possibility to yield an irreducible. As in the answer to Exercise 1, we must have $\alpha = a + bi$ where a is ± 1 and b is ± 2 or where a is ± 2 and b is ± 1. A bit of trial and error shows that $4 + 3i = (1 + 2i)(2 - i)$, and $1 + 2i$ and $2 - i$ are irreducible since they have the prime 5 as norm.

4. Proceeding as in the answer to Exercise 1, we find that $N(\alpha)$ must be a divisor of 85, and must be either 5 or 17.

   If $N(\alpha) = 5$, then $\alpha = a + bi$ where $a^2 + b^2 = 5$, so either $a = \pm 1$ and $b = \pm 2$ or $a = \pm 2$ and $b = \pm 1$. We compute

   $$\frac{6 - 7i}{1 + 2i} = \frac{6 - 7i}{1 + 2i} \cdot \frac{1 - 2i}{1 - 2i} = \frac{-8 - 19i}{5}$$

   and find that the answer is not in $\mathbb{Z}[i]$. There is no use trying $\pm 1(1 + 2i)$ of $\pm i(1 + 2i)$, so we try $1 - 2i$. We obtain

$$\frac{6 - 7i}{1 - 2i} = \frac{6 - 7i}{1 - 2i}\cdot\frac{1 + 2i}{1 + 2i} = \frac{20 + 5i}{5} = 4 + i.$$

Thus $6 - 7i = (1 - 2i)(4 + i)$, and $1 - 2i$ and $4 + i$ are irreducibles since their norms are the primes 5 and 17 respectively.

5. We have $6 = 2\cdot 3 = (-1 + \sqrt{-5})(-1 - \sqrt{-5})$. The numbers 2 and 3 are both irreducible in $\mathbf{Z}[i]$ since the equations $a^2 + 5b^2 = 2$ and $a^2 + 5b^2 = 3$ have no solutions in integers. (See Example 3 in the text.) If $-1 + \sqrt{-5}$ were not irreducible, then it would be a product $\alpha\beta$ where neither α nor β is a unit and $N(\alpha\beta) = N(\alpha)N(\beta) = 6$. This means that we would have to have $N(\alpha) = 2$ or 3, which is impossible as we have just seen.

   Since the only units in $\mathbf{Z}[\sqrt{-5}]$ are ± 1, we have two essentially different factorizations of 6.

6. We compute α/β:

$$\frac{7 + 2i}{3 - 4i} = \frac{7 + 2i}{3 - 4i}\cdot\frac{3 + 4i}{3 + 4i} = \frac{13 + 34i}{25}.$$

   Now we take the integer 1 closest to 13/25 and the integer 1 closest to 34/25, and let $\sigma = 1 + 1i = 1 + i$. Then we compute

$$\begin{aligned}\rho = \alpha - \beta\sigma &= (7 + 2i) - (3 - 4i)(1 + i)\\ &= (7 + 2i) - (7 - i) = 3i.\end{aligned}$$

   Then $\alpha = \sigma\beta + \rho$ where $N(\rho) = N(3i) = 9 < 25 = N(3 - 4i) = N(\beta)$.

7. We let $\alpha = 5 - 15i$ and $\beta = 8 + 6i$, and compute σ and ρ as in the answer to Exercise 6:

   $\dfrac{5 - 15i}{8 + 6i} = \dfrac{5 - 15i}{8 + 6i}\cdot\dfrac{8 - 6i}{8 - 6i} = \dfrac{-50 - 150i}{100}$. We take $\sigma = -i$, and obtain $\rho = \alpha - \sigma B = (5 - 15i) - (-i)(8 + 6i) = -1 - 7i$, so

$$5 - 15i = (8 + 6i)(-i) + (-1 - 7i).$$

   Continuing the Euclidean algorithm, we now take $\alpha = 8 + 6i$ and $\beta = -1 - 7i$, and obtain

$$\frac{8 + 6i}{-1 -7i} = \frac{8 + 6i}{-1 - 7i}\cdot\frac{-1 + 7i}{-1 + 7i} = \frac{-50 + 50i}{50} = -1 + i.$$

   Since $-1 + i \in \mathbf{Z}[i]$, we are done, and a gcd of $5 - 15i$ and $8 + 6i$ is $-1 - 7i$. Of course, the other gcd's are obtained by multiplying by the units -1, $\pm i$, so $1 + 7i$, $-7 + i$, and $7 - i$ are also acceptable answers.

8. T T T F T T T F T T

9. Suppose that $\pi = \alpha\beta$. Then $N(\pi) = N(\alpha)N(\beta)$. Since $N(\pi)$ is the minimal norm > 1, one of $N(\alpha)$ and $N(\beta)$ must be $N(\pi)$ and the other must be 1. Thus either α of β has norm 1, and is thus a unit by hypothesis. Therefore π is an irreducible in D.

10. a) We know that in $\mathbf{Z}[i]$, the units are precisely the elements ± 1, $\pm i$ of norm 1. By Theorem 7.9, every element of $\mathbf{Z}[i]$ having as norm a prime in $\mathbf{Z}$ is an irreducible. Since $N(1 + i) = 1^2 + 1^2 = 2$, we see that $1 + i$ is an irreducible. The equation $2 = -i(1 + i)^2$ thus gives the desired factorization of 2.

b) Every odd prime in $\mathbf{Z}$ is congruent to either 1 or 3 modulo 4. If $p \equiv 1 \pmod 4$, then Theorem 7.10 shows that $p = a^2 + b^2 = (a + ib)(a - ib)$ where neither $a + ib$ nor $a - ib$ is a unit since they each have norm $a^2 + b^2 = p > 1$, so p is not an irreducible.

Conversely, if p is not an irreducible, then $p = (a + bi)(c + di)$ in $\mathbf{Z}[i]$ where neither factor is a unit, so that both $a + bi$ and $c + di$ have norm greater than 1. Taking the norm of both sides of the equation, we obtain $p^2 = (a^2 + b^2)(c^2 + d^2)$, so we must have $p = a^2 + b^2 = c^2 + d^2$. Theorem 7.10 then shows that we must have $p \equiv 1 \pmod 4$.

We have shown that an odd prime p is not irreducible if and only if $p \equiv 1 \pmod 4$ so an odd prime p is irreducible if and only if $p \equiv 3 \pmod 4$.

11. *Property 1:* Let $\alpha = a + bi$. Then $N(\alpha) = a^2 + b^2$. As a sum of squares, $a^2 + b^2 \geq 0$

Property 2: Continuing the argument for Property 1, we see that $a^2 + b^2 = 0$ if and only if $a = b = 0$, so $N(\alpha) = 0$ if and only if $\alpha = 0$.

Property 3: Let $\beta = c + di$. Then $\alpha\beta = (a + bi)(c + di) = (ac - bd) + (ad + bc)i$, so

$$N(\alpha\beta) = (ac - bd)^2 + (ad + bc)^2$$
$$= a^2c^2 - 2abcd + b^2d^2 + a^2d^2 + 2abcd + b^2c^2$$

$= a^2c^2 + b^2d^2 + a^2d^2 + b^2c^2$

$= (a^2 + b^2)(c^2 + d^2) = N(\alpha)N(\beta).$

12. *Property 1:* Let $\alpha = a + b\sqrt{-5}$. Then $N(\alpha) = a^2 + 5b^2$. As a sum of squares, $a^2 + (\sqrt{5}b)^2 \geq 0$
Property 2: Continuing the argument for Property 1, we see that $a^2 + 5b^2 = 0$ if and only if $a = b = 0$, so $N(\alpha) = 0$ if and only if $\alpha = 0$.
Property 3: (See also Exercise 17.) Let $\beta = c + d\sqrt{-5}$. Then $\alpha\beta = (a + b\sqrt{-5})(c + d\sqrt{-5}) = (ac - 5bd) + (ad + bc)\sqrt{-5}$, so

$N(\alpha\beta) = (ac - 5bd)^2 + 5(ad + bc)^2$

$= a^2c^2 - 10abcd + 25b^2d^2 + 5a^2d^2 + 10abcd + 5b^2c^2$

$= a^2c^2 + 25b^2d^2 + 5a^2d^2 + 5b^2c^2$

$= (a^2 + 5b^2)(c^2 + 5d^2) = N(\alpha)N(\beta).$

13. Let $\alpha \in D$. We give a proof by induction on $N(\alpha)$, starting with $N(\alpha) = 2$, that α has a factorization into irreducibles. Let $N(\alpha) = 2$. Then α is itself an irreducible by Theorem 7.9, and we are done.

Suppose that every element of norm > 1 but < k has a factorization into irreducibles, and let $N(\alpha) = k$. If α is an irreducible, then we are done. Otherwise, $\alpha = \beta\gamma$ where neither β nor γ is a unit, so $N(\beta) > 1$ and $N(\gamma) > 1$. From $N(\beta\gamma) = N(\beta)N(\gamma) = N(\alpha) = k$, we then see that $1 < N(\beta) < k$ and $1 < N(\gamma) < k$, so by the induction assumption, both β and γ have factorizations into a product of irreducibles. The product of these two factorizations then provides a factorization of α into irreducibles.

14. We work in the complex field $\mathbb{C}$. First note that the proof that $N(\alpha)N(\beta) = N(\alpha\beta)$ in Exercise 11 made no use of the fact that we were working with numbers $a + bi$ in $\mathbb{Z}[i]$. The proof is valid for numbers $c + di$ in $\mathbb{C}$, so if we define $N(a + bi) = a^2 + b^2 = |a + bi|^2$ for any $(a + bi) \in \mathbb{C}$, this norm continues to have the multiplicative property.

By construction of σ, we see that $|r - q_1| \leq 1/2$ and $|s - q_2| \leq 1/2$. Thus $N(\frac{\alpha}{\beta} - \sigma) = N((r+si) - (q_1 + q_2i)) \leq (1/2)^2 + (1/2)^2 = 1/4 + 1/4 = 1/2$. Thus we obtain

$N(\rho) = N(\alpha - \beta\sigma) = N(\beta(\frac{\alpha}{\beta} - \sigma)) = N(\beta)N(\frac{\alpha}{\beta} - \sigma) \leq N(\beta)\frac{1}{2},$

so we do indeed have $N(\rho) < N(\beta)$ as claimed.

15. $\dfrac{16 + 7i}{10 - 5i} = \dfrac{16 + 7i}{10 - 5i}\cdot\dfrac{10 + 5i}{10 + 5i} = \dfrac{125 + 150i}{125} = 1 + \dfrac{6}{5}i$ so we let $\sigma = 1 + i$. Then $16 + 7i = (10 - 5i)((1 + i) + (1 + 2i)$.

$\dfrac{10 - 5i}{1 + 2i} = \dfrac{10 - 5i}{1 + 2i}\cdot\dfrac{1 - 2i}{1 - 2i} = \dfrac{0 - 25i}{5} = -5i$ so $10 - 5i =$

$(1 + 2i)(-5i)$. Thus $1 + 2i$ is a gcd of $16 + 7i$ and $10 + 5i$. Other possible answers are $-1 - 2i$, $-2 + i$, and $2 - i$.

16. a) Let $\gamma + \langle\alpha\rangle$ be a coset of $\mathbf{Z}[i]/\langle\alpha\rangle$. By the division algorith, $\gamma = \alpha\sigma + \rho$ where either $\rho = 0$ or $N(\rho) < N(\alpha)$. Then $\gamma + \langle\alpha\rangle = (\rho + \sigma\alpha) + \langle\alpha\rangle$. Now $\sigma\alpha \in \langle\alpha\rangle$, so $\gamma + \langle\alpha\rangle = \rho + \langle\alpha\rangle$. Thus every coset of $\langle\alpha\rangle$ contains a representative of norm less than $N(\alpha)$. Since there are only a finite number of elements of $\mathbf{Z}[i]$ having norm less than $N(\alpha)$, we see that $\mathbf{Z}[i]/\langle\alpha\rangle$ is a finite ring.

b) Let π be an irreducible in $\mathbf{Z}[i]$, and let $\langle\mu\rangle$ be an ideal of $\mathbf{Z}[i]$ such that $\langle\pi\rangle \subseteq \langle\mu\rangle$. (Remember that $\mathbf{Z}[i]$ is a PID, so every ideal is principal.) Then $\pi \in \langle\mu\rangle$ so $\pi = \mu\beta$. Since π is an irreducible, either μ is a unit, in which case $\langle\mu\rangle = \mathbf{Z}[i]$, or β is a unit, in which case $\mu = \pi\beta^{-1}$ so $\mu \in \langle\pi\rangle$ and $\langle\mu\rangle = \langle\pi\rangle$. We have shown that $\langle\pi\rangle$ is a maximal ideal of $\mathbf{Z}[i]$, so $\mathbf{Z}[i]/\langle\pi\rangle$ is a field.

c) i) Each coset contains a unique representative of the form $a + bi$ where a and b are both in the set $\{0, 1, 2\}$. Thus there are 9 elements in all, and the ring has characteristic 3 since $1 + 1 + 1 = 0$.

ii) By part (a), each coset contains a representative of norm less than $N(1 + i) = 2$. The only elements $\neq 0$ of $\mathbf{Z}[i]$ of norm less than 2 are ± 1 and $\pm i$. Since $i = -1 + (1 + i)$ and $-i = 1 - (1 + i)$, we see that $1 + \langle 1+i\rangle$ and $-1 + \langle 1+i\rangle$ are the only cosets. Thus the order of the ring is 2, and the characteristic is 2.

iii) By part (a), each coset contains a representative of norm less than $N(1 + 2i) = 5$. The only elements $\neq 0$ of $\mathbf{Z}[i]$ of norm less than 5 are of the form $a + bi$ where a and b are in the set $\{0, 1, -1\}$ or where one of a and b is ± 2 and the other is zero. These elements are $1, -1, i, -i, 1 + i, 1 - i, -1 + i, -1 - i, 2, -2$, and $2i, -2i$. Since

$$
\begin{aligned}
i &= 2 + (1+2i)i, & 1 + i &= -2 + (1+2i)(1-i),\\
-i &= -2 + (1+2i)(-i), & 1 - i &= -1 + (1+2i)(-i),\\
2i &= -1 + (1+2i), & -1 + i &= 1 + (1+2i)(i),\\
-2i &= 1 + (1+2i)(-1), & -1 - i &= 2 + (1+2i)(-1+i),
\end{aligned}
$$

we see that every coset contains either 0, 1, -1, 2, or -2 as a representative. The ring has 5 elements and is of characteristic 5.

17. a) *Property 1:* Since $n > 0$, we see that $a^2 + nb^2 \geq 0$.

Property 2: Since $n > 0$, we see that $a^2 + nb^2 = 0$ if and only if $a = b = 0$.

Property 3: Let $\alpha = a + b\sqrt{-n}$ and $\beta = c + d\sqrt{-n}$. Then $N(\alpha\beta) = N((ac-bdn) + (ad+bc)\sqrt{-n}) = (ac-bdn)^2 + n(ad+bc)^2 = a^2c^2 - 2abcdn + b^2d^2n^2 + a^2d^2n + 2abcdn + b^2c^2n = a^2c^2 + b^2d^2n^2 + a^2d^2n + b^2c^2n = (a^2 + nb^2)(c^2 + nd^2) = N(\alpha)N(\beta)$. [Of course it also follows from the fact that $|\alpha\beta|^2 = |\alpha|^2|\beta|^2$ for all $\alpha, \beta \in \mathbb{C}$.]

b) Theorem 7.9 shows that if $\alpha \in \mathbb{Z}[\sqrt{-n}]$ is a unit, then $N(\alpha) = 1$. Conversely, suppose that $N(\alpha) = 1$. Now $a^2 + nb^2 = 1$ where $n \in \mathbb{Z}^+$ if and only if either $a = \pm 1$ and $b = 0$, or $a = 0$ and $n = 1$ and $b = \pm 1$. In the former case, $\alpha = \pm 1$, and of course 1 and -1 are units. In the latter case with $n = 1$, we are in the Gaussian integers which are a Euclidean domain, and Theorem 7.5 tells us that the elements of norm (valuation) 1 are indeed units.

c) By parts (a) and (b), we have a multiplicative norm on $\mathbb{Z}[\sqrt{-n}]$ such that the elements of norm 1 are precisely the units. By Exercise 13, every nonzero nonunit has a factorization into irreducibles.

Note that the hypothesis that n is square free was not used in this exercise. Since $a + b\sqrt{-m^2n} = a + (bm)\sqrt{-n}$, we see that the square-free assumption is really no loss of generality. The assumption that $n > 0$ was used in both part (a) and part (b). The square-free assumption is used in the following exercise, however.

18. a) *Property 1:* Of course $|a^2 - nb^2| \geq 0$.

Property 2: If $|a^2 - nb^2| = 0$, then $a^2 = nb^2$. If $b = 0$, then $a = 0$. If $b \neq 0$, then $n = (a/b)^2$, contradicting the hypothesis that n is square free. Thus $a = 0$ and $b = 0$.

Property 3: Let $\alpha = a + b\sqrt{n}$ and $\beta = c + d\sqrt{n}$. Then

$$\begin{aligned} N(\alpha\beta) &= N((ac+bdn) + (ad+bc)\sqrt{n}) \\ &= |(ac+bdn)^2 - n(ad+bc)^2| \\ &= |(a^2c^2+2acbdn+b^2d^2n^2 - a^2d^2n-2abcdn-b^2c^2n| \\ &= |(a^2c^2 + b^2d^2n^2 - a^2d^2n - b^2c^2n| \\ &= |(a^2-nb^2)(c^2-nd^2| = |a^2-nb^2||c^2-nd^2| = N(\alpha)N(\beta). \end{aligned}$$

b) As an integral domain with a multiplicative norm, the norm of every unit is 1 by Theorem 7.9. Now suppose that $\alpha = a + b\sqrt{n}$ has norm 1, so that $|a^2 - nb^2| = 1$. Then $\frac{1}{\alpha} =$

$$\frac{1}{a + b\sqrt{n}} = \frac{a - b\sqrt{n}}{a - b\sqrt{n}} = \frac{a - b\sqrt{n}}{a^2 - nb^2} = \pm(a - b\sqrt{n})$$ and

$a + (-b)\sqrt{n} \in \mathbb{Z}[\sqrt{n}]$ so α is a unit.

c) The reasoning here is identical with that of part (c) in Exercise 17.

19. Given α and β in one of these integral domains, we proceed to construct $\sigma = q_1 + q_2\sqrt{*}$ and $\rho = \alpha - \beta\sigma$ as described in the hint to Exercise 14, where now $*$ may be either 2, -2, or 3. Namely, working in $\mathbb{C}$, we compute $\frac{\alpha}{\beta} = \frac{a + b\sqrt{*}}{c + d\sqrt{*}} \cdot \frac{c - d\sqrt{*}}{c - d\sqrt{*}} =$ $r + s\sqrt{*}$ for $r, s \in \mathbb{Q}$. Again, we choose q_1 and q_2 to be integers in $\mathbb{Z}$ as close as possible to r and s respectively.

By construction of σ, we see that $|r - q_1| \le 1/2$ and $|s - q_2| \le 1/2$. Thus $N(\frac{\alpha}{\beta} - \sigma) = N((r+s\sqrt{*}) - (q_1 + q_2\sqrt{*})) \le$

$$\left[\begin{array}{l} (1/2)^2 + 2(1/2)^2 \le 1/4 + 2(1/4) = 3/4 \text{ if } * = -2, \\ |0^2 - 3(1/2)^2| = 3/4 \text{ if } * = 2 \text{ or } * = 3, \end{array}\right.$$

where in the second case, we took the values for $*$, $|r - q_1|$ and $|s - q_2|$ which result in the largest possible value for the norm. Thus we obtain

$$N(\rho) = N(\alpha - \beta\sigma) = N(\beta(\frac{\alpha}{\beta} - \sigma)) = N(\beta)N(\frac{\alpha}{\beta} - \sigma) \le N(\beta)\frac{3}{4},$$

so we do indeed have $N(\rho) < N(\beta)$ as claimed.

CHAPTER 8

EXTENSION FIELDS

SECTION 8.1 - Introduction to Extension Fields

1. Let $\alpha = 1 + \sqrt{2}$. Then $(\alpha - 1)^2 = 2$ so $\alpha^2 - 2\alpha - 1 = 0$. Thus α is a zero of $x^2 - 2x - 1$ in $\mathbf{Q}[x]$.

2. Let $\alpha = \sqrt{2} + \sqrt{3}$. Then $\alpha^2 = 2 + 2\sqrt{6} + 3$ so $\alpha^2 - 5 = 2\sqrt{6}$. Squaring again, we obtain $\alpha^4 - 10\alpha^2 + 1 = 0$, so α is a zero of $x^4 - 10x^2 + 1$ in $\mathbf{Q}[x]$.

3. Let $\alpha = 1 + i$. Then $(\alpha - 1)^2 = -1$ so $\alpha^2 - 2\alpha + 2 = 0$. Thus α is a zero of $x^2 - 2x + 2$ in $\mathbf{Q}[x]$.

4. Let $\alpha = \sqrt{1 + \sqrt[3]{2}}$. Then $\alpha^2 = 1 + \sqrt[3]{2}$ so $\alpha^2 - 1 = \sqrt[3]{2}$. Cubing, we obtain $\alpha^6 - 3\alpha^4 + 3\alpha^2 - 3 = 0$, so α is a zero of $x^6 - 3x^4 + 3x^2 - 3$ in $\mathbf{Q}[x]$.

5. Let $\alpha = \sqrt{\sqrt[3]{2} - i}$. Then $\alpha^2 + i = \sqrt[3]{2}$. Cubing, we obtain $\alpha^6 + 3\alpha^4 i - 3\alpha^2 - i = 2$, so $\alpha^6 - 3\alpha^2 - 2 = (1 - 3\alpha^4)i$. Squaring, we obtain $\alpha^{12} - 6\alpha^8 - 4\alpha^6 + 9\alpha^4 + 12\alpha^2 + 4 = -1 + 6\alpha^4 - 9\alpha^8$. Thus $\alpha^{12} + 3\alpha^8 - 4\alpha^6 + 3\alpha^4 + 12\alpha^2 + 5 = 0$, so α is a zero of $x^{12} + 3x^8 - 4x^6 + 3x^4 + 12x^2 + 5$ in $\mathbf{Q}[x]$.

6. Let $\alpha = \sqrt{3 - \sqrt{6}}$. Then $\alpha^2 - 3 = -\sqrt{6}$. Squaring again, we obtain $\alpha^4 - 6\alpha^2 + 3 = 0$, so α is a zero of $f(x) = x^4 - 6x^2 + 3$ in $\mathbf{Q}[x]$. Now $f(x)$ is monic and is irreducible by Eisenstein with $p = 3$. Thus $\deg(\alpha,\mathbf{Q}) = 4$ and $\operatorname{irr}(\alpha,\mathbf{Q}) = f(x)$.

7. Let $\alpha = \sqrt{(1/3) + \sqrt{7}}$. Then $\alpha^2 - 1/3 = \sqrt{7}$. Squaring again, we obtain $\alpha^4 - (2/3)\alpha^2 - (62/9) = 0$, or $9\alpha^4 - 6\alpha^2 - 62 = 0$. Let $f(x) = 9x^4 - 6x^2 - 62$. Then $f(x)$ is irreducible by

Eisenstein with $p = 2$. Thus $\deg(\alpha,\mathbf{Q}) = 4$ and $\mathrm{irr}(\alpha,\mathbf{Q}) = (1/9)f(x)$.

8. Let $\alpha = \sqrt{2} + i$. Then $\alpha^2 = 2 - 2\sqrt{2}i - 1$ so $\alpha^2 - 1 = 2\sqrt{2}i$. Squaring again, we obtain $\alpha^4 - 2\alpha^2 + 1 = -8$, so $\alpha^4 - 2\alpha^2 + 9 = 0$. Let $f(x) = x^4 - 2x^2 + 9$. One can show that $f(x)$ is irreducible by the technique of Example 6 in Section 5.6. Thus $\deg(\alpha,\mathbf{Q}) = 4$ and $\mathrm{irr}(\alpha,\mathbf{Q}) = f(x)$.

9. i is algebraic over $\mathbf{Q}$ since it is a zero of $x^2 + 1$ in $\mathbf{Q}[x]$. $\mathrm{Deg}(i,\mathbf{Q}) = 2$.

10. Let $\alpha = 1 + i$. Then $\alpha - 1 = i$ so $\alpha^2 - 2\alpha + 2 = 0$. Since α is not in $\mathbf{R}$, we see that α is algebraic over $\mathbf{R}$ of degree 2.

11. The text told us that π is transcendental over $\mathbf{Q}$, behaving just like an indeterminant. Thus $\sqrt{\pi}$ is also transcendental over $\mathbf{Q}$. [It is easy to see that if a polynomial expression in $\sqrt{\pi}$ is zero, then a polynomial in π is zero. Namely, starting with $f(\sqrt{\pi}) = 0$, move all odd-degree terms to the right-hand side, factor $\sqrt{\pi}$ out from them, and then square both sides.]

12. $\sqrt{\pi} \in \mathbf{R}$ so it is algebraic over $\mathbf{R}$ of degree 1. It is a zero of $x - \sqrt{\pi}$ in $\mathbf{R}[x]$.

13. $\sqrt{\pi}$ is algebraic over $\mathbf{Q}(\pi)$ of degree 2. It is not in $\mathbf{Q}(\pi)$. Note that $\sqrt{x}$ is not in $\mathbf{Q}(x)$, but it is a zero of $y^2 - x$ in $(\mathbf{Q}(x))[y]$.

14. π^2 is transcendental over $\mathbf{Q}$ for the text told us that π is transcendental over $\mathbf{Q}$, and a polynomial expression in π^2 equal to zero and having rational coefficients would immediately give rise to a polynomial expression in π equal to zero, all with coefficients in $\mathbf{Q}$.

15. $\pi^2 \in \mathbf{Q}(\pi)$ so it is algebraic over $\mathbf{Q}(\pi)$ of degree 1. It is a zero of $x - \pi^2$ in $(\mathbf{Q}(\pi))[x]$.

16. π^2 is algebraic over $\mathbf{Q}(\pi^3)$ of degree 3. It is not in $\mathbf{Q}(\pi^3)$, (note that x^2 is not a polynomial in x^3,) but it is a zero of

$x^3 - (\pi^3)^2 = x^3 - \pi^6$ in $(\mathbf{Q}(\pi^3))[x]$.

17.
$$\begin{array}{r|l} & x + (1+\alpha) \\ \hline x - \alpha & x^2 + x + 1 \\ & x^2 - \alpha x \\ \hline & (1+\alpha)x + 1 \\ & (1+\alpha)x - \alpha^2 - \alpha \\ \hline & \alpha^2 + \alpha + 1 = 2\cdot(\alpha+1) = 0. \end{array}$$

$x^2 + x + 1 = (x - \alpha)(x + \alpha + 1)$.

18. a) Let $f(x) = x^2 + 1$. Then $f(0) = 1$, $f(1) = 2$, and $f(-1) = 2$ so $f(x)$ has no zeros in $\mathbf{Z}_2[x]$ and is thus irreducible.

b)

+	0	1	2	α	2α	$1+\alpha$	$1+2\alpha$	$2+\alpha$	$2+2\alpha$
0	0	1	2	α	2α	$1+\alpha$	$1+2\alpha$	$2+\alpha$	$2+2\alpha$
1	1	2	0	$1+\alpha$	$1+2\alpha$	$2+\alpha$	$2+2\alpha$	α	2α
2	2	0	1	$2+\alpha$	$2+2\alpha$	α	2α	$1+\alpha$	$1+2\alpha$
α	α	$1+\alpha$	$2+\alpha$	2α	0	$1+2\alpha$	1	$2+2\alpha$	2
2α	2α	$1+2\alpha$	$2+2\alpha$	0	α	1	$1+\alpha$	2	$2+\alpha$
$1+\alpha$	$1+\alpha$	$2+\alpha$	α	$1+2\alpha$	1	$2+2\alpha$	2	2α	0
$1+2\alpha$	$1+2\alpha$	$2+2\alpha$	2α	1	$1+\alpha$	2	$2+\alpha$	0	α
$2+\alpha$	$2+\alpha$	α	$1+\alpha$	$2+2\alpha$	2	2α	0	$1+2\alpha$	1
$2+2\alpha$	$2+2\alpha$	2α	$1+2\alpha$	2	$2+\alpha$	0	α	1	$1+\alpha$

$\cdot$	0	1	2	α	2α	$1+\alpha$	$1+2\alpha$	$2+\alpha$	$2+2\alpha$
0	0	0	0	0	0	0	0	0	0
1	0	1	2	α	2α	$1+\alpha$	$1+2\alpha$	$2+\alpha$	$2+2\alpha$
2	0	2	1	2α	α	$2+2\alpha$	$2+\alpha$	$1+2\alpha$	$1+\alpha$
α	0	α	2α	2	1	$2+\alpha$	$1+\alpha$	$2+2\alpha$	$1+2\alpha$
2α	0	2α	α	1	2	$1+2\alpha$	$2+2\alpha$	$1+\alpha$	$2+\alpha$
$1+\alpha$	0	$1+\alpha$	$2+2\alpha$	$2+\alpha$	$1+2\alpha$	2α	2	1	α
$1+2\alpha$	0	$1+2\alpha$	$2+\alpha$	$1+\alpha$	$2+2\alpha$	2	α	2α	1
$2+\alpha$	0	$2+\alpha$	$1+2\alpha$	$2+2\alpha$	$1+\alpha$	1	2α	α	2
$2+2\alpha$	0	$2+2\alpha$	$1+\alpha$	$1+2\alpha$	$2+\alpha$	α	1	2	2α

19. T T T T F T F T F T

20. a) $\mathbf{Q}(\pi^3)$ (Other answers are possible.)

b) $\mathbf{Q}(e^2)$ (Other answers are possible.)

21. a) Let $f(x) = x^3 + x^2 + 1$. Then $f(0) = 1$ and $f(1) = 0$ so $f(x)$ has no zeros in $\mathbf{Z}_2$ and is thus irreducible.

b) Long division:

$$\begin{array}{r|l} & x^2 + (1+\alpha)x + (\alpha^2+\alpha) \\ \hline x - \alpha & x^3 + x^2 + 1 \\ & \underline{x^3 - \alpha x^2} \\ & (1+\alpha)x^2 \\ & \underline{(1+\alpha)x^2 - (\alpha+\alpha^2)x} \\ & (\alpha^2+\alpha)x + 1 \\ & \underline{(\alpha^2+\alpha)x - 1} \\ & 0 \end{array}$$

$$\alpha^3 = \alpha^2 + 1$$
$$-\alpha^3 - \alpha^2 = -\alpha^2 - 1 - \alpha^2 = -1$$

Try α^2 as a zero of $q(x) = x^2 + (1+\alpha)x + (\alpha^2+\alpha)$:

$\alpha^4 + (1+\alpha)\alpha^2 + (\alpha^2+\alpha) = \alpha(\alpha^2+1) + \alpha^2+(\alpha^2+1) + (\alpha^2+\alpha)$

$= (\alpha^2+1)+\alpha + \alpha^2+(\alpha^2+1) + (\alpha^2+\alpha) = 2\cdot(\alpha^2+1) + 2\cdot\alpha^2 + 2\cdot\alpha = 0$

so α^2 is a zero of $q(x)$. We do a long division.

$$\begin{array}{r|l} & x + (\alpha^2+\alpha+1) \\ \hline x - \alpha^2 & x^2 + (1+\alpha)x + (\alpha^2+\alpha) \\ & \underline{x^2 - \alpha^2 x} \\ & (\alpha^2+\alpha+1)x + (\alpha^2+\alpha) \\ & \underline{(\alpha^2+\alpha+1)x + (\alpha^2+\alpha)} \\ & 0 \end{array}$$

$$\alpha^2(\alpha^2+\alpha+1) = \alpha\alpha^3 + \alpha^3 + \alpha^2$$
$$= \alpha(\alpha^2+1) + (\alpha^2+1) + \alpha^2$$
$$= (\alpha^2+1)+\alpha + (\alpha^2+1) + \alpha^2$$
$$= \alpha^2+\alpha$$

Thus in $(\mathbb{Z}_2(\alpha))[x]$,

$$x^3 + x^2 + 1 = [x - \alpha][x - \alpha^2][x - (\alpha^2 + \alpha + 1)].$$

22. $\langle \mathbb{Z}_2(\alpha), +\rangle$ is an abelian group of order 8 with the property that $a + a = 0$ for all a in the group. Thus the group must the group must be isomorphic to $\mathbb{Z}_2 \times \mathbb{Z}_2 \times \mathbb{Z}_2$. $\langle \mathbb{Z}_2(\alpha)^*, \cdot\rangle$ is an abelian group of order 7, which must be cyclic and isomorphic to $\mathbb{Z}_7$.

23. It is the monic polynomial in $\phi[x]$ of *minimal* degree having α as a zero.

24. Every element of $F(\beta)$ can be expressed as a quotient of polynomials in β with coefficients in F. Since α is algebraic over $F(\beta)$, there is a polynomial in α with coefficients in $F(\beta)$ which is equal to zero. By multiplying this equation by the polynomial in β which is the product of

the denominators of the coefficients in this equation, we obtain a polynomial in α equal to zero and having as coefficients polynomials in β. Now a polynomial in α with coefficients that are polynomials in β can be formally rewritten as a polynomial in β with coefficients that are polynomials in α. (Recall that $(F[x])[y] \simeq (F[y])[x]$.) This polynomial expression is still zero, which shows that β is algebraic over $F(\alpha)$.

25. Theorem 8.4 shows that every element of $F(\alpha)$ can be *uniquely* expressed in the form

$$b_0 + b_1\alpha + b_2\alpha^2 + \cdots + b_{n-1}\alpha^{n-1}.$$

Since F has q elements, there are q choices for b_0, then q choices for b_1, etc. Thus there are q^n such expressions altogether. The *uniqueness* property shows that different expressions correspond to distinct elements of $F(\alpha)$, which must therefore have q^n elements.

26. a) Let $f(x) = x^3 + x^2 + 2$. Then $f(0) = 2$, $f(1) = 1$, and $f(-1) = 2$ so $f(x)$ has no zeros in $\mathbb{Z}_3$ and thus is irreducible over $\mathbb{Z}_3[x]$.

b) Exercise 25 shows that $\mathbb{Z}_3(\alpha)$ has $3^3 = 27$ elements. Thus there is a field of 27 elements.

27. a) If $p \neq 2$, then $1 \neq -1$ in $\mathbb{Z}_p$, so $1^2 = (p - 1)^2$. Thus the squaring function mapping $\mathbb{Z}_p \longrightarrow \mathbb{Z}_p$ is not one to one; in fact, its image can have at most $p - 1$ elements. Thus some element of $\mathbb{Z}_p$ is not a square if $p \neq 2$.

b) We saw in Example 9 that there exists a finite field of four elements. Let p be an odd prime. By part (a), there exists $a \in \mathbb{Z}_p$ such that $x^2 - a$ has no zeros in $\mathbb{Z}_p$. This means that $x^2 - a$ is irreducible. Let α be a zero of $x^2 - a$ in an extension field of $\mathbb{Z}_p$. By Exercise 25, $\mathbb{Z}_p(\alpha)$ has p^2 elements.

28. Let $\beta \in F(\alpha)$. Then β is equal to a quotient $r(\alpha)/s(\alpha)$ of polynomials in α with coefficients in F. Suppose that $f(\beta) =$

0 where $f(x) \in F[x]$ and is of degree n. Multiplying the equation $f(\beta) = 0$ by $s(\alpha)^n$, we obtain a polynomial in α with coefficients in F which is equal to zero. But then, α is algebraic over F, which is contrary to hypothesis. Therefore there is no nonzero polynomial expression $f(\beta) = 0$, that is, β is transcendental over F.

29. We know that $x^3 - 2$ is irreducible in $\mathbf{Q}[x]$ by the Eisenstein criterion with $p = 2$. Therefore $\sqrt[3]{2}$ is algebraic of degree 3 over $\mathbf{Q}$. By Theorem 8.4, the field $\mathbf{Q}(\sqrt[3]{2})$ consists of all elements of $\mathbf{R}$ of the form $a + b(\sqrt[3]{2}) + c(\sqrt[3]{2})^2$, and distinct values of a, b, and c give distinct elements of $\mathbf{R}$. The given set in the problem consists of precisely these elements of $\mathbf{R}$, so the given set is the field $\mathbf{Q}(\sqrt[3]{2})$.

30. We keep using Theorem 8.4 and Exercise 25. Now $x^3 + x + 1$ has no zeros in $\mathbf{Z}_2[x]$ and is therefore irreducible. If α is a zero of this polynomial, then $\mathbf{Z}_2(\alpha)$ has $2^3 = 8$ elements by Exercise 25.

Similarly, let α be a zero of the irreducible polynomial $x^4 + x + 1$ in $\mathbf{Z}_2[x]$. Then $\mathbf{Z}_2(\alpha)$ has $2^4 = 16$ elements.

Finally, let α be a zero of the irreducible polynomial $x^2 - 2$ in $\mathbf{Z}_5[x]$. Then $\mathbf{Z}_5(\alpha)$ has 5^2 elements.

31. Following the hint, we let F^* be the multiplicative group of nonzero elements of F. We are given that F is finite; suppose that F has m elements. Then F^* has $m - 1$ elements. Since the order of an element of a finite group divides the order of the group, we see that for all $a \in F^*$ we have $a^{m-1} = 1$. Thus every $a \in F^*$ is a zero of the polynomial $x^{m-1} - 1$. Of course, 0 is a zero of x. Thus every $\alpha \in F$ is algebraic over the prime field $\mathbf{Z}_p$ of F_p, for the polynomial $x^{m-1} - 1$ lies in $\mathbf{Z}_p[x]$ for all primes p.

32. Let E be a finite field with prime subfield $\mathbf{Z}_p$. If $E = \mathbf{Z}_p$, then the order of E is p and we are done. Otherwise, let $\alpha_1 \in E$ where α_1 is not in $\mathbf{Z}_p$. Exercise 31 shows that α_1 is

algebraic over the field $\mathbb{Z}_p$. Let $F_1 = \mathbb{Z}_p(\alpha_1)$. By Exercise 25, the field F_1 has order p^{n_1} where n_1 is the degree of α_1 over $\mathbb{Z}_p$. If $F_1 = E$, we are done. Otherwise, we find $\alpha_2 \in E$ where α_2 is not in F_1 and form $F_2 = F_1(\alpha_2)$, obtaining a field of order $p^{n_1 n_2}$ where n_2 is the degree of α_2 over F_1. We continue this process, constructing fields F_i of order $p^{n_1 n_2 \cdots n_i}$. Since E is a finite field, this process must eventually terminate with a field $F_r = E$. Thus E has order $p^{n_1 n_2 \cdots n_r}$ which is a power of p as asserted.

SECTION 8.2 - Vector Spaces

1. (See the text answer.)

2. Suppose that $a(1,1,0) + b(1,0,1) + c(0,1,1) = (d,e,f)$. Then $a + b = d$, $a + c = e$, and $b + c = f$. Subtracting the first two equations, we obtain $b - c = d - e$. Adding this to the last equation, we obtain $2b = f + d - e$, so $b = (f + d - e)/2$. Then $a = (d + e - f)/2$ and $c = (e + f - d)/2$. This shows that the given vectors span $\mathbb{R}^3$. Setting $d = e = f = 0$, we see that we must then have $a = b = c = 0$ so the vectors are also independent, and hence are a basis for $\mathbb{R}^3$.

3. We claim the vectors are dependent, and therefore cannot form a basis. If $a(-1,1,2) + b(2,-3,1) + c(10,-14,0) = (0,0,0)$, then

$$\begin{aligned} -a + 2b + 10c &= 0 \\ a - 3b - 14c &= 0 \\ 2a + b \quad\quad &= 0. \end{aligned}$$

Adding the first two equation, we find that $-b - 4c = 0$. Adding twice the first equation to the last, we find that $5b + 20c = 0$, which is essentially the same equation. We set $c = 1$ so $b = -4$ and $a = 2$, and see that $2(-1,1,2) + (-4)(2,-3,1) + 1(10,-14,0) = (0,0,0)$, so the vectors are indeed dependent.

Section 8.2

4. Since $\sqrt{2}$ is a zero of irreducible $x^2 - 2$ of degree 2, Theorem 8.8 shows that a basis is $\{1, \sqrt{2}\}$.

5. Since $\sqrt{2}$ is in $\mathbb{R}$ and is a zero of $x - \sqrt{2}$ of degree 1, Theorem 8.8 shows that a basis is $\{1\}$.

6. Since $\sqrt[3]{2}$ is a zero of irreducible $x^3 - 2$ of degree 3, Theorem 8.8 shows that a basis is $\{1, \sqrt[2]{2}, (\sqrt[3]{2})^2\}$.

7. Since $\mathbb{C} = \mathbb{R}[i]$ where i is a zero of irreducible $x^2 + 1$ of degree 2, Theorem 8.8 shows that a basis is $\{1, i\}$.

8. Since i is a zero of irreducible $x^2 + 1$ of degree 2, Theorem 8.8 shows that a basis is $\{1, i\}$.

9. Since $\sqrt[4]{2}$ is a zero of irreducible $x^4 - 2$ of degree 4, Theorem 8.8 shows that a basis is $\{1, \sqrt[4]{2}, \sqrt{2}, (\sqrt[4]{2})^3\}$.

10. Recall that α is a zero of $x^2 + x + 1$, so $\mathbb{Z}_2(\alpha)$ is a 2-dimensional vector space over $\mathbb{Z}_2$. Thus the three elements $1, 1 + \alpha, (1 + \alpha)^2 = 1 + 2\cdot\alpha + \alpha^2 = 1 + \alpha + 1 = \alpha$ must be independent. By inspection, we see that

$$1(1) + 1(1 + \alpha) + 1(1 + \alpha)^2 = 1 + (1 + \alpha) + \alpha = 0,$$

so α is a zero of $x^2 + x + 1$, which is thus not only $\mathrm{irr}(\alpha, \mathbb{Z}_2)$ but also $\mathrm{irr}(1 + \alpha, \mathbb{Z}_2)$. Of course, we should have known this since the other zero besides α must lie in $\mathbb{Z}_2(\alpha)$ since $x^2 + x + 1$ has to factor into linear factors there, and $1 + \alpha$ is the only possibility for the other zero.

11. T F T T F F F T T T

12. a) A subspace of V is a subset W of V that is closed under vector addition and under scalar multiplication, and is itself a vector space under these two operations.
    b) Let $\{W_i \mid i \in I\}$ be a collection of subspaces of V. Since $\langle W_i, +\rangle$ is an abelian group, Exercise 26 of Section 3.3 shows that $\bigcap_{i\in I} W_i$ is again an abelian group. Let $a \in F$ and let $\alpha \in \bigcap_{i\in I} W_i$. Then $\alpha \in W_i$ for each $i \in I$, so

$a\alpha \in W_i$ for each $i \in I$ since each W_i is a vector space. Hence $a\alpha \in \bigcap_{i \in I} W_i$ so the intersection is an abelian group closed under multiplication by elements of F. All the other axioms for a vector space (distributive laws, etc.) certainly hold in this intersection, since they hold for all elements in V.

13. a) The subspace of a vector space V generated by a subset S is the intersection of all subspaces of V that contain S.

b) Clearly, the sum of two finite linear combinations of elements of S is again a finite linear combination of elements of S. Also, a scalar times a finite linear combination is again a finite linear combination:

$$a(b_1\alpha_1 + \cdots + b_n\alpha_n) = (ab_1)\alpha_1 + \cdots + (ab_n)\alpha_n.$$

Since $0 = 0\alpha$ for $\alpha \in S$, we see that 0 is a finite linear combination of elements of S. Multiplying by the scalar -1, we see that an additive inverse of such a linear combination is again a finite linear combination of elements of S. Therefore the set of finite linear combinations of elements of S is a vector space, and is clearly the smallest vector space that contains S.

This result is analogous to the case of Theorem 1.11 for abelian groups, although we do have to check here that scalar multiplication is well behaved.

14. The direct sum of vector spaces $V_1, V_2, \cdots, V_n$ over the same field F is $\{(\alpha_1, \alpha_2, \cdots, \alpha_n) \mid \alpha_i \in V_i\}$ with addition and scalar multiplication defined by

$$(\alpha_1, \alpha_2, \cdots, \alpha_n) + (\alpha_1', \alpha_2', \cdots, \alpha_n') = (\alpha_1 + \alpha_1', \alpha_2 + \alpha_2', \cdots, \alpha_n + \alpha_n')$$

and $a(\alpha_1, \alpha_2, \cdots, \alpha_n) = (a\alpha_1, a\alpha_2, \cdots, a\alpha_n)$.

Since addition and multiplication are defined by performing the operations in each component, and since the vectors appearing in each component form a vector space over F, it is clear that this direct sum is again a vector space.

15. Let F be any field and let $F^n = \{(a_1, a_2, \cdots, a_n) \mid a_i \in F\}$. Then F^n is a vector space with addition and multiplication of n-tuples defined by performing those operations in each component. (It is the direct sum of F with itself n times,

as defined in Exercise 14.) A basis for F^n is $\{(1,0,0,\cdots,0), (0,1,0,\cdots, 0), \cdots, (0,0,0,\cdots,1)\}$.

16. Let V and V' be vector spaces of the same field F. A map $\phi: V \longrightarrow V'$ is an isomorphism if ϕ is a one-to-one map, $\phi(V) = V'$, and furthermore

$$\phi(\alpha + \beta) = \phi(\alpha) + \phi(\beta) \quad \text{and} \quad \phi(a\alpha) = a\phi(\alpha)$$

for all α, $\beta \in V$ and all $a \in F$.

17. Since each vector in V can be expressed as a linear combination of the β_i, we see that $\{\beta_i \mid i = 1, 2, \cdots, n\}$ generates V. Now $0 = 0\beta_1 + 0\beta_2 + \cdots + 0\beta_n$. By hypothesis, this is the *unique* linear combination of the β_i that yields 0, so the vectors are independent. Therefore, $\{\beta_i \mid i = 1, 2, \cdots, n\}$ is a basis for V.

18. a) The system can be rewritten as

$$X_1\alpha_1 + X_2\alpha_2 + \cdots + X_n\alpha_n = \beta \qquad (1)$$

since the ith component of the vector on the left side of the equation (1) is $a_{i1}X_1 + a_{i2}X_2 + \cdots + a_{in}X_n$ and the ith component of β is β_i. Eq.(1) shows that the system has a solution if and only if β is a finite linear combination of the vectors α_j for $j = 1, 2, \cdots, n$. By Exercise 13, this means that the system has a solution if and only if β lies in the subspace generated by the vectors $\alpha_1, \alpha_2, \cdots, \alpha_n$.

b) If $\{\alpha_j \mid j = 1, \cdots, n\}$ is a basis for F^n, then Exercise 17 shows that each $\beta \in F^n$ can be expressed *uniquely* as a linear combination of the vectors $\alpha_1, \cdots, \alpha_n$. By part (a), this means that the system has a unique solution.

19. Let $\{\gamma_1, \gamma_2, \cdots, \gamma_n\}$ be a basis for V. Let $\phi: F^n \longrightarrow V$ be defined by $\phi(a_1, a_2, \cdots, a_n) = a_1\gamma_1 + a_2\gamma_2 + \cdots + a_n\gamma_n$

Since addition and multiplication in F^n is by components, it is obvious that $\phi(\alpha + \beta) = \phi(\alpha) + \phi(\beta)$ and $\phi(a\alpha) = a\phi(\alpha)$ for for α, $\beta \in F^n$ and $a \in F$. Since $\gamma_1, \gamma_2, \cdots, \gamma_n$ form a basis for V, every vector in V can be expressed as a linear combination of these vectors, so ϕ maps F^n *onto* V. By Exercise 17, the expression for a vector in V as a linear combination of the vectors γ_i is *unique*, so ϕ is *one to one*. Thus ϕ is an isomorphism.

20. a) Let $\alpha \in V$. Since $\{\beta_1, \beta_2, \cdots, \beta_n\}$ is a basis for V, we know that there are scalars a_i such that $\alpha = a_1\beta_1 + a_2\beta_2 + \cdots + a_n\beta_n$. By the conditions for a linear transformation, we then have

$$\phi(\alpha) = a_1\alpha(\beta_1) + a_2\phi(\beta_2) + \cdots + a_n\phi(\beta_n).$$

This shows that as soon as the values $\phi(\beta_i)$ are known for $i = 1, 2, \cdots, n$, then ϕ is completely determined.

b) Let $\phi: V \longrightarrow V'$ be defined as follows: For $\alpha \in V$, express α as a linear combination

$$\alpha = a_1\beta_1 + a_2\beta_2 + \cdots + a_n\beta_n. \tag{2}$$

This can be done since $\{\beta_1, \beta_2, \cdots, \beta_n\}$ is a basis for V. Define $\phi(\alpha) = a_1\beta_1' + a_2\beta_2' + \cdots + a_n\beta_n'$. Since the expression (2) for α is unique by Exercise 17, we see that ϕ is well defined, and of course, $\phi(\beta_i) = \beta_i'$ for $i = 1, 2, \cdots, n$. Since addition and scalar multiplication of linear combinations of the β_i and the β_i' are both achieved by adding and multiplying respectively the coefficients in the linear combinations, we see at once that ϕ satisfies the required properties for a linear transformation. Part (a) shows that this transformation is completely determined by the vectors β_i', that is, the linear transformation is unique.

21. a) A linear transformation of vector spaces is analogous to a homomorphism of groups.

b) The kernel or nullspace of ϕ is $\text{Ker}(\phi) = \phi^{-1}(0) = \{\alpha \in V \mid \phi(\alpha) = 0\}$. Considering just the additive groups of V and V', group theory shows that $\text{Ker}(\phi)$ is an additive group. Let $\alpha \in \text{Ker}(\phi)$. Then $\phi(a\alpha) = a\phi(\alpha) = a0 = 0$, so $\text{Ker}(\phi)$ is closed under scalar multiplication by scalars $a \in F$. Hence $\text{Ker}(\phi)$ is a subspace of V.

c) ϕ is an isomorphism of V with V' if ϕ is one to one (equivalently, if $\text{Ker}(\phi) = \{0\}$,) and if ϕ maps V onto V'.

22. Let V/S be the factor group $\langle V,+\rangle/\langle S,+\rangle$, which is abelian since V is abelian. Define scalar multiplication on V/S by $a(\alpha + S) = a\alpha + S$ for $a \in F$, $(\alpha + S) \in V/S$. Since $a(\alpha + \sigma) = a\alpha + a\sigma$ and since $a\sigma \in S$ for σ in the subspace S, we see that this scalar multiplication is well defined, independent of the choice of representative in the coset $\alpha + S$. Since addition and scalar multiplication in V/S is computed in terms of representatives in V and since V is a vector space, we see that addition and scalar multiplication in V/S satisfy the axioms for a vector space.

23. a) By group theory, we know that $\langle \phi[V],+\rangle$ is a subgroup of V'. Let $\alpha \in V$ and $a \in F$. Then $a\phi(\alpha) = \phi(a\alpha)$ shows that $\phi[V]$ is closed under multiplication by scalars in F. Thus $\phi[V]$ is a subspace of V'.

b) Let $\{\alpha_1, \alpha_2, \cdots, \alpha_r\}$ be a basis for $\text{Ker}(\phi)$. By Theorem 8.7, this set can be enlarged to a basis $\{\alpha_1, \alpha_2, \cdots, \alpha_r, \beta_1, \beta_2, \cdots, \beta_m\}$ for V. Let $\gamma \in V$. Then $\gamma = a_1\alpha_1 + \cdots + a_r\alpha_r + b_1\beta_1 + \cdots + b_m\beta_m$ for scalars $a_i, b_j \in F$. Since $\phi(\alpha_i) = 0$ for $i = 1, \cdots, r$ we see that that $\phi(\gamma) = b_1\phi(\beta_1) + \cdots + b_m\phi(\beta_m)$. Thus $\{\phi(\beta_1), \cdots, \phi(\beta_m)\}$ spans $\phi[V]$. We claim that this set is independent, and hence is actually a basis for $\phi[V]$. Suppose that $c_1\phi(\beta_1) + \cdots + c_m\phi(\beta_m) = 0$ for some scalars $c_j \in F$. Then $\phi(c_1\beta_1 + \cdots + c_m\beta_m) = 0$, so $(c_1\beta_1 + \cdots + c_m\beta_m) \in \text{Ker}(\phi)$, and hence

$$c_1\beta_1 + \cdots + c_m\beta_m = d_1\alpha_1 + \cdots + d_r\alpha_r$$

for some scalars d_i. Moving everything to the lefthand side of this equation, we obtain a linear combination of of vectors α_i and β_j which is equal to 0. Since the α_i and β_j form a basis for V, they are independent so all the coefficients d_i and c_j must be zero. The fact that the c_j must be zero shows that $\{\beta_1, \cdots, \beta_m\}$ is independent, and and thus is a basis for $\phi[V]$. By our construction $\dim(\text{Ker}(\phi)) = r$, $\dim(V) = r + m$, and we have shown that dim($\phi[V]$) = m. Thus dim($\phi[V]$) = m = ($r + m$) - r = dim(V) - dim(Ker(ϕ)).

SECTION 8.3 - Algebraic Extensions

1. $\{1, \sqrt{2}\}$ since irr($\sqrt{2}$,$\mathbf{Q}$) = $x^2 - 2$. Degree 2

2. $\{1, \sqrt{2}, \sqrt{3}, \sqrt{6}\}$ as shown in Example 2. Degree 4

3. We form products from the basis of Exercise 2 and the basis $\{1, \sqrt{5}\}$ for $\mathbf{Q}(\sqrt{2},\sqrt{3},\sqrt{5})$ over $\mathbf{Q}(\sqrt{2},\sqrt{3})$, obtaining $\{1, \sqrt{2}, \sqrt{3}, \sqrt{5}, \sqrt{6}, \sqrt{10}, \sqrt{15}, \sqrt{30}\}$. Degree 8.

4. We form products from the bases $\{1, \sqrt{3}\}$ of $\mathbf{Q}(\sqrt{3})$ over $\mathbf{Q}$ and $\{1, \sqrt[3]{2}, (\sqrt[3]{2})^2\}$ for $\mathbf{Q}(\sqrt{3}, \sqrt[3]{2})$ over $\mathbf{Q}(\sqrt{3})$ obtaining $\{1, \sqrt[3]{2}, (\sqrt[3]{2})^2, \sqrt{3}, \sqrt{3}(\sqrt[3]{2}), \sqrt{3}(\sqrt[3]{2})^2\}$. Degree 6

5. We form products from the bases $\{1, \sqrt{2}\}$ of $\mathbf{Q}(\sqrt{2})$ over $\mathbf{Q}$ and $\{1, \sqrt[3]{2}, (\sqrt[3]{2})^2\}$ for $\mathbf{Q}(\sqrt{2}, \sqrt[3]{2})$ over $\mathbf{Q}(\sqrt{2})$ obtaining $\{1, \sqrt[3]{2}, (\sqrt[3]{2})^2, \sqrt{2}, \sqrt{2}(\sqrt[3]{2}), \sqrt{2}(\sqrt[3]{2})^2\}$. It is easy to see that $\mathbf{Q}(\sqrt{2}, \sqrt[3]{2}) = \mathbf{Q}(\sqrt[6]{2})$ since $2^{1/6} = 2^{7/6}/2 = 2^{3/6}2^{4/6}/2 = 2^{1/2}(2^{1/3})^2/2$, so another basis is $\{1, 2^{1/6}, 2^{2/6}, 2^{3/6}, 2^{4/6}, 2^{5/6}\}$. Degree 6

6. As in Example 2, we see that deg($\sqrt{2} + \sqrt{3}$, $\mathbf{Q}$) = 4, so $\mathbf{Q}(\sqrt{2} + \sqrt{3}) = Q(\sqrt{2}, \sqrt{3})$ and the basis can be the same as in Exercise 2.

7. Since $\sqrt{2}\sqrt{3} = \sqrt{6}$, we see the field is $\mathbf{Q}\sqrt{6}$ which has as basis over $\mathbf{Q}$ the set $\{1, \sqrt{6}\}$. Degree 2

8. We form products from the bases $\{1, \sqrt{2}\}$ of $\mathbf{Q}(\sqrt{2})$ over $\mathbf{Q}$ and $\{1, \sqrt[3]{5}, (\sqrt[3]{5})^2\}$ for $\mathbf{Q}(\sqrt{2}, \sqrt[3]{5})$ over $\mathbf{Q}(\sqrt{2})$ obtaining $\{1, \sqrt[3]{5}, (\sqrt[3]{5})^2, \sqrt{2}, \sqrt{2}(\sqrt[3]{5}), \sqrt{2}(\sqrt[3]{5})^2\}$. Degree 6

9. Since $\sqrt[3]{6}/\sqrt[3]{2} = \sqrt[3]{3}$ and $\sqrt[3]{24} = 2(\sqrt[3]{3})$, we see that $\mathbf{Q}(\sqrt[3]{2}, \sqrt[3]{6}, \sqrt[3]{24}) = \mathbf{Q}(\sqrt[3]{2}, \sqrt[3]{3})$, so we can take products from

the bases $\{1, \sqrt[3]{2}, (\sqrt[3]{2})^2\}$ and $\{1, \sqrt[3]{3}, (\sqrt[3]{3})^2\}$ for $\mathbf{Q}(\sqrt[3]{2})$ over $\mathbf{Q}$ and $\mathbf{Q}(\sqrt[3]{2}, \sqrt[3]{3})$ over $\mathbf{Q}(\sqrt[3]{2})$ respectively, obtaining $\{1, \sqrt[3]{2}, \sqrt[3]{4}, \sqrt[3]{3}, \sqrt[3]{6}, \sqrt[3]{12}, \sqrt[3]{9}, \sqrt[3]{18}, \sqrt[3]{36}\}$. Degree 9

10. Since $\mathbf{Q}(\sqrt{2},\sqrt{6}) = \mathbf{Q}(\sqrt{2},\sqrt{3})$, we can take as basis over $\mathbf{Q}(\sqrt{3})$ the set $\{1, \sqrt{2}\}$. Degree 2

11. Example 2 shows that $\mathbf{Q}(\sqrt{2} + \sqrt{3}) = \mathbf{Q}(\sqrt{2}, \sqrt{3})$ so we can take as basis over $\mathbf{Q}(\sqrt{3})$ the set $\{1, \sqrt{2}\}$. Degree 2

12. $\mathbf{Q}(\sqrt{2} + \sqrt{3}) = \mathbf{Q}(\sqrt{2},\sqrt{3})$ so $\{1\}$ is a basis. Degree 1

13. Since $\sqrt{6} + \sqrt{10} = \sqrt{2}(\sqrt{3} + \sqrt{5})$ we see that $\mathbf{Q}(\sqrt{2}, \sqrt{6} + \sqrt{10}) = \mathbf{Q}(\sqrt{2}, \sqrt{3} + \sqrt{5})$ so a basis over $\mathbf{Q}(\sqrt{3} + \sqrt{5})$ is $\{1, \sqrt{2}\}$. Degree 2

14. Let $E = \mathbf{Q}(\sqrt{2})$ and let $F = \mathbf{Q}$. The algebraic closure of $\mathbf{Q}$ in $\mathbf{Q}(\sqrt{2})$ is $\mathbf{Q}(\sqrt{2})$ since it is an algebraic extension of $\mathbf{Q}$. However, $\mathbf{Q}(\sqrt{2})$ is not algebraically closed, since the polynomial $x^2 + 1$ has no zeros in $\mathbf{Q}(\sqrt{2})$.

15. T F T F F T F F F F

16. If $b \neq 0$, then $a + bi \in \mathbf{C}$ but $a + bi$ is not in $\mathbf{R}$. By Theorem 8.9, $a + bi$ is algebraic over $\mathbf{R}$. Then $[\mathbf{C}: \mathbf{R}] = [\mathbf{C}: \mathbf{R}(a + bi)][\mathbf{R}(a + bi): \mathbf{R}] = 2$ and since $a + bi$ is not in $\mathbf{R}$, we must have $[\mathbf{R}(a + bi): \mathbf{R}] = 2$, so $[\mathbf{C}: \mathbf{R}(a + bi)] = 1$. Thus $\mathbf{C} = \mathbf{R}(a + bi)$.

17. Let α be any element in E that is not in F. Then by Theorem 8.10, $[E: F] = [E: F(\alpha)][F(\alpha): F] = p$ for some prime p. Since α is not in F, we know that $[F(\alpha): F] > 1$, so we must have $[F(\alpha): F] = p$ and therefore $[E: F(\alpha)] = 1$. As we remarked after Definition 8.13, this shows that $E = F(\alpha)$, which is what we wished to show.

18. If $x^2 - 3$ were reducible over $\mathbf{Q}(\sqrt[3]{2})$, then it would factor into linear factors over $\mathbf{Q}(\sqrt[3]{2})$, so $\sqrt{3}$ would lie in the field $\mathbf{Q}(\sqrt[3]{2})$, and we would have $\mathbf{Q}(\sqrt{3}) \leq \mathbf{Q}(\sqrt[3]{2})$. But then by Theorem 8.10,

$$[\mathbf{Q}(\sqrt[3]{2}): \mathbf{Q}] = [\mathbf{Q}(\sqrt[3]{2}): \mathbf{Q}(\sqrt{3})][\mathbf{Q}(\sqrt{3}): \mathbf{Q}].$$

This equation is impossible since $[\mathbf{Q}(\sqrt[3]{2}): \mathbf{Q}] = 3$ while $[\mathbf{Q}(\sqrt{3}): \mathbf{Q}] = 2$.

19. Corollary 1 of Theorem 8.10 shows that the degree of an extension obtained by successively adjoining square roots must be 2^n for $n \in \mathbf{Z}^+$. Since $x^{14} - 3x^2 + 12$ is irreducible over $\mathbf{Q}$ by Eisenstein with $p = 3$, and since $[\mathbf{Q}(\alpha): \mathbf{Q}] = 14$ for any zero α of this polynonmial, and since 14 is not a divisor of 2^n for any $n \in \mathbf{Z}$, we see see that α cannot lie in any field obtained by adjoining just square roots. Therefore α cannot be expressed as a rational function of square roots, square roots of rational functions of square roots, etc.

20. We need only show that for each $\alpha \in D$, $\alpha \neq 0$, its multiplicative inverse $1/\alpha$ is in D also. Since E is a finite extension of F, we know α is algebraic over F. Let $\deg(\alpha, F) = n$. Then by Theorem 8.8, we have $F(\alpha) =$ $\{a_0 + a_1\alpha + a_2\alpha^2 + \cdots + a_n\alpha^n \mid a_i \in F,\ i = 0,\cdots,n\}$. In particular, $1/\alpha \in F(\alpha)$ so $1/\alpha$ is a polynomial in α with coefficients in F, and hence is in D.

21. Obviously $\mathbf{Q}(\sqrt{3} + \sqrt{7}) \subseteq \mathbf{Q}(\sqrt{3}, \sqrt{7})$. Let $\alpha = \sqrt{3} + \sqrt{7}$. Then $\alpha^2 = 10 + 2\sqrt{21}$ so $\sqrt{21} \in \mathbf{Q}(\sqrt{3} + \sqrt{7})$. Then $\sqrt{21}(\sqrt{3} + \sqrt{7}) =$ $(3\sqrt{7} + 7\sqrt{3}) \in \mathbf{Q}(\sqrt{3} + \sqrt{7})$. Hence $(3\sqrt{7} + 7\sqrt{3}) - 7(\sqrt{3} + \sqrt{7}) =$ $2\sqrt{7}$ is in $\mathbf{Q}(\sqrt{3} + \sqrt{7})$, so this field contains $\sqrt{7}$ and also $(\sqrt{3} + \sqrt{7}) - \sqrt{7} = \sqrt{3}$. Therefore $\mathbf{Q}(\sqrt{3}, \sqrt{7}) \subseteq \mathbf{Q}(\sqrt{3} + \sqrt{7})$, so $\mathbf{Q}(\sqrt{3}, \sqrt{7}) = \mathbf{Q}(\sqrt{3} + \sqrt{7})$. [One can also make an argument like that in Example 2 of the text, finding $\operatorname{irr}(\sqrt{3} + \sqrt{7}, \mathbf{Q})$ and showing that it is of degree 4. Then $[\mathbf{Q}(\sqrt{3}, \sqrt{7}): \mathbf{Q}(\sqrt{3} + \sqrt{7})] = 1$, so the fields are equal.]

22. If $a = b$ the result is clear; we assume $a \neq b$. Obviously $\mathbf{Q}(\sqrt{a} + \sqrt{b}) \subseteq \mathbf{Q}(\sqrt{a}, \sqrt{b})$. Let $\alpha = \sqrt{a} + \sqrt{b}$. Then $\alpha^2 =$ $a^2 + b^2 + 2\sqrt{ab}$ so $\sqrt{ab} \in \mathbf{Q}(\sqrt{a} + \sqrt{b})$. Then $\sqrt{ab}(\sqrt{a} + \sqrt{b}) =$ $(a\sqrt{b} + b\sqrt{a}) \in \mathbf{Q}(\sqrt{a} + \sqrt{b})$. Hence $(a\sqrt{b} + b\sqrt{a}) - b(\sqrt{a} + \sqrt{b})$ $(a - b)\sqrt{b}$ is in $\mathbf{Q}(\sqrt{a} + \sqrt{b})$. Thus $\sqrt{b} \in \mathbf{Q}(\sqrt{a} + \sqrt{b})$ and hence so is $\sqrt{a} = (\sqrt{a} + \sqrt{b}) - \sqrt{b}$. Therefore $\mathbf{Q}(\sqrt{a}, \sqrt{b}) \subseteq \mathbf{Q}(\sqrt{a} + \sqrt{b})$ also, so $\mathbf{Q}(\sqrt{a}, \sqrt{b}) = \mathbf{Q}(\sqrt{a} + \sqrt{b})$.

Section 8.3

23. If a zero α of $p(x)$ were in E, then since $p(x)$ is irreducible over F, we would have $[F(\alpha)\colon F] = \deg(p(x))$, and $[F(\alpha)\colon F]$ would be a divisor of $[E\colon F]$ by Theorem 8.10. By hypothesis, this is not the case. Therefore $p(x)$ has no zeros in E.

24. If $F(\alpha^2) \neq F(\alpha)$, then $F(\alpha)$ must be an extension of $F(\alpha^2)$ of degree 2, since α is a zero of $x^2 - \alpha^2$. By Theorem 8.10, we would then have $2 = [F(\alpha)\colon F(\alpha^2)]$ divides $[F(\alpha)\colon F]$, which is impossible since $[F(\alpha)\colon F]$ is odd. Therefore $F(\alpha^2) = F(\alpha)$, so $\deg(\mathrm{irr}(\alpha^2, F)) = \deg(\mathrm{irr}(\alpha, F)) = [F(\alpha)\colon F]$ which is odd.

25. Suppose K is algebraic over F. Then every element of K is a zero of a nonzero polynomial in $F[x]$, and hence in $E[x]$. This shows that K is algebraic over E. Of course E is algebraic over F, since each element of E is also an element of K.

    Conversely, suppose that K is algebraic over E and that E is algebraic over F. Let $\alpha \in K$. We must show that α is is algebraic over F. Now α is a zero of some polynomial $a_0 + a_1x + a_2x^2 + \cdots + a_nx^n$ in $E[x]$. Since E is algebraic over F, the a_i are algebraic over F for $i = 0, 1, \cdots, n$. Hence $F(a_0, a_1, a_2, \cdots, a_n)$ is an extension of F of some finite degree m by Theorem 8.11. Since α is algebraic over E of degree $r \le n$, Theorem 8.10 shows that $F(a_0, a_1, a_2, \cdots, a_n, \alpha)$ is a finite extension of F of degree $\le mr$. By Theorem 8.9, α is algebraic over F.

26. If α were algebraic over $\overline{F}_E$, then $\overline{F}_E(\alpha)$ would be algebraic over $\overline{F}_E$ and by definition, $\overline{F}_E$ is algebraic over F. By Exercise 25, then $\overline{F}_E(\alpha)$ is algebraic over F, so in particular α is algebraic over F. But then $\alpha \in \overline{F}_E$ contrary to hypothesis. Thus α is transcendental over $\overline{F}_E$.

27. Let $f(x)$ be a nonconstant polynomial in $\overline{F}_E[x]$. We must show that $f(x)$ has a zero in $\overline{F}_E$. Now $f(x) \in E[x]$ and E is algebraically closed by hypothesis, so $f(x)$ has a zero α in

E. By Exercise 26, if α is not in $\overline{F}_E$, then α is transcendental over $\overline{F}_E$. But by construction, α is a zero of $f(x) \in \overline{F}_E[x]$, so this is impossible. Hence $\alpha \in \overline{F}_E$, which shows that $\overline{F}_E$ is algebraically closed.

28. Let $\alpha \in E$ and let $p(x) = \text{irr}(\alpha,F)$ have degree n. Now $p(x)$ factors in $\overline{F}[x]$ into $(x - \alpha_1)(x - \alpha_2)\cdots(x - \alpha_n)$. Since by hypothesis all zeros of $p(x)$ in $\overline{F}$ are also in E, we see that this same factorization is also valid in $E[x]$. Hence $p(\alpha) = (\alpha - \alpha_1)(\alpha - \alpha_2)\cdots(\alpha - \alpha_n) = 0$ so $\alpha = \alpha_i$ for some i. This shows that $F \leq E \leq \overline{F}$. Since by definition $\overline{F}$ contains only elements that are algebraic over F and E contains all of these, we see that $E = \overline{F}$ and is therefore algebraically closed.

29. If F is a finite field of odd characteristic, then $1 \neq -1$ in F. Since $1^2 = (-1)^2 = 1$, the squares of elements of F can run through at most $|F| - 1$ elements of F, so there is some $a \in F$ that is not a square. The polynomial $x^2 - a$ then has no zeros in F, so F is not algebraically closed.

30. For all $n \in \mathbb{Z}$, $n \geq 2$, the polynomial $x^n - 2$ is irreducible in $\mathbb{Q}[x]$ by Eisenstein with $p = 2$. This shows that $\mathbb{Q}$ has finite extensions contained in $\mathbb{C}$ of arbitrarily high degree. If $\overline{\mathbb{Q}}_{\mathbb{C}}$ were a finite extension of $\mathbb{Q}$ of degree r, then there would be no algebraic extensions of $\mathbb{Q}$ in $\mathbb{C}$ of degree greater than r. Thus the algebraic closure $\overline{\mathbb{Q}}_{\mathbb{C}}$ of $\mathbb{Q}$ in $\mathbb{C}$ cannot be a finite extension of $\mathbb{Q}$.

31. Since $[\mathbb{C}: \mathbb{R}] = 2$ and $\mathbb{C}$ is an algebraic closure of $\mathbb{R}$, it must be that every irreducible $p(x)$ in $\mathbb{R}[x]$ of degree > 1 is actually of degree 2. Since $\mathbb{C} = \mathbb{R}(\alpha)$ where α is a zero of any such polynomial, we know by the construction in Theorem 8.1 that $\mathbb{C} \simeq \mathbb{R}[x]/\langle p(x)\rangle$. Now let E be any finite extension of $\mathbb{R}$. If $E \neq \mathbb{R}$, then let β be an in E but not in $\mathbb{R}$. Then $p(x) = \text{irr}(\beta, \mathbb{R})$ has degree 2 since we have seen that there are no irreducible polynomials in $\mathbb{R}[x]$ of greater degree. The construction in Theorem 8.1 shows that $\mathbb{R}(\beta) \simeq \mathbb{R}[x]/\langle p(x)\rangle$ and hence $\mathbb{R}(\beta) \simeq \mathbb{C}$. Since $\mathbb{C}$ is algebraically closed, $\mathbb{R}(\beta)$ is algebraically closed also, and admits no proper algebraic

extensions. Since E is an algebraic extension of $\mathbb{R}(\beta)$, we must have $E = \mathbb{R}(\beta)$, so $E \simeq \mathbb{C}$.

32. If R contains no nontrivial proper ideals, then $\{0\}$ is the only proper ideal, and is a maximal ideal, and of course it is contained in itself so we are done.

Suppose R contains a nontrivial proper ideal N which of course does not contain the unity 1 of R. The set S of ideals of R that do not contain 1 is partially ordered by inclusion. Let $T = \{N_i \mid i \in I\}$ be a chain of S. We claim that $U = \bigcup_{i\in I} N_i$ is an element of S that is an upper bound of T. Let x, $y \in U$. Then $x \in N_j$ and $y \in N_k$ for some j, $k \in I$. Since T is a chain, one of these ideals is contained in the other, say $N_j \subseteq N_k$. Then x, $y \in N_k$ which is an ideal, so $x \pm y$, 0, rx, and xr are all in N_k and hence in U. This shows that U is an ideal. Clearly $N_i \subseteq U$ for all $i \in I$, and 1 is not in U since 1 is not in N_i for any $i \in I$. Thus $U \in S$ and is an upper bound for T, so the hypotheses of Zorn's lemma are satisfied.

Let M be a maximal element of S; such an element of S exists by Zorn's lemma. Since $M \in S$, we see that M is an ideal of R, and does not contain 1 so $M \neq R$. Suppose that L is an ideal of R such that $M \subseteq L \subseteq R$. If $L \in S$, then $M = L$ since M is a maximal element of S under set inclusion. Otherwise, $1 \in L$ so $L = R$. Thus M is a maximal ideal of R.

SECTION 8.4 - Geometric Constructions

1. T T T F T F T T T F f) It is true that every such real number is constructible, but we have not shown this.

2. If a regular 9-gon could be constructed, then an angle of $360°/9 = 40°$ could be constructed, and then bisected to construct an angle of 20°. The proof of Theorem 8.20 shows, however, that an angle of 20° is not constructible.

3. One can construct an angle of 30° if and only if one can construct $\cos 30° = \sqrt{3}/2$. Since $\sqrt{3}$ is constructible and quotients of constructible numbers are constructible, an angle of 30° is constructible.

4. Since $|\overline{OA}| = |\overline{OP}| = 1$, $\angle OAP = \angle APO = (180° - 36°)/2 = 72°$. Then $\angle QAP = 36°$ so triangle OAP is similar to triangle APQ. Now $|\overline{AP}| = |\overline{AQ}| = |\overline{OQ}| = r$, so $|\overline{QP}| = 1 - r$. Taking ratios of corrresponding sides, we obtain $|\overline{AP}|/|\overline{QP}| = |\overline{OA}|/|\overline{AP}|$ so $r/(1 - r) = 1/r$. Thus $r^2 = 1 - r$ so $r^2 + r - 1 = 0$. By the quadratic formula, we find that $r = \frac{-1 + \sqrt{5}}{2}$ which is a constructible number. Thus we can construct an angle of 36° by taking a line segment $\overline{OP}$ of length 1, drawing circles of radii 1 and r from its endpoints, and finding a point A of intersection of the two circles. Then $\angle AOP$ measures 36°. Thus a regular 10-gon with central angles of 36° is constructible. A regular pentagon is obtained by starting at vertex 1 of a regular 10-gon and drawing line segments to vertex 3, then to vertex 5, then to vertex 7, then to vertex 9, and then to vertex 1.

5. A regular 20-gon is constructible since we can bisect the constructible angle of 36° (see Exercise 4) to obtain an angle of 18° = 360°/20.

6. Since we can construct an angle of 72° = 2(36°) by Exercise 4, and since 60° is a constructible angle, we can construct an angle of 72° - 60° = 12° = 360°/30. Therefore a regular 30-gon can be constructed.

7. Exercise 6 shows that a 12° angle can be constructed, so a 24° = 2(12°) = 72°/3 angle can be constructed. Thus an angle of 72° can be trisected.

8. Exercise 6 shows that an angle of 12° can be constructed so an angle of 24° = 2(12°) = 360°/15 can be constructed. Hence a regular 15-gon can be constructed.

SECTION 8.5 - Finite Fields

1. Since $4096 = 2^{12}$ is a power of a prime, a finite field of order 4096 does exist.

2. Since $3127 = 53 \cdot 59$ is not a power of a prime, no finite field of order 3127 exists.

Section 8.5

3. Since $68921 = 41^3$ is a power of a prime, a finite field of order 68921 does exist.

4. $GF(9)^*$ is a cyclic group under multiplication of order 8 and has $\phi(8) = 4$ generators, so there are 4 primitive 8th roots of unity.

5. $GF(19)^*$ is a cyclic group under multiplication of order 18 and has $\phi(18) = 6$ generators, so there are 6 primitive 18th roots of unity.

6. $GF(30)^*$ is a cyclic group under multiplication of order 30. Its cyclic subgroup of order 15 has $\phi(15) = 8$ generators, so there are 8 primitive 15th roots of unity.

7. $GF(23)^*$ is a cyclic group under multiplication of order 22. Since 10 is not a divisor of 22, it contains no elements of order 10, so GF(23) contains no primitive 10th roots of unity.

8. T F T F T F T T F T

9. Since both the given polynomials are irreducible over $\mathbb{Z}_2$, both $\mathbb{Z}_2(\alpha)$ and $\mathbb{Z}_2(\beta)$ are extensions of $\mathbb{Z}_2$ of degree 3 and thus are subfields of $\overline{\mathbb{Z}}_2$ containing $2^3 = 8$ elements. By Theorem 8.22, both of these fields must consist precisely of the zeros in $\overline{\mathbb{Z}}_2$ of the polynomial $x^8 - x$. Thus the fields are the same.

10. Let $p(x)$ be irreducible of degree m in $\mathbb{Z}_p[x]$. Let K be the finite extension of $\mathbb{Z}_p$ obtained by adjoining *all* the zeros of $p(x)$ in $\overline{\mathbb{Z}}_p$. Then K is a finite field of order p^n for some positive integer n, and consists precisely of all zeros of $x^{p^n} - x$ in $\overline{\mathbb{Z}}_p$. Now $p(x)$ factors into linear factors in $K[x]$, and these linear factors are among the linear factors of $x^{p^n} - x$ in $K[x]$. Thus $p(x)$ is a divisor of $x^{p^n} - x$.

11. Since $\alpha \in F$, we have $\mathbb{Z}_p(\alpha) \subseteq F$. But since α is a generator of the multiplicative group F^*, we see that $\mathbb{Z}_p(\alpha) = F$. Since

$|F| = p^n$, the degree of α over $\mathbb{Z}_p$ must be n.

12. Let F be a finite field of p^n elements containing (up to isomorphism) the prime field $\mathbb{Z}_p$. Let m be a divisor of n, so that $n = mq$. Let $\overline{F} = \overline{\mathbb{Z}}_p$ be an algebraic closure of F. If $\alpha \in \overline{\mathbb{Z}}_p$ and $\alpha^{p^m} = \alpha$, then $\alpha^{p^n} = \alpha^{p^{mq}} = (\alpha^{p^m})^{p^{m(q-1)}} = \alpha^{p^{m(q-1)}}$ $= (\alpha^{p^m})^{p^{m(q-2)}} = \alpha^{p^{m(q-2)}} = \cdots = \alpha^{p^m} = \alpha$. By Theorem 8.22, the zeros of $x^{p^m} - x$ in $\overline{\mathbb{Z}}_p$ form the *unique* subfield of $\overline{\mathbb{Z}}_p$ of order p^m. Our computation shows that the elements in this subfield are also zeros of $x^{p^n} - x$, and consequently all lie in the field F, which by Theorem 8.22 consists of all zeros of $x^{p^n} - x$ in $\overline{\mathbb{Z}}_p$.

13. Let F be the extension of $\mathbb{Z}_p$ of degree n, consisting of all zeros of $x^{p^n} - x$ by Theorem 8.22. Each $\alpha \in F$ is algebraic over $\mathbb{Z}_p$ and has degree that divides n by Theorem 8.10. Thus each $\alpha \in F$ is a zero of a monic irreducible polynomial of a degree dividing n. Conversely, a zero β of an irreducible monic polynomial having degree m dividing n lies in a field $\mathbb{Z}_p(\beta)$ of p^m elements that is contained in F by Exercise 12. Thus the elements of F are precisely the zeros of all monic irreducible polynomials in $\mathbb{Z}_p[x]$ of degree dividing n. But by Theorem 8.22, the elements of F consist precisely of all zeros of $x^{p^n} - x$ in $\overline{\mathbb{Z}}_p$. Looking at their factorizations in $\overline{\mathbb{Z}}_p$ into linear factors, we see that $x^{p^n} - x$ is the product of all monic polynomials in $\mathbb{Z}_p[x]$ of degree d dividing n.

14. a) $x^2 \equiv a \pmod{p}$ has a solution in $\mathbb{Z}$ if and only if $x^2 - b$ has a solution in $\mathbb{Z}_p$ where b is the remainder of a modulo p. Now $\mathbb{Z}_p^*$ is cyclic of order $p - 1$. The elements b of a

cyclic group that are squares are those that are even powers of a generator, and these are precisely the elements satisfying $b^{(p-1)/2} = 1$. Thus we see that $x^2 \equiv a \pmod{p}$, where a is not congruent to zero modulo p, has a solution in $\mathbb{Z}$ if and only if $a^{(p-1)/2} \equiv 1 \pmod{p}$.

b) We know that $x^2 - 6$ is irreducible in $\mathbb{Z}_{17}[x]$ if and only if it has no zero in $\mathbb{Z}_{17}$, so $6 \neq b^2$ for $b \in \mathbb{Z}_{17}$. By part (a), we must determine whether $6^{(17 - 1)/2} = 6^8$ is congruent to 1 modulo 17. Computing in $\mathbb{Z}_{17}$, we have $6^2 = 2$, $6^4 = 2^2 = 4$, and $6^8 = 4^2 = 16$, so 6 is not a square in $\mathbb{Z}_{17}$ and $x^2 - 6$ is irreducible.

15. Let F and F' be two fields of order p^n. Both of their prime fields are isomorphic to $\mathbb{Z}_p$ in a natural way. By Corollary 2 of Theorem 8.23, both F and F' are simple extensions of their prime subfields. Let $p(x)$ be a monic irreducible polynomial of degree n with coefficients in the prime field P of F such that $F = P(\alpha)$ where α is a zero of $p(x)$. Let $p'(x)$ be the corresponding polynomial in $P'[x]$ under the natural isomorphism of $P[x]$ with $P'[x]$. By Theorem 8.22 a zero α' of $p'(x)$ in $\overline{F'}$ actually must lie in F', and then $F' = P'(\alpha)$. The construction in Theorem 8.1 shows that $P(\alpha) \simeq P[x]/\langle p(x)\rangle$ and $P'(\alpha) \simeq P'[x]/\langle p'(x)\rangle$. But the natural isomorphism of $P[x]$ with $P'[x]$ carrying $p(x)$ into $p'(x)$ gives an isomorphism of $P[x]/\langle p(x)\rangle$ with $P'[x]/\langle p'(x)\rangle$. Thus F and F' are isomorphic.

SECTION 8.6 - Additional Algebraic Structures

1. $\langle G, O, \cdot_G, \cdot_{OG}\rangle$. Other notations are possible.

2. $\langle M, R, +_M, +_R, \cdot_R, \cdot_{RM}\rangle$. Other notations are possible.

3. $\langle V, F, +_V, \cdot_V, +_F, \cdot_F, \cdot_{FG}\rangle$. Other notations are possible.

4. $\mathbb{Z}_2 \times \{0\}$ is not a characteristic subgroup of $\mathbb{Z}_2 \times \mathbb{Z}_2$ since

$\phi: \mathbb{Z}_2 \times \mathbb{Z}_2 \to \mathbb{Z}_2 \times \mathbb{Z}_2$ where $\phi(a,b) = (b,a)$ is an automorphism carrying $\mathbb{Z}_2 \times \{0\}$ onto $\{0\} \times \mathbb{Z}_2$.

5. (See the text answer for the definition.) Since a O-homomorphism is, in particular, a group homomorphism of G into G', we know that $\mathrm{Ker}(\phi)$ is a subgroup of G. It remains only to show that $\mathrm{Ker}(\phi)$ is closed under multiplication by elements of O. Let $a \in O$ and $\alpha \in \mathrm{Ker}(\phi)$ so that $\phi(\alpha) = 1$, the identity of G. We claim that $a1 = 1$. Condition 2 for a group with operators shows that $a1 = a[(1)(1)] = (a1)(a1)$, and group cancellation proves that $a1 = 1$. Then $\phi(a\alpha) = a\phi(\alpha) = a1 = 1$, so $a\alpha \in \mathrm{Ker}(\phi)$ and we are done.

6. A submodule N of a (left) R-module M is a subset of M that is a (left) R-module under induced operations of addition in M and external multiplication by R.

7. (See the text answer.)

8. We know that an intersection of subgroups of $\langle G, \cdot \rangle$ is again a subgroup of $\langle G, \cdot \rangle$. It remains just to show that an intersection of O-subgroups H_i for $i \in I$ is again closed under multiplication by elements of O. Let $a \in O$ and $\alpha \in \bigcap_{i \in I} H_i$. Then $\alpha \in H_i$ for $i \in I$. Since each H_i is an O-subgroup, we know that $a\alpha \in H_i$ for $i \in I$. Hence $a\alpha \in \bigcap_{i \in I} H_i$.

9. Let H be an admissible normal subgroup of G. Neglecting multiplication by O, we know that G/H is a group of cosets of H with well-defined group multiplication by multiplying representatives in G. It remains to define multiplication of a coset by elements of O and to show that G/H satifies the conditions for an O-group.

   Let $a \in O$ and let $\alpha H \in G/H$. Define $a(\alpha H) = (a\alpha)H$. Let $\beta \in H$, so that $\alpha\beta$ is another representative of αH. By Condition 2 for an O-group, $a(\alpha\beta) = (a\alpha)(a\beta)$. But $a\beta \in H$ since H is an admissible normal subgroup, so $a(\alpha\beta) \in (a\alpha)H$ and our multiplication of cosets in G/H by elements of O is well defined, and of course Condition 1 for an O-group is satisfied.

   For $a \in O$ and $\alpha H, \gamma H \in G/H$, we have $a[(\alpha H)(\gamma H)] = a[(\alpha\gamma)H] = [a(\alpha\gamma)]H = [(a\alpha)(a\gamma)]H = [(a\alpha)H][(a\gamma)H]$ so Condition 2 for an O-group is satisfied.

10. The proof is identical with that for Theorem 8.5. Just read R for F and M for V in the proof of that theorem.

Section 8.6

11. Let $a, b \in L_\alpha$ so that $a\alpha = b\alpha = 0$. Then by Condition 3 for an R-module, $(a + b)\alpha = a\alpha + b\alpha = 0 + 0 = 0$, so $a + b \in L_\alpha$ and L_α is closed under addition. For any $r \in R$, Condition 4 of an R-module shows that $(ra)\alpha = r(a\alpha) = r0$, and $r0 = 0$ by Exercise 10. Thus $ra \in L_\alpha$ which is thus closed under left multiplication by elements of R. Since $0\alpha = 0$ and $(-a)\alpha = -(a\alpha) = -0 = 0$ by Exercise 10, we see that L_α contains the additive identity and additive inverses, so it is an additive subgroup of R closed under left multiplication by elements of R, that is, L_α is a left ideal of R.

12. We know that $M_n(F)$ is an abelian group under addition. Let $A = (a_{ij})$, $B = (b_{ij})$ and $C = (c_{ij})$ be in $M_n(F)$, where a_{ij} is the element in the ith row and jth column of A with similar notations for B and C. We show the conditions of Definition 8.20 for an algebra.

Condition 1: For $r \in F$, we have $(rA)B = D = (d_{ij})$ where

$$d_{ij} = \sum_{k=1}^{n} (ra_{ik})b_{kj} = \sum_{k=1}^{n} a_{ik}(rb_{kj}) = r\sum_{k=1}^{n} a_{ik}b_{kj},$$ showing

that we also have $D = A(rB)$ and $D = r(AB)$. Thus $(rA)B = A(rB) = r(AB)$ and Condition 1 is satisfied.

Condition 2: We have $(A + B)C = E$ where

$$e_{ij} = \sum_{k=1}^{n} (a_{ik}+b_{ik})c_{kj} = \sum_{k=1}^{n} a_{ik}c_{kj} + \sum_{k=1}^{n} b_{ik}c_{kj}.$$

This equation in F shows that $(A + B)C = AC + BC$.

Condition 3: We have $A(B + C) = H$ where

$$h_{ij} = \sum_{k=1}^{n} a_{ik}(b_{kj}+c_{kj}) = \sum_{k=1}^{n} a_{ik}b_{kj} + \sum_{k=1}^{n} a_{ik}c_{kj}.$$

This equation in F shows that $A(B + C) = AB + AC$. [Actually, $M_n(F)$ is an associative algebra.]

A basis for $M_n(F)$ is clearly the set of all $n\times n$ matrices having a single 1 as only nonzero entry. Since there are n^2 possible positions for this entry 1, we see that $M_n(F)$ has a basis of n^2 elements, and hence dimension n^2.

13. Let α, $\gamma \in V$, and suppose that $\alpha = \sum_{r=1}^{n} a_r\beta_r$ and $\gamma = \sum_{s=1}^{n} c_s\beta_s$.

Using all three conditions listed for an algebra, we see that

$$\alpha\gamma = \sum_{r=1}^{n} a_r\beta_r\left(\sum_{s=1}^{n} c_s\beta_s\right) = \sum_{r,s=1}^{n} (a_r\beta_r)(c_s\beta_s) = \sum_{r,s=1}^{n} (a_rc_s)\beta_r\beta_s$$

showing that $\alpha\gamma$ is completely determined if all the n^2 products $\beta_r\beta_s$ for r, $s = 1, 2, \cdots, n$ are known.

14. Continuing with the notation from Exercise 13, let also $\delta = \sum_{t=1}^{n} d_t\beta_t$. As in Exercise 13, using all three properties of an algebra, we find that $(\alpha\gamma)\delta = \sum_{r,s,t=1}^{n} (a_rc_sd_t)[(\beta_t\beta_s)\beta_t]$ while $\alpha(\gamma\delta) = \sum_{r,s,t=1}^{n} (a_rc_sd_t)[\beta_r(\beta_s\beta_r)]$. Thus we see that if $\beta_r(\beta_s\beta_t) = (\beta_r\beta_s)\beta_t$ for r, s, $t = 1, 2, \cdots, n$, then V is associative. If V is associative, then of course these products of the β_i, in particular, must satisfy this associative property.

15. Exercise 13 shows that there is at most one algebra satisfying these conditions, for it shows that multiplication on V is completely determined once the products $\beta_r\beta_s$ are known for r, $s = 1, 2, \cdots, n$. It remains to show that a multiplication can be defined using these structure constants c_{rst} such that the three conditions for an algebra in Definition 8.20 are satisfied.

Let $\alpha = \sum_{r=1}^{n} a_r\beta_r$ and $\delta = \sum_{s=1}^{n} d_s\beta_s$ be elements of V. We define $\alpha\delta = \sum_{t=1}^{n}\left(\sum_{r,s=1}^{n} a_rd_sc_{rst}\right)\beta_t$. For $v \in F$, the associativity and commutativity of multiplication in F shows that $(va_r)d_sc_{rst} = v(a_rd_sc_{rst}) = a_r(vd_s)c_{rst}$, so we see that $(v\alpha)\delta = v(\alpha\delta) = \alpha(v\delta)$ and Condition 1 for an algebra is satisfied. Let $\mu = \sum_{i=1}^{n} m_i\beta_i$ also be in V. The distributive laws in F show that $(a_r + d_r)m_sc_{rst} = a_rm_sc_{rst} + d_rm_sc_{rst}$ so $(\alpha + \delta)\mu = \alpha\mu + \delta\mu$, and that $a_r(d_s + m_s)c_{rst} = a_rd_sc_{rst} + a_rm_sc_{rst}$ so $\alpha(\delta + \mu) = \alpha\delta + \alpha\mu$. Thus Conditions 2 and 3 for an algebra also hold.

CHAPTER 9

AUTOMORPHISMS AND GALOIS THEORY

SECTION 9.1 - Automorphisms of Fields

1. $\sqrt{2}$ and $-\sqrt{2}$

2. $\sqrt{2}$, for $\sqrt{2} \in \mathbb{R}$

3. $3 + \sqrt{2}$ and $3 - \sqrt{2}$

4. $\sqrt{2} - \sqrt{3}$, $\sqrt{2} + \sqrt{3}$, $-\sqrt{2} - \sqrt{3}$, and $-\sqrt{2} + \sqrt{3}$

5. $\sqrt{2} + i$, $\sqrt{2} - i$, $-\sqrt{2} + i$, and $-\sqrt{2} - i$

6. $\sqrt{2} + i$ and $\sqrt{2} - i$ since $\sqrt{2} \in \mathbb{R}$

7. $\sqrt{1 + \sqrt{2}}$, $\sqrt{1 - \sqrt{2}}$, $-\sqrt{1 + \sqrt{2}}$, and $-\sqrt{1 - \sqrt{2}}$

8. $\sqrt{1 + \sqrt{2}}$ and $\sqrt{1 - \sqrt{2}}$ since $\sqrt{2} \in \mathbb{Q}(\sqrt{2})$

9. $\tau_2(\sqrt{3}) = \sqrt{3}$

10. $\tau_2(\sqrt{2} + \sqrt{5}) = -\sqrt{2} + \sqrt{5}$

11. $(\tau_3\tau_2)(\sqrt{2} + 3\sqrt{5}) = \tau_3[\tau_2(\sqrt{2} + 3\sqrt{5})] = \tau_3(-\sqrt{2} + 3\sqrt{5}) = -\sqrt{2} + 3\sqrt{5}$

12. $(\tau_5\tau_3)\left(\dfrac{\sqrt{2} - 3\sqrt{5}}{2\sqrt{3} - \sqrt{2}}\right) = \tau_5\left[\tau_3\left(\dfrac{\sqrt{2} - 3\sqrt{5}}{2\sqrt{3} - \sqrt{2}}\right)\right] = \tau_5\left(\dfrac{\sqrt{2} - 3\sqrt{5}}{-2\sqrt{3} - \sqrt{2}}\right)$

$$= \frac{\sqrt{2} + 3\sqrt{5}}{-2\sqrt{3} - \sqrt{2}}$$

13. $(\tau_5{}^2\tau_3\tau_2)(\sqrt{2} + \sqrt{45}) = \tau_5{}^2\tau_3\left[\tau_2(\sqrt{2} + 3\sqrt{5})\right]$

$$= \tau_5{}^2\left[\tau_3(-\sqrt{2} + 3\sqrt{5})\right] = \tau_5{}^2(-\sqrt{2} + 3\sqrt{5}) = \tau_5(-\sqrt{2} - 3\sqrt{5})$$

$$= -\sqrt{2} + 3\sqrt{5} = -\sqrt{2} + \sqrt{45}$$

14. $\tau_3\left[\tau_5(\sqrt{2} - \sqrt{3}) + (\tau_2\tau_5)(\sqrt{30})\right] = \tau_3\left[(\sqrt{2}-\sqrt{3}) + \tau_2(\tau_5(\sqrt{2}\sqrt{3}\sqrt{5}))\right]$

$$= \tau_3\left[(\sqrt{2} - \sqrt{3}) + \tau_2(-\sqrt{2}\sqrt{3}\sqrt{5})\right] = (\sqrt{2} + \sqrt{3}) + \tau_3(\sqrt{2}\sqrt{3}\sqrt{5})$$

$= \sqrt{2} + \sqrt{3} - \sqrt{2}\sqrt{3}\sqrt{5} = \sqrt{2} + \sqrt{3} - \sqrt{30}$

15. a) $\mathbf{Q}$ b) $\mathbf{Q}\sqrt{6}$ c) $\mathbf{Q}$

16. $\mathbf{Q}(\sqrt{2}, \sqrt{5})$ 17. $\mathbf{Q}(\sqrt{2}, \sqrt{3}, \sqrt{5})$ 18. $\mathbf{Q}(\sqrt{5})$

19. $\mathbf{Q}(\sqrt{3}, \sqrt{10})$ 20. $\mathbf{Q}(\sqrt{6}, \sqrt{10})$ 21. $\mathbf{Q}$

22. a) We have $\tau_2^{\,2}(\sqrt{2}) = \tau_2(\tau_2(\sqrt{2})) = \tau_2(-\sqrt{2}) = \sqrt{2}$ while of course $\tau_2^{\,2}(\sqrt{3}) = \sqrt{3}$ and $\tau_2^{\,2}(\sqrt{5}) = \sqrt{5}$, so $\tau_2^{\,2}$ leaves $\sqrt{2}$, $\sqrt{3}$, and $\sqrt{5}$ all fixed and is thus the identity automorphism of $\mathbf{Q}(\sqrt{2}, \sqrt{3}, \sqrt{5})$. A similar argument shows that τ_3 and τ_5 are also the identity automorphism, so these elements are all of order 2.

b) $H = \{\iota, \tau_2, \tau_3, \tau_5, \tau_2\tau_3, \tau_2\tau_5, \tau_3\tau_5, \tau_2\tau_3\tau_5\}$

	ι	τ_2	τ_3	τ_5	$\tau_2\tau_3$	$\tau_2\tau_5$	$\tau_3\tau_5$	$\tau_2\tau_3\tau_5$
ι	ι	τ_2	τ_3	τ_5	$\tau_2\tau_3$	$\tau_2\tau_5$	$\tau_3\tau_5$	$\tau_2\tau_3\tau_5$
τ_2	τ_2	ι	$\tau_2\tau_3$	$\tau_2\tau_5$	τ_3	τ_5	$\tau_2\tau_3\tau_5$	$\tau_3\tau_5$
τ_3	τ_3	$\tau_2\tau_3$	ι	$\tau_3\tau_5$	τ_2	$\tau_2\tau_3\tau_5$	τ_5	$\tau_2\tau_5$
τ_5	τ_5	$\tau_2\tau_5$	$\tau_3\tau_5$	ι	$\tau_2\tau_3\tau_5$	τ_2	τ_3	$\tau_2\tau_3$
$\tau_2\tau_3$	$\tau_2\tau_3$	τ_3	τ_2	$\tau_2\tau_3\tau_5$	ι	$\tau_3\tau_5$	$\tau_2\tau_5$	τ_5
$\tau_2\tau_5$	$\tau_2\tau_5$	τ_5	$\tau_2\tau_3\tau_5$	τ_5	$\tau_3\tau_5$	ι	$\tau_2\tau_3$	τ_3
$\tau_3\tau_5$	$\tau_3\tau_5$	$\tau_2\tau_3\tau_5$	τ_5	τ_3	$\tau_2\tau_5$	$\tau_2\tau_3$	ι	τ_2
$\tau_2\tau_3\tau_5$	$\tau_2\tau_3\tau_5$	$\tau_3\tau_5$	$\tau_2\tau_5$	$\tau_2\tau_3$	τ_5	τ_3	τ_2	ι

c) An automorphism in $G(E/\mathbf{Q})$ is completely determined by its values on $\sqrt{2}$, $\sqrt{3}$, and $\sqrt{5}$. Each of these is either left alone or mapped into its negative. Thus there are two possibilities for the value of $\sigma \in G(E/\mathbf{Q})$ on $\sqrt{2}$, two possibilties for $\sigma(\sqrt{3})$, and two possibilites for $\sigma(\sqrt{5})$ giving a total of $2 \cdot 2 \cdot 2 = 8$ automorphisms in all. Since $|H| = 8$, we see that $H = G(E/\mathbf{Q})$.

23. a) $\beta = 3 - \sqrt{2}$

b) Since $\psi_{\alpha,\beta}(\sqrt{2}) = \psi_{\alpha,\beta}[-3 + (3 + \sqrt{2})]$

$= \psi_{\alpha,\beta}(-3) + \psi_{\alpha,\beta}(3 + \sqrt{2}) = -3 + (3 - \sqrt{2}) = -\sqrt{2}$

$= \psi_{\sqrt{2},-\sqrt{2}}(\sqrt{2})$, we see that $\psi_{\alpha,\beta}$ and $\psi_{\sqrt{2},-\sqrt{2}}$ are the same maps.

24. $\sigma_2(0) = 0^2 = 0$, $\sigma^2(1) = 1^2 = 1$, $\sigma_2(\alpha) = \alpha^2 = \alpha + 1$,

$\sigma_2(\alpha + 1) = (\alpha + 1)^2 = \alpha^2 + 2\cdot\alpha + 1 = \alpha^2 + 1 = \alpha + 1 + 1 = \alpha$.

$\mathbb{Z}_2(\alpha)_{\{\sigma_2\}} = \{0, 1\} = \mathbb{Z}_2$

25. Using the table for this field on page 184 of this manual, we find that $\sigma_3(0) = 0^3 = 0$, $\sigma_3(1) = 1^3 = 1$, $\sigma_3(2) = (2^3) = 2$,

$\sigma_3(\alpha) = \alpha^3 = 2\alpha$, $\sigma_3(2\alpha) = (2\alpha)^3 = \alpha$, $\sigma_3(1+\alpha) = (1+\alpha)^3 = 1+2\alpha$,

$\sigma_3(1+2\alpha) = (1+2\alpha)^3 = 1+\alpha$, $\sigma_3(2+\alpha) = (2+\alpha)^3 = 2+2\alpha$,

$\sigma_3(2+2\alpha) = (2+2\alpha)^3 = 2+\alpha$. Thus $\mathbb{Z}_3(\alpha)_{\{\sigma_3\}} = \mathbb{Z}_3$.

26. $\sigma_2\colon \mathbb{Z}_2(x) \longrightarrow \mathbb{Z}_2(x)$, where x is an indeterminant, is not an automorphism since the image is $\mathbb{Z}_2(x^2)$. Thus σ_2 is not onto $\mathbb{Z}_2(x)$, but rather maps $\mathbb{Z}_2(x)$ one to one onto a proper subfield of itself.

27. F F T T F T T T T T

28. By Corollary 1 of Theorem 9.1, such an isomorphism must map α onto one of its conjugates over F. Since $\deg(\alpha, F) = n$, there are at most n conjugates of α in $\overline{F}$, for a polynomial of degree n has at most n zeros in a field. On the other hand, Corollary 1 of Theorem 9.1 asserts that there is exactly one such isomorphism for each conjugate of α over F, so the number of such isomorphisms is equal to the number of conjugates of α over F, which is $\leq n$.

29. We proceed by induction on n. For $n = 1$, Corollary 1 of Theorem 9.1 shows that σ is completely determined by $\sigma(\alpha_1)$, which must be a conjugate of α_1 over F. Suppose that the statement is true for $n \leq k$, and let $n = k + 1$. Suppose that σ is known on $\alpha_1, \alpha_2, \cdots, \alpha_k, \alpha_{k+1}$. Let $r =$

$\deg(\alpha_{k+1}, F(\alpha_1, \cdots, \alpha_k))$. Then each element β in $F(\alpha_1, \cdots, \alpha_k, \alpha_{k+1})$ can be written as

$$\beta = \gamma_0 + \gamma_1\alpha_{k+1} + \gamma_2\alpha_{k+1}^2 + \cdots + \gamma_{r-1}\alpha_{k+1}^{r-1}$$

where $\gamma_i \in F(\alpha_1, \cdots, \alpha_k)$ for $i = 0, \cdots, r - 1$ according to Theorem 8.8. By our induction assumption, we know $\sigma(\gamma_i)$ for for $i = 0, \cdots, r - 1$, and we are assuming that we also know $\sigma(\alpha_{k+1})$. The expression for β and the fact that σ is an automorphism shows that we known $\sigma(\beta)$. This completes our proof by induction.

30. By Corollary 1 of Theorem 7.1, σ maps each zero of $\text{irr}(\alpha,F)$ onto a zero of this same polynomial. Since σ is an automorphism, it is a one-to-one map of E onto E. By counting, it must map the set of zeros in E of this polynomial onto itself, so it is a permutation of this set.

31. Since $S \subseteq H$, it is clear that $E_H \subseteq E_S$. Let $\alpha \in E_S$, so that $\sigma_i(\alpha) = \alpha$ for $i \in I$. Then $\sigma_i^{-1}(\alpha) = \alpha$ also. It follows at once that $\sigma_i^{\,n}(\alpha) = \alpha$ for $i \in I$ and $n \in \mathbb{Z}$. Theorem 1.11 shows that every element of H is a product of a finite number of such powers of the σ_i. Since products of automorphisms are computed by function composition, it follows that α is left fixed by each element in H. Therefore $\alpha \in E_H$, so $E_S \subseteq E_H$ and therefore $E_S = E_H$.

32. a) Suppose that $\zeta^i = \zeta^j$ for $i < j \le p - 1$. Then $\zeta^{j-i} = 1$ and ζ would be a zero of $x^{j-i} - 1$ which is of degree less than $p - 1$, contradicting the fact that $\Phi_p(x)$ is irreducible. Thus these powers ζ^i for $0 \le i \le p - 1$ are distinct. Since $\zeta^i \ne 1$ for $i \le p - 1$ but $(\zeta^i)^p = (\zeta^p)^i = 1^i = 1$, we see that ζ^i is a zero of $x^p - 1$ that is different from 1 for $1 \le i \le p - 1$, so these distinct powers of ζ must account for all $p - 1$ zeros of $\Phi_p(x)$.

b) Let $\sigma, \tau \in G(\mathbf{Q}(\zeta)/\mathbf{Q})$. Suppose that $\sigma(\zeta) = \zeta^i$ and that $\tau(\zeta) = \zeta^j$. Then $(\sigma\tau)(\zeta) = \sigma(\tau(\zeta)) = \sigma(\zeta^j) = [\sigma(\zeta)]^j =$

$(\zeta^i)^j = \zeta^{ij} = \zeta^{ji} = (\zeta^j)^i = [\tau(\zeta)]^i = \tau(\zeta^i) = \tau(\sigma(\zeta)) = (\tau\sigma)(\zeta)$. Since $(\sigma\tau)(\zeta) = (\tau\sigma)(\zeta)$, Corollary 2 of Theorem 9.1 shows that $\sigma\tau = \tau\sigma$. Thus $G(\mathbf{Q}(\zeta)/\mathbf{Q})$ is abelian.

c) We know that $B = \{1, \zeta, \zeta^2, \cdots, \zeta^{p-2}\}$ is a basis for $\mathbf{Q}(\zeta)$ over $\mathbf{Q}$. Let $\beta \in \mathbf{Q}(\zeta)$. We can write

$$\frac{\beta}{\zeta} = a_0 + a_1\zeta + a_2\zeta^2 + \cdots + a_{p-2}\zeta^{p-2}$$

for $a_i \in \mathbf{Q}$. Multiplying by ζ, we see that we have

$$\beta = a_0\zeta + a_1\zeta^2 + a_2\zeta^3 + \cdots + a_{p-2}\zeta^{p-1} \qquad (1)$$

so these powers of ζ do span $\mathbf{Q}(\zeta)$. They are independent since a linear combination of them equal to zero yields a linear combination of elements in B equal to zero upon division by ζ. Thus the set $\{\zeta, \zeta^2, \cdots, \zeta^{p-1}\}$ is a basis for $\mathbf{Q}(\zeta)/\mathbf{Q}$.

By Theorem 9.1, and part (a), there exists an automorphism σ_i in $G(\mathbf{Q}(\zeta)/\mathbf{Q})$ such that $\sigma_i(\zeta) = \zeta^i$ for $i = 1, 2, \cdots, p - 1$. Thus if β in Eq.(1) is left fixed by all such σ_i, we must have $a_0 = a_1 = \cdots = a_{p-2}$ so $\beta = a_0(\zeta + \zeta^2 + \cdots + \zeta^{p-1}) = -a_0$ since ζ is a zero of $\Phi_p(x)$. Thus the elements of $\mathbf{Q}(\zeta)$ left fixed lie in $\mathbf{Q}$, so $\mathbf{Q}$ is the fixed field of $G(\mathbf{Q}(\zeta),\mathbf{Q})$.

33. Yes. If α and β are transcendentals over F, then $\phi: F(\alpha) \longrightarrow F(\beta)$, where $\phi(a) = a$ for each $a \in F$, and $\phi(\frac{f(\alpha)}{g(\beta)}) = \frac{f(\beta)}{g(\beta)}$ for $f(x)$, $g(x) \in F[x]$ and $g(x) \neq 0$, is an isomorphism. Both $F(\alpha)$ and $F(\beta)$ are isomorphic to $F(x)$ as we saw in Case II under the heading SIMPLE EXTENSIONS in Section 8.1.

34. In the notation of Exercise 33, taking $\alpha = x$, we must find all β transcendental over F such that $F(\alpha) = F(x) = F(\beta)$. This means that not only must β be a quotient of polymomials in x that does not lie in the field F, but also, we must be able to solve and express x as a quotient of polynomials in β. This is only possible if β is a quotient $\frac{ax + b}{cx + d}$ of linear polynomials in $F[x]$, and for this quotient not to be in F, we must require that $ad - bc \neq 0$.

35. a) Let σ be an automorphism of E and let $\alpha \in E$. Then $\sigma(\alpha^2) = \sigma(\alpha\alpha) = \sigma(\alpha)\sigma(\alpha) = \sigma(\alpha)^2$, so σ indeed carries squares into squares.

b) Since the positive numbers in $\mathbf{R}$ are precisely the squares in $\mathbf{R}$, this follows at once from part (a).

c) From $a < b$, we deduce that $b - a > 0$. By part (b), we see that $\sigma(b - a) = \sigma(b) - \sigma(a) > 0$, so $\sigma(a) < \sigma(b)$.

d) Let $a \in \mathbf{R}$, and find sequences $\{r_i\}$ and $\{s_i\}$ of rational numbers, both converging to a, and satisfying

$$r_i < r_{i+1} < a < s_{i+1} < s_i$$

for $i \in \mathbf{Z}^+$. By part (c), we see that

$$\sigma(r_i) < \sigma(r_{i+1}) < \sigma(a) < \sigma(s_{i+1}) < \sigma(s_i). \qquad (2)$$

The automorphism σ of $\mathbf{R}$ must leave the prime field $\mathbf{Q}$ fixed, since $\sigma(1) = 1$. Thus inequality (2) becomes

$$r_i < r_{i+1} < \sigma(a) < s_{i+1} < s_i$$

for all $i \in \mathbf{Z}^+$. Since $\{r_i\}$ and $\{s_i\}$ converge to a, we see that $\sigma(a) = a$, so σ is the identity automorphism.

SECTION 9.2 - The Isomorphism Extension Theorem

1. (See the text answer.)

2. τ_1 given by $\tau_1(\sqrt{2}) = \sqrt{2}$, $\tau_1(\sqrt{3}) = -\sqrt{3}$, $\tau_1(\sqrt{5}) = \sqrt{5}$;

   τ_2 given by $\tau_2(\sqrt{2}) = \sqrt{2}$, $\tau_2(\sqrt{3}) = \sqrt{3}$, $\tau_2(\sqrt{5}) = -\sqrt{5}$;

3. (See the text answer.)

4. The identity map of $\mathbf{Q}(\sqrt[3]{2})$ onto itself;
   τ_1 given by $\tau_1(\alpha_1) = \alpha_2$ that is, $\psi_{\alpha_1, \alpha_2}$; $\tau_2 = \psi_{\alpha_1, \alpha_3}$

5. (See the text answer.)

6. τ_1 given by $\tau_1(i) = i$, $\tau_1(\sqrt{3}) = -\sqrt{3}$, $\tau_1(\alpha_1) = \alpha_1$

   τ_2 given by $\tau_2(i) = i$, $\tau_2(\sqrt{3}) = -\sqrt{3}$, $\tau_2(\alpha_2) = \alpha_2$

τ_3 given by $\tau_3(i) = i$, $\tau_3(\sqrt{3}) = -\sqrt{3}$, $\tau_3(\alpha_3) = \alpha_3$

τ_4 given by $\tau_4(i) = -i$, $\tau_4(\sqrt{3}) = -\sqrt{3}$, $\tau_4(\alpha_4) = \alpha_1$

τ_5 given by $\tau_5(i) = -i$, $\tau_5(\sqrt{3}) = -\sqrt{3}$, $\tau_5(\alpha_5) = \alpha_2$

τ_6 given by $\tau_6(i) = -i$, $\tau_6(\sqrt{3}) = -\sqrt{3}$, $\tau_6(\alpha_6) = \alpha_3$

7. (See the text answer.)

8. F T F T F T T T T F

9. Since $\sigma: K \longrightarrow K$ is an isomorphism, the map $\sigma^{-1}: \sigma[K] \longrightarrow K$ is an isomorphism. Since K is algebraically closed and is algebraic over $\sigma[K]$, Theorem 9.6 shows that σ^{-1} has an extension to an isomorphism τ mapping $\sigma[K]$ onto a subfield of K. But σ^{-1} is already onto K, and since τ must be a one-to-one map, we see that it cannot be defined on any elements of K not already in $\sigma[K]$. Thus we must have $K = \sigma[K]$, so σ is an automorphism of K.

10. Let E be an algebraic extension of F and let τ be an isomorphism of E onto a subfield of $\overline{F}$ that leaves F fixed. Since E is an algebraic extension of F, the field $\overline{F}$ is an algebraic extension of E and is an algebraic closure of E. By Theorem 9.6, τ can be extended to an isomorphism σ of $\overline{F}$ onto a subfield of $\overline{F}$. By Exercise 9, such an isomorphism σ is an automorphism of $\overline{F}$.

11. By Theorem 9.6, the identity map of F onto F has an extension to an isomorphism τ mapping E onto a subfield of $\overline{F}$. By Theorem 9.6, τ can be extended to an isomorphism σ mapping $\overline{E}$ onto a subfield of $\overline{F}$. Then σ^{-1} is an isomorphism mapping $\sigma[\overline{E}]$ onto $\overline{F}$. By Theorem 9.6, σ^{-1} can be extended to an isomorphism of $\overline{F}$ onto a subfield of $\overline{E}$. Since σ^{-1} is already onto $\overline{E}$ and its extension must be one to one, we see that the domain of σ^{-1} must already by $\overline{F}$. Thus $\sigma[\overline{E}] = \overline{F}$ and σ is an isomorphism of $\overline{E}$ onto $\overline{F}$.

12. We know that π is transcendental over $\mathbf{Q}$. Therefore, $\sqrt{\pi}$ must be transcendental over $\mathbf{Q}$, for if it were algebraic, then $\pi = (\sqrt{\pi})^2$ would be algebraic since algebraic numbers form a closed set under field operations. Thus the map $\tau: \mathbf{Q}(\sqrt{\pi}) \to \mathbf{Q}(x)$ where $\tau(a) = a$ for $a \in \mathbf{Q}$ and $\tau(\sqrt{\pi}) = x$ is an isomorphism. Theorem 9.6 shows that τ can be extended to an isomorphism σ mapping $\overline{\mathbf{Q}(\sqrt{\pi})}$ onto a subfield of $\overline{\mathbf{Q}(x)}$. Then σ^{-1} is an isomorphism mapping $\sigma[\overline{\mathbf{Q}(\sqrt{\pi})}]$ onto a subfield of $\overline{\mathbf{Q}(\sqrt{\pi})}$ which can be extended to an isomorphism of $\overline{\mathbf{Q}(x)}$ onto a subfield of $\overline{\mathbf{Q}(\sqrt{\pi})}$. But since σ^{-1} is already onto $\overline{\mathbf{Q}(\sqrt{\pi})}$, we see that σ must actually onto $\overline{\mathbf{Q}(x)}$, so σ provides the required isomorphism of $\overline{\mathbf{Q}(\sqrt{\pi})}$ with $\overline{\mathbf{Q}(x)}$.

13. Let E be a finite extension of F. Then by Theorem 8.11, $E = F(\alpha_1, \alpha_2, \cdots, \alpha_n)$ where each α_i is algebraic over F. Now suppose that $L = F(\alpha_1, \alpha_2, \cdots, \alpha_{k+1})$ and $K = F(\alpha_1, \alpha_2, \cdots, \alpha_k)$. Every isomorphism of L onto a subfield $\overline{F}$ and leaving F fixed can be viewed as an extension of an isomorphism of K onto a subfield of $\overline{F}$. The extension of such an isomorphism τ of K to an isomorphism σ of L onto a subfield of $\overline{F}$ is completely determined by $\sigma(\alpha_{k+1})$. Let $p(x)$ be the irreducible polynomial for α_{k+1} over K, and let $q(x)$ be the polynomial in $\tau[K][x]$ obtained by applying τ to each of the coefficients of $p(x)$. Since $p(\alpha_{k+1}) = 0$, we must have $q(\sigma(\alpha_{k+1})) = 0$, so the number of choices for $\sigma(\alpha_{k+1})$ is at most $\deg(q(x)) = \deg(p(x)) = [L: K]$. Thus $\{L: K\} \le [L: K]$, that is,

$$\{F(\alpha_1, \cdots, \alpha_{k+1}): F(\alpha_1, \cdots, \alpha_k)\} \le [F(\alpha_1, \cdots, \alpha_{k+1}): F(\alpha_1, \cdots, \alpha_k)]. \quad (1)$$

We have such an inequality (1) for each $k = 1, 2, \cdots, n - 1$. Using the multiplicative properties of the index and of the degree (the corollary of Theorem 9.7 and Corollary 1 of

Theorem 8.10), we obtain upon multiplication of these $n - 1$ inequalities the desired result $\{E: F\} \le [E: F]$.

SECTION 9.3 - Splitting Fields

1. 2

2. Degree 2; $x^4 - 1 = (x - 1)(x + 1)(x^2 + 1)$.

3. $2 \cdot 2 = 4$

4. Degree 6; replace 2 by 3 in Example 4.

5. Degree 2; $x^3 - 1 = (x - 1)(x^2 + x + 1)$.

6. Degree $2 \cdot 6 = 12$; see Example 4 for the splitting field of $x^3 - 2$.

7. Order 1; $\mathbf{Q}(\sqrt[3]{2}) < \mathbb{R}$ and the other conjugates of $\sqrt[3]{2}$ do not lie in $\mathbb{R}$ so they yield isomorphisms into $\mathbb{C}$ rather than automorphisms of $\mathbf{Q}(\sqrt[3]{2})$. See also Exercise 35 of Section 9.1.

8. Order $3 \cdot 2 = 6$; this is a splitting field by Example 4. An automorphism can carry $\sqrt[3]{2}$ into any one of its three conjugates and $i\sqrt{3}$ into any one of its two conjugates.

9. Degree 2; This is the splitting field of $x^2 + 3$ over $\mathbf{Q}(\sqrt[3]{2})$.

10. Theorem 8.22 shows that the only field of order 8 in $\overline{\mathbb{Z}_2}$ is the splitting field of $x^8 - x$ over $\mathbb{Z}_2$. Since a field of order 8 can be obtained by adjoining to $\mathbb{Z}_2$ a root of any cubic polynomial that is irreducible in $\mathbb{Z}_2[x]$, it must be that all roots of every irreducible cubic lie in this unique subfield of order 8 in $\overline{\mathbb{Z}_2}$.

11. $1 \le [E: F] \le n!$. The example $E = F = \mathbf{Q}$ and $f(x) = x^2 - 1$ shows that the lower bound 1 cannot be improved unless we are told that $f(x)$ is irreducible over F. Example 4 shows that the upper bound $n!$ cannot be improved.

12. T F T T T T F F T T

13. (See the text answer.)

14. a) Not necessarily. For example, $6 = |G(\mathbf{Q}(\sqrt[3]{2}, i\sqrt{3})/\mathbf{Q})| \ne |G(\mathbf{Q}(\sqrt[3]{2}, i\sqrt{3})/\mathbf{Q}(\sqrt[3]{2}))| \cdot |G(\mathbf{Q}(\sqrt[3]{2})/\mathbf{Q})| = 2 \cdot 1 = 2$.

b) Yes. Each field is a splitting field of the one immediately under it. If E is a splitting field over F then $|G(E/F)| = \{E: F\}$, and the index is multiplicative by the corollary of Theorem 9.7.

15. Let E be the splitting field of a set S of polynomials in $F[x]$. If $E = F$, then E is the splitting field of x over F. If $E \neq F$, then find a polynomial $f_1(x)$ in S that does not split in F, and form its splitting field, which is a subfield a subfield E_1 of E where $[E_1: F] > 1$. If $E = E_1$, then E is the splitting field of $f_1(x)$ over F. If $E \neq E_1$, find a polynomial $f_2(x)$ in S that does not split in E_1, and form its splitting field $E_2 \leq E$ where $[E_2: E_1] > 1$. If $E = E_2$, then E is the splitting field of $f_1(x)f_2(x)$ over F. If $E \neq E_2$, then continue the construction in the obvious way. Since by hypothesis E is a *finite* extension of F, this process must eventually terminate with some $E_r = E$, which is then the splitting field of the product $g(x) = f_1(x)f_2(x)\cdots f_r(x)$ over F.

16. Find $\alpha \in E$ that is not in F. Now α is algebraic over F, and must be of degree 2 since $[E: F] = 2$ and $[F(\alpha): F] = \deg(\alpha, F)$. Thus $\operatorname{irr}(\alpha, F) = x^2 + bx + c$ for some $b, c \in F$. Since $\alpha \in E$, this polynomial factors in $E[x]$ into a product $(x - \alpha)(x - \beta)$, so the other root β of $\operatorname{irr}(\alpha,F)$ lies in E also. Thus E is the splitting field of $\operatorname{irr}(\alpha,F)$.

17. Let E be a splitting field over F. Let α be in E but not in F. By Corollary 1 of Theorem 9.8, the polynomial $\operatorname{irr}(\alpha,F)$ splits in E since it has a zero α in E. Thus E contains all conjugates of α over F.
Conversely, suppose that E contains all conjugates of $\alpha \in E$ over F, where $F \leq E \leq \overline{F}$. Since an automorphism σ of $\overline{F}$ leaving F fixed carries every element of $\overline{F}$ into one of its conjugates over F, we see that $\sigma(\alpha) \in E$. Thus σ induces a one to one map of E into E. Since the same is true of σ^{-1}, we see that σ maps E *onto* E and thus induces an automorphism of E leaving F fixed. Theorem 9.8 shows that under these conditions, E is a splitting field of F.

18. Since $\mathbb{Q}(\sqrt[3]{2})$ lies in $\mathbb{R}$ and the other two conjugates of $\sqrt[3]{2}$ do not lie in $\mathbb{R}$, we see that no map of $\sqrt[3]{2}$ into any conjugate other than $\sqrt[3]{2}$ itself can give rise to an automorphism of $\mathbb{Q}(\sqrt[3]{2})$; the other two maps give rise to isomorphisms of $\mathbb{Q}(\sqrt[3]{2})$ onto a subfield of $\bar{\mathbb{Q}}$. Since any automorphism of $\mathbb{Q}(\sqrt[3]{2})$ must leave the prime field $\mathbb{Q}$ fixed, we see that the identity is the only automorphism of $\mathbb{Q}(\sqrt[3]{2})$. [For an alternate argument, see Exercise 35 of Section 9.1.]

19. The conjugates of $\sqrt[3]{2}$ over $\mathbb{Q}(i\sqrt{3})$ are

$$\sqrt[3]{2},\quad \sqrt[3]{2}\,\frac{-1 + i\sqrt{3}}{2},\quad \text{and}\quad \sqrt[3]{2}\,\frac{-1 - i\sqrt{3}}{2}.$$

Maps of $\sqrt[3]{2}$ into each of them give rise to the only three automorphisms of $G(\mathbb{Q}(\sqrt[3]{2}, i\sqrt{3})/\mathbb{Q}(i\sqrt{3}))$. Let σ be the automorphism such that $\sigma(\sqrt[3]{2}) = \sqrt[3]{2}\,\frac{-1 + i\sqrt{3}}{2}$. Then σ must be a generator of this group of order 3, since σ is not the identity map, and every group of order 3 is cyclic. Thus the automorphism group is isomorphic to $\mathbb{Z}_3$.

20. a) Each automorphism of E leaving F fixed is a one-to-one map that carries each zero of $f(x)$ into one of its conjugates, which must be a zero of an irreducible factor of $f(x)$ and hence is also a zero of $f(x)$. Thus each automorphism gives rise to a one-to-one map of the set of zeros of $f(x)$ onto itself, that is, a permutation of the zeros of $f(x)$.

b) Since E is the splitting field of $f(x)$ over F, we know that $E = F(\alpha_1, \alpha_2, \cdots, \alpha_n)$ where $\alpha_1, \alpha_2, \cdots, \alpha_n$ are the zeros of $f(x)$. As Exercise 29 of Section 9.1 shows, an automorphism σ of E leaving F fixed is completely determined by the values $\sigma(\alpha_1), \sigma(\alpha_2), \cdots, \sigma_n(\alpha)$, that is, by the permutation of the zeros of $f(x)$ given by σ.

c) We associate with each $\sigma \in G(E/F)$ its permutation of the zeros of $f(x)$ in E. Since multiplication $\sigma\tau$ in $G(E/F)$ is function composition and since multiplication of the permutations of zeros is again composition of these same functions, with domain restricted to the zeros of $f(x)$, we see that $G(E/F)$ is isomorphic to a subgroup of the group of all permutations of the zeros of $f(x)$.

21. a) $|G(E/\mathbb{Q})| = 2\cdot 3 = 6$, for $\{\mathbb{Q}(i\sqrt{3}): \mathbb{Q}\} = 2$ since $\text{irr}(i\sqrt{3}, \mathbb{Q}) = x^2 + 3$ and $\{\mathbb{Q}(\sqrt[3]{2}): \mathbb{Q}(i\sqrt{3})\} = 3$ since $\text{irr}(\sqrt[3]{2}, \mathbb{Q}(i\sqrt{3})) = x^3 - 2$. The index is multiplicative by the Corollary of Theorem 9.7.

b) Since E is the splitting field of $x^3 - 2$ over $\mathbb{Q}$, Exercise 20 shows that $G(E/\mathbb{Q})$ is isomorphic to a subgroup of the group of all permutations of the three zeros of $x^3 - 2$ in E. Since the group of all permutations of three objects has order 6 and $|G(E/\mathbb{Q})| = 6$ by part (a), we see that $G(E/\mathbb{Q})$ is isomorphic to the full symmetric group on three letters, that is, to S_3.

22. We have $x^p - 1 = (x - 1)(x^{p-1} + x^{p-2} + \cdots + x + 1)$, and the corollary of Theorem 5.21 shows that the second of these factors, the cyclotomic polynomial $\Phi_p(x)$, is irreducible over $\mathbb{Q}$. Let ζ be a zero of $\Phi_p(x)$ in its splitting field over $\mathbb{Q}$. Exercise 32a of Section 9.1 shows that then $\zeta, \zeta^2, \zeta^3, \cdots, \zeta^{p-1}$ are distinct and are all zeros of $\Phi_p(x)$. Thus all zeros of $\Phi_p(x)$ lie in the simple extension $\mathbb{Q}(\zeta)$, so $\mathbb{Q}(\zeta)$ is the splitting field of $x^p - 1$ and of course has degree $p - 1$ over $\mathbb{Q}$ since $\Phi_p(x) = \text{irr}(\zeta, \mathbb{Q})$ has degree $p - 1$.

23. By Corollary 2 of Theorem 9.6, there exists an isomorphism $\phi: \overline{F} \longrightarrow \overline{F}'$ leaving each element of F fixed. Since the coefficients of $f(x) \in F[x]$ are all left fixed by ϕ, we see that ϕ carries each zero of $f(x)$ in $\overline{F}$ into a zero of $f(x)$ in $\overline{F}'$. Since the zeros of $f(x)$ in $\overline{F}$ generate its splitting field E in $\overline{F}$, we see that $\phi[E]$ is contained in the splitting field E' of $f(x)$ in $\overline{F}'$. But the same argument can be made for ϕ^{-1}; we must have $\phi^{-1}[E'] \subseteq E$. Thus ϕ maps E onto E', so these two splitting fields of $f(x)$ are isomorphic.

SECTION 9.4 - Separable Extensions

1. Since $\sqrt[3]{2}\sqrt{2} = 2^{1/3}2^{1/2} = 2^{5/6}$, we see that $\sqrt[6]{2} = 2/(\sqrt[3]{2}\sqrt{2})$ so $\mathbf{Q}(\sqrt[6]{2}) \subseteq \mathbf{Q}(\sqrt[3]{2}, \sqrt{2})$. Since $\sqrt[3]{2} = (\sqrt[6]{2})^2$ and $\sqrt{2} = (\sqrt[6]{2})^3$, we see that $\mathbf{Q}(\sqrt[3]{2}, \sqrt{2}) \subseteq \mathbf{Q}(\sqrt[6]{2})$, so $\mathbf{Q}(\sqrt[6]{2}) = \mathbf{Q}(\sqrt[3]{2}, \sqrt{2})$. We can take $\alpha = \sqrt[6]{2}$.

2. Since $(\sqrt[4]{2})^3(\sqrt[6]{2}) = 2^{3/4}2^{1/6} = 2^{9/12}2^{2/12} = 2^{11/12}$, we see that $\mathbf{Q}(\sqrt[12]{2}) \subseteq \mathbf{Q}(\sqrt[4]{2}, \sqrt[6]{2})$. Since $\sqrt[4]{2} = (\sqrt[12]{2})^3$ and $\sqrt[6]{2} = (\sqrt[12]{2})^2$, we see that $\mathbf{Q}(\sqrt[4]{2}, \sqrt[6]{2}) \subseteq \mathbf{Q}(\sqrt[12]{2})$, so $\mathbf{Q}(\sqrt[12]{2}) = \mathbf{Q}(\sqrt[4]{2}, \sqrt[6]{2})$. We can take $\alpha = \sqrt[12]{2}$.

3. We try $\alpha = \sqrt{2} + \sqrt{3}$. Then $\alpha^2 = 5 + 2\sqrt{2}\sqrt{3}$ and $\alpha^3 = 11\sqrt{2} + 9\sqrt{3}$. Since $\sqrt{2} = \dfrac{\alpha^3 - 9\alpha}{2}$ and $\sqrt{3} = \dfrac{11\alpha - \alpha^3}{2}$, we see that $\mathbf{Q}(\sqrt{2} + \sqrt{3}) = \mathbf{Q}(\sqrt{2}, \sqrt{3})$.

4. Of course $\mathbf{Q}(i\cdot\sqrt[3]{2}) \subseteq \mathbf{Q}(i, \sqrt[3]{2})$. Since $i = -(i\cdot\sqrt[3]{2})^3/2$ and $\sqrt[3]{2} = -2/(i\cdot\sqrt[3]{2})^2$, we see that $\mathbf{Q}(i, \sqrt[3]{2}) \subseteq \mathbf{Q}(i\cdot\sqrt[3]{2})$. Thus $\mathbf{Q}(i, \sqrt[3]{2}) = \mathbf{Q}(i\cdot\sqrt[3]{2})$ so we can take $\alpha = i\cdot\sqrt[3]{2}$.

5. (See the text answer.)

6. F T T F F T T T T T

7. We are given that α is separable over F, so by definition, $F(\alpha)$ is a separable extension over F. Since β is separable over F, it follows that β is separable over $F(\alpha)$ because $q(x) = \text{irr}(\beta, F(\alpha))$ divides $\text{irr}(\beta, F)$ so β is a zero of $q(x)$ of multiplicity 1. Therefore $F(\alpha,\beta)$ is a separable extension of F by Theorem 9.11. The corollary of Theorem 9.11 then asserts that each element of $F(\alpha,\beta)$ is separable over F. In particular, $\alpha \pm \beta$, $\alpha\beta$, and α/β if $\beta \neq 0$ are all separable over F.

8. We know that $[\mathbf{Z}_p(y): \mathbf{Z}_p(y^p)]$ is at most p. If we can show that $\{1, y, y^2, \cdots, y^{p-1}\}$ is an independent set over $\mathbf{Z}_p(y^p)$, then by Theorem 8.7, this set could be enlarged to a basis for $\mathbf{Z}_p(y)$ over $\mathbf{Z}_p(y^p)$. But since a basis can have at most p

elements, it would already be a basis, and $[\mathbb{Z}_p(y): \mathbb{Z}_p(y^p)] = p$, showing that $\mathrm{irr}(y, \mathbb{Z}_p(y^p))$ would have degree p and must therefore be $x^p - y^p$. Thus our problem is reduced to showing that $S = \{1, y, y^2, \cdots, y^{p-1}\}$ is an independent set over $\mathbb{Z}_p(y^p)$.

Suppose that

$$\frac{r_0(y^p)}{s_0(y^p)}(1) + \frac{r_1(y^p)}{s_1(y^p)}(y) + \frac{r_2(y^p)}{s_2(y^p)}(y^2) + \cdots + \frac{r_{p-1}(y^p)}{s_{p-1}(y^p)}(y^{p-1}) = 0$$

where $r_i(y^p)$, $s_i(y^p) \in \mathbb{Z}_p[y^p]$ for $i = 0, 1, 2, \cdots, p - 1$. We want to show that all these coefficients in $\mathbb{Z}_p(y^p)$ must be zero. Clearing denominators, we see that it is no loss of generality to assume that all $s_i(y^p) = 1$ for $i = 0, 1, 2, \cdots, p - 1$. Now the powers of y appearing in $r_i(y^p)(y^i)$ are all congruent to i modulo p, and consequently no terms in this expression can be combined with any terms of $r_j(y^p)(y^j)$ for $j \neq i$. Since y is an indeterminant, we then see that this linear combination of elements in S can be zero only if all the coefficients $r_i(y^p)$ are zero, so S is an independent set over $\mathbb{Z}_p(y^p)$, and we are done.

9. Let E be an algebraic extension of a perfect field F and let K be a finite extension of E. To show that E is perfect, we must show that K is a separable extension of E. Let α be an element of K. Since $[K: E]$ is finite, α is algebraic over K. Since K is algebraic over F, then α is algebraic over F by Exercise 25 of Section 8.3. Since F is perfect, α is a zero of $\mathrm{irr}(\alpha, F)$ of multiplicity 1. Since $\mathrm{irr}(\alpha, E)$ divides $\mathrm{irr}(\alpha, F)$, we see that α is a zero of $\mathrm{irr}(\alpha, E)$ of multiplicity 1, so α is separable over E by the italicized remark preceding Theorem 9.11. Thus each $\alpha \in K$ is separable over E, so K is separable over E by the corollary of Theorem 9.11.

10. Since K is algebraic over E and E is algebraic over F, we have K algebraic over F by Exercise 25 of Section 8.3. Let $\beta \in K$ and let $\beta_0, \beta_1, \cdots, \beta_n$ be the coefficients in E of $\text{irr}(\beta, E)$. Since β is a zero of $\text{irr}(\beta, E)$ of algebraic multiplity 1, we see that $F(\beta_0, \beta_1, \cdots, \beta_n, \beta)$ is a separable extension of $F(\beta_0, \beta_1, \cdots, \beta_n)$, which in turn is a separable extension of F by the corollary of Theorem 9.11. Thus we are back to a tower of finite extensions, and deduce from Theorem 9.11 that $F(\beta_0, \beta_1, \cdots, \beta_n, \beta)$ is a separable extension of F. In particular, β is separable over F. This shows that every element of K is separable over F, so by definition, K is separable over F.

11. Exercise 7 shows that the set S of all elements in E that are separable over F is closed under addition and multiplication. Of course 0 and 1 are separable over F, so Exercise 7 further shows that S contains additive inverses and reciprocals of nonzero elements. Therefore S is a subfield of E.

12. a) We know that the nonzero elements of E form a cyclic group E^* of order $p^n - 1$ under multiplication, so all elements of E are zeros of $x^{p^n} - x$. (See Section 8.6.) Thus for $\alpha \in E$, we have

$$\begin{aligned}\sigma_p{}^n(\alpha) &= \sigma_p{}^{n-1}(\sigma_p(\alpha)) = \sigma_p{}^{n-1}(\alpha^p) = \sigma_n{}^{n-2}(\sigma_p(\alpha^p)) \\ &= \sigma_p{}^{n-2}(\sigma_p(\alpha))^p = \sigma_p{}^{n-2}(\alpha^p)^p = \sigma_p{}^{n-2}(\alpha^{p^2}) = \cdots \\ &= \alpha^{p^n} = \alpha\end{aligned}$$

so $\sigma_p{}^n$ is the identity automorphism. If α is a generator of the group E^*, then $\alpha^{p^i} \neq \alpha$ for $i < n$, so we see that n is indeed the order of σ_p.

b) Section 8.6 shows that E is an extension of $\mathbb{Z}_p$ of order n, and is the splitting field of any irreducible polynomial of degree n in $\mathbb{Z}_p[x]$. Since E is a separable extension of the finite perfect field $\mathbb{Z}_p[x]$, we see that $|G(E/F)| = \{E: F\} = [E: F] = n$. Since $\sigma_p \in G(E/F)$ has order n, we see that $G(E/F)$ is cyclic of order n.

Section 9.4

13. a) Let $f(x) = \sum_{i=1}^{\infty} a_i x^i$ and $g(x) = \sum_{i=1}^{\infty} b_i x^i$. Then

$$D(f(x) + g(x)) = D\left[\sum_{i=1}^{\infty} (a_i+b_i)x^i\right] = \sum_{i=1}^{\infty} (i\cdot 1)(a_i+b_i)x^i$$

$$= \sum_{i=1}^{\infty} (i\cdot 1)a_i x^i + \sum_{i=1}^{\infty} (i\cdot 1)b_i x^i$$

$$= D(f(x)) + D(g(x)).$$

Thus D is a homomorphism of $\langle F[x], +\rangle$.

b) $\mathrm{Ker}(D) = F$ if F has characteristic zero.

c) $\mathrm{Ker}(D) = F[x^p]$ if F has characteristic p.

14. a) Let $f(x) = \sum_{i=1}^{\infty} a_i x^i$. Then $D(af(x)) = D\left[\sum_{i=1}^{i} aa_i x^i\right]$

$$= \sum_{i=1}^{\infty} (i\cdot 1)aa_i x^i = a \sum_{i=1}^{\infty} (i\cdot 1)a_i x^i = a\, D(f(x)).$$

b) We proceed by induction on $n = \deg(f(x)g(x))$. If $n = 0$ then $f(x)$, $g(x)$, $f(x)g(x) \in F$ and $D(f(x)) = D(g(x)) = D(f(x)g(x)) = 0$ by parts (b) and (c) of Exercise 13. Suppose the formula is true for $n < k$, and let us prove it for $n = k > 0$. Write $f(x) = h(x) + a_r x^r$ where $a_r x^r$ is the term of highest degree in $f(x)$, and similarly write $g(x)$ as $g(x) = q(x) + b_s x^s$. Then $f(x)g(x) =$

$h(x)q(x) + h(x)b_s x^s + a_r x^r q(x) + a_r b_s x^{r+s}$. Of these four terms, all are of degree less than $k = r + s$ except for the last term, so by part (a) of Exercise 13 and our induction hypothesis, we have

$$\begin{aligned} D(f(x)g(x)) &= h(x)q'(x) + h'(x)q(x) \\ &\quad + h(x)(s\cdot 1)b_s x^{s-1} + h'(x)b_s x^s \\ &\quad + a_r x^r q'(x) + (r\cdot 1)a_r x^{r-1} q(x) \\ &\quad + [(r+s)\cdot 1)]a_r b_s x^{r+s-1} \end{aligned}$$

$$
\begin{aligned}
&= h(x)g'(x) + f'(x)q(x) \\
&\quad + h'(x)b_s x^s + a_r x^r q'(x) \\
&\quad + (s\cdot 1)a_r b_s x^{r+s-1} + (r\cdot 1)a_r b_s x^{r+s-1} \\
&= h(x)g'(x) + f'(x)q(x) \\
&\quad + a_r x^r[q'(x) + (s\cdot 1)b_s x^{s-1}] \\
&\quad + [h'(x) + (r\cdot 1)a_r x^{r-1}]b_s x^s \\
&= h(x)g'(x) + f'(x)q(x) \\
&\quad + a_r x^r g'(x) + f'(x)b_s x^s \\
&= [h(x) + a_r x^r]g'(x) \\
&\quad + f'(x)[q(x) + b_s x^s] \\
&= f(x)g'(x) + f'(x)g(x).
\end{aligned}
$$

15. Let $f(x) = (x - \alpha)^{\nu}g(x)$ where $g(\alpha) \neq 0$ and $\nu \geq 1$ since $f(\alpha) = 0$. Then by Exercise 13, we have in $\overline{F}[x]$

$$f'(x) = (x - \alpha)^{\nu}g'(x) + \nu(x - \alpha)^{\nu-1}g(x). \tag{1}$$

Remembering that $\nu \geq 1$ and that $g(\alpha) \neq 0$, we see that $f'(\alpha) = 0$ if and only if $\nu > 1$, that is, if and only if α is a zero of $f(x)$ of multiplicity > 1.

16. Let $f(x)$ be an irreducible polynomial in $F[x]$. Suppose that α is a zero of $f(x)$ in $\overline{F}$. Since $f(x) \in F[x]$ has minimal degree among all polynomials having α as a zero, we see that $f'(\alpha) \neq 0$, since the degree of $f'(x)$ is always one less than the degree of $f(x)$ in the characteristic 0 case. By Exercise 15, α is a zero of $f(x)$ of multiplicity 1. By Theorem 9.9, all zeros of $f(x)$ have this same multiplicity 1, so $f(x)$ is separable.

17. Let α be a zero of $q(x)$ in the algebraic closure $\overline{F}$. The argument in Exercise 16 shows that $q'(\alpha) \neq 0$ unless $q'(x)$ should be the zero polynomial. Now $q'(x) = 0$ if and only if each exponent of each term of $q(x)$ is divisible by p. If this is not the case, then $q'(\alpha) \neq 0$ so α has multiplicity 1 by Exercise 15, and so do other zeros of $q(x)$ by Theorem 9.9. Thus $q(x)$ is a separable polynomial. This proves the "only if" part of the exercise.

If every exponent in $q(x)$ is divisible by p, let $g(x)$ be the polynomial obtained from $q(x)$ by dividing each exponent by p. Then $\alpha \in \overline{F}$ is a zero of $q(x)$ if and only if α^p is a zero of $g(x)$. Let $g(x)$ factor into $(x - \alpha^p)h(x)$ in $\overline{F}[x]$. Then $q(x) = (x^p - \alpha^p)h(x^p) = (x - \alpha)^p h(x^p)$ in $\overline{F}[x]$, showing showing that α is a zero of $q(x)$ of algebraic multiplicity at least p.

18. The polynomials $f(x)$ and $f'(x)$ have a common nonconstant factor in $\overline{F}[x]$ if and only if they have a common zero in $\overline{F}$, since a zero of the common nonconstant factor must be a zero of each polynomial, and a common zero α gives rise to a common factor $x - \alpha$. By Exercise 15, this is equivalent to saying that $f(x)$ has a zero of multiplicity greater than 1. Thus there is no nonconstant factor of $f(x)$ and $f'(x)$ if and only if $f(x)$ has no zero of multiplicity greater than 1.

19. Suppose that 1 is a gcd of $f(x)$ and $f'(x)$ in $F[x]$. By Theorem 7.6, $1 = h(x)f(x) + g(x)f'(x)$ for some polynomials $h(x), g(x) \in F[x]$. Viewing this equation in $\overline{F}[x]$, we see $F[x]$, we see that every common factor of $f(x)$ and $f'(x)$ in $\overline{F}[x]$ must divide 1, so the only such common factors are elements of $\overline{F}$ and 1 is a gcd of $f(x)$ and $f'(x)$ in $\overline{F}[x]$ also. in $F[x]$ also. Thus $f(x)$ and $f'(x)$ have no common nonconstant factor in $F[x]$ if and only if they have no common nonconstant factor in $\overline{F}[x]$. By Exercise 18, this is equivalent to $f(x)$ having no zero in $\overline{F}$ of multiplicity greater than 1.

20. Compute a gcd of $f(x)$ and $f'(x)$ using the Euclidean algorithm. Then $f(x)$ has a zero of multiplicity > 1 if and only if this gcd is of degree > 0.

SECTION 9.5 - Totally Inseparable Extensions

1. The separable closure is $\mathbb{Z}_3(y^3, z^9)$ since $(y^3)^4 = u$ and $(z^9)^2 = v$ and 3 does not divide 4 or 2. $\mathbb{Z}_3(y,z)$ is clearly totally inseparable over $\mathbb{Z}_3(y^3, z^9)$.

2. Clearly the separable closure contains y^3. Therefore it must

contain $(y^2z^{18})^3/(y^3)^2 = z^{54}$, and hence must contain z^{27}. Since it must also contain y^2z^{18}, we see that the separable closure is $\mathbf{Z}_3(y^3,\ y^2z^{18},\ z^{27})$. Clearly $\mathbf{Z}_3(y,z)$ is totally inseparable over this field.

3. The totally inseparable closure is $\mathbf{Z}_3(y^4,\ z^2)$.

4. The totally inseparable closure must contain y^4, and hence $(y^2z^{18})^2/y^4 = z^{36}$, so it must also contain z^4. Of course it must also contain y^2z^{18} and therefore $y^2z^{18}/(z^4)^4 = y^2z^2$. We see that the totally inseparable closure is $\mathbf{Z}_3(y^4,\ y^2z^2,\ z^4)$. Note that $(y^2z^2)^{27} = (y^{12})^4(y^2z^{18})^3$, and that $(z^4)^{27} = (y^2z^{18})^6/y^{12}$.

5. F T F F F F T F T T

6. If E is a separable extension of F, then there are no elements of E totally inseparable over F so the totally inseparable closure of F in E is just F, which is a subfield of E.

Suppose that E does contain some elements totally inseparable over F, and let K be the union of F with the set of all such totally inseparable elements. We need only show that for $\alpha,\ \beta \in K$, the elements $\alpha \pm \beta$, $\alpha\beta$, and $1/\alpha$ if $\alpha \neq 0$, are in either in F or are totally inseparable over F. Suppose that α is not in F, but is totally inseparable over F, so that $\alpha^{p^r} \in F$. Then for any element b in F, we have $(\alpha \pm b)^{p^r} = \alpha^{p^r} \pm b^{p^r}$ and this sum or difference is in F. Also $(a\alpha)^{p^r} = a^{p^r}\alpha^{p^r}$ is in F, and $(1/\alpha)^{p^r} = 1/\alpha^{p^r}$ is in F if $\alpha \neq 0$. This shows that for α in K but not in F and for b in F, the elements $\alpha \pm b$, αb, and $1/\alpha$ are in K. The other case we have to worry about is where α and β are both totally inseparable over F, that is they are both in K, but neither one is in F. Then $\alpha^{p^r} \in F$ and $\beta^{p^s} \in F$ for some $r,\ s \in \mathbf{Z}^+$. Suppose that $s \geq r$. Then $(\alpha \pm \beta)^{p^s} = \alpha^{p^s} \pm \beta^{p^s} = (\alpha^{p^r})^{p^{s-r}} \pm \beta^{p^s}$ is in F, and $(\alpha\beta)^{p^s} = (\alpha^{p^r})^{p^{s-r}}\beta^{p^s}$ is in F.

Thus $\alpha \pm \beta$ and $\alpha\beta$ are either already in F or are totally inseparable over F. This shows that K is closed under the field operations of addition, subtraction, multiplication, and contains multiplicative inverses of nonzero elements.

7. Suppose F is perfect. If $x^p - a$ has no zero in F for some $a \in F$, then $F(\sqrt[p]{a})$ is a proper extension of F and is totally inseparable over F, contradicting the hypothesis that F is is perfect. Thus $x^p - a$ has a zero in F for every $a \in F$, that is, $F^p = F$.

   Conversely, suppose that $F^p = F$ and let $f(x)$ be an irreducible polynomial in $F[x]$. We must show that $f(x)$ is a separable polynomial. Let E be the separable closure of F in the splitting field K of $f(x)$ in $\overline{F}$. Let $[E: F] = n$. Now the map $\sigma_p: E \longrightarrow E^p$ is an isomorphism. Since $F^p = F$ and σ_p is is one to one, no $\alpha \in E$ that is not in F is carried into F. Since σ_p is an isomorphism, the extension E of degree n over F is carried into an extension E^p of $F^p = F$ of degree n. Since $E^p \leq E$, we see that $n = [E: F] = [E: E^p][E^p: F] = [E: E^p]n$, so $[E: E^p] = 1$ and $E = E^p$. But then E has no totally inseparable extension, for an element α of such an extension must satisfy $\alpha^{p^r} \in E$, α not in E. But since $E^p = E$ the polynomial $x^{p^r} - \beta$ has a solution γ in E, so that $x^{p^r} - \beta = x^{p^r} - \gamma^{p^r} = (x - \gamma)^{p^r}$, showing that γ is the only zero of this polynomial. Thus we must have $E = K$, so the splitting field of $f(x)$ is a separable extension of F, and therefore $f(x)$ is a separable polynomial.

8. The solution of Exercise 7 showed that if $F^p = F$, then $E^p = E$. Conversely, suppose that $E^p = E$. Let $n = [E: F]$. Since σ_p is an isomorphism, it must be that $[E^p: F^p] = n$. Of course $F^p \leq F$. Then we have $n = [E^p: F^p] = [E: F^p] = [E: F][F: F^p] = n[F: F^p]$, so $[F: F^p] = 1$ and $F = F^p$.

SECTION 9.6 - Galois Theory

1. $2 \cdot 2 \cdot 2 = 8$
2. $|G(K/\mathbf{Q})| = [K: \mathbf{Q}] = 8$, since K is normal over $\mathbf{Q}$.
3. $|\lambda\mathbf{Q}| = |G(K/\mathbf{Q})| = 8$
4. $|\lambda(\mathbf{Q}(\sqrt{2},\sqrt{3}))| = [K: \mathbf{Q}(\sqrt{2},\sqrt{3})] = 2$
5. $|\lambda(\mathbf{Q}(\sqrt{6}))| = [K: \mathbf{Q}(\sqrt{6})] = 4$
6. $|\lambda(\mathbf{Q}(\sqrt{30}))| = [K: \mathbf{Q}(\sqrt{30})] = 4$
7. $|\lambda(\mathbf{Q}(\sqrt{2} + \sqrt{6}))| = [K: \mathbf{Q}(\sqrt{2} + \sqrt{6})] = 2$
8. $|\lambda(K)| = [K: K] = 1$
9. $x^4 - 1 = (x^2 - 1)(x^2 + 1) = (x - 1)(x + 1)(x^2 + 1)$ so the splitting field of $x^4 - 1$ over $\mathbf{Q}$ is the same as the splitting field of $x^2 + 1$ over $\mathbf{Q}$. This splitting field is $\mathbf{Q}(i)$. It is of degree 2, and its Galois group is cyclic of order 2 with generator σ where $\sigma(i) = -i$.
10. Since $729 = 9^3$, Theorem 9.19 shows that the Galois group of $GF(729)$ over $GF(9)$ is cyclic of order 3, generated by σ_9 where $\sigma_9(\alpha) = \alpha^9$ for $\alpha \in GF(729)$.
11. (See the text answer.)
12. $x^4 - 5x^2 + 6 = (x^2 - 2)(x^2 - 3)$ so the splitting field is $\mathbf{Q}(\sqrt{2}, \sqrt{3})$. Its Galois group over $\mathbf{Q}$ is isomorphic to the Klein 4-group $\mathbb{Z}_2 \times \mathbb{Z}_2$. See Example 1 for a description of the action of each element on $\sqrt{2}$ and on $\sqrt{3}$.
13. $x^3 - 1 = (x - 1)(x^2 + x + 1)$. Since a primitive cube root of unity is $(-1 + i\sqrt{3})/2$, we see that its splitting field over $\mathbf{Q}$ is $\mathbf{Q}(i\sqrt{3})$. The Galois group is cyclic of order 2 and is generated by σ where $\sigma(i\sqrt{3}) = -i\sqrt{3}$.
14. Let $F = \mathbf{Q}$, $K_1 = \mathbf{Q}(\sqrt{2})$ and $K_2 = \mathbf{Q}(i)$. The fields are not isomorphic since the additive inverse of unity is a square in

K_2 but is not a square in K_1. However, the Galois groups over $\mathbf{Q}$ are isomorphic, for they are both cyclic of order 2.

15. F F T T T F F T F T

16. Since $G(K/E) \le G(K/F)$ and $G(K/F)$ is abelian, we see that $G(K/E)$ is abelian, for a subgroup of an abelian group is abelian. Since $G(E/F) \simeq G(K/F)/G(K/E)$ and $G(K/F)$ is abelian, we see that $G(E/F)$ is abelian, for a factor group of an abelian group, where multiplication is done by choosing representatives, must again be abelian.

17. To show that $N_{K/F}(\alpha) \in F$, we need only show that it is left fixed by each $\tau \in G(K/F)$. From the given formula and the fact that τ is an automorphism, we have $\tau(N_{K/F}(\alpha)) =$

$$\prod_{\sigma \in G(K/F)} (\tau\sigma)(\alpha).$$ But as σ runs through the elements of $G(K/F)$, $\tau\sigma$ again runs through all elements, since $G(K/F)$ is a group. Thus only the order of the factors in the product is changed, and since multiplication in K is commutative, the product is unchanged. Thus $\tau(N_{K/F}(\alpha)) = N_{K/F}(\alpha)$ so $N_{K/F}(\alpha) \in F$. Precisely the same argument shows that $Tr_{K/F}(\alpha) \in F$, only this time it is the order of the summands in the sum that gets changed when computing $\tau(Tr_{K/F}(\alpha))$.

18. a) $N_{K/\mathbf{Q}}(\sqrt{2}) = (\sqrt{2})(\sqrt{2})(-\sqrt{2})(-\sqrt{2}) = 4$, since two of the elements of the Galois group leave $\sqrt{2}$ fixed, and two carry it into $-\sqrt{2}$. The computations shown for parts (b) - (h) are based similarly on the action of the Galois group on $\sqrt{2}$ and on $\sqrt{3}$.

b) $N_{K/\mathbf{Q}}(\sqrt{2}+\sqrt{3}) = (\sqrt{2}+\sqrt{3})(\sqrt{2}-\sqrt{3})(-\sqrt{2}+\sqrt{3})(-\sqrt{2}-\sqrt{3}) = (-1)^2 = 1$

c) $N_{K/\mathbf{Q}}(\sqrt{6}) = (\sqrt{6})(-\sqrt{6})(-\sqrt{6})(\sqrt{6}) = 36$

d) $N_{K/\mathbf{Q}}(2) = (2)(2)(2)(2) = 16$

e) $Tr_{K/\mathbf{Q}}(\sqrt{2}) = \sqrt{2} + \sqrt{2} + (-\sqrt{2}) + (-\sqrt{2}) = 0$

f) $Tr_{K/\mathbf{Q}}(\sqrt{2}+\sqrt{3}) = (\sqrt{2}+\sqrt{3}) + (-\sqrt{2}+\sqrt{3}) + (\sqrt{2}-\sqrt{3}) + (-\sqrt{2}-\sqrt{3}) = 0$

g) $Tr_{K/\mathbf{Q}}(\sqrt{6}) = \sqrt{6} + (-\sqrt{6}) + (-\sqrt{6}) + \sqrt{6} = 0$

h) $Tr_{K/\mathbf{Q}}(2) = 2 + 2 + 2 + 2 = 8$

19. Let $f(x) = \text{irr}(\alpha,F)$. Since $K = F(\alpha)$ is normal over F, it is a splitting field of $f(x)$. Let the factorization of $f(x)$ in $K[x]$ be

$$f(x) = (x - \alpha_1)(x - \alpha_2) \cdots (x - \alpha_n) \qquad (1)$$

where $\alpha = \alpha_1$ Now $|G(K/F)| = n$, and α is carried onto each α_i for $i = 1, 2, \cdots, n$ by precisely one element of $G(K/F)$. Thus $N_{K/F}(\alpha) = \alpha_1\alpha_2\cdots\alpha_n$ If we multiply the linear factors in (1) together, we see that the constant term a_0 is $(-1)^n\alpha_1\alpha_2\cdots\alpha_n = (-1)^n N_{K/F}(\alpha)$. Similarly, we see that $Tr_{K/F}(\alpha) = \alpha_1 + \alpha_2 + \cdots + \alpha_n$. If we pick up the coefficient of x^{n-1} in $f(x)$ by multiplying the linear factors in (1), we find that this coefficient is $a_{n-1} = -\alpha_1 - \alpha_2 - \cdots - \alpha_3 = -Tr_{K/F}(\alpha)$.

20. Let $\alpha_1, \alpha_2, \cdots, \alpha_r$ be the distinct zeros of $f(x)$ in $\overline{F}$ and form the splitting field $K = F(\alpha_1, \alpha_2, \cdots, \alpha_r)$ of $f(x)$ in $\overline{F}$. Note that $r \le n$ since $f(x)$ has at most n distinct zeros. Since all irreducible factors of $f(x)$ are separable, we see that K is normal over F. Now each $\sigma \in G(K/F)$ provides a permutation of $S = \{\alpha_1, \alpha_2, \cdots, \alpha_r\}$ and distinct elements of $G(K/F)$ correspond to distinct permutations of S since an automorphism of K leaving F fixed is uniquely determined by its values on the elements of S. Since permutation multiplication and multiplication in $G(K/F)$ are both function composition, we see that $G(K/F)$ is ismorphic to a subgroup of the group of all permutations of S, which is isomorphic to a subgroup of S_r. By the Theorem of Lagrange, it follows that $|G(K/F)|$ divides r! which in turn divides $n!$ since $r \le n$.

21. The solution to Exercise 20 shows that the group of $f(x)$ over can be regarded as a group of permutations of the set $S = \{\alpha_1, \alpha_2, \cdots, \alpha_r\}$ of distinct zeros of $f(x)$ in $\overline{F}$.

Section 9.6

22. a) Exercise 15 of Section 9.4 shows that $x^n - 1$ has no zeros of multiplicity greater than 1 as long as $(n \cdot 1)$ is not equal to zero in F. Thus the splitting field of $x^n - 1$ over F is a normal extension. If ζ is a primitive nth root of unity, then $1, \zeta, \zeta^2, \cdots, \zeta^{n-1}$ are distinct elements, and are all zeros of $x^n - 1$. Thus the splitting field of $x^n - 1$ over F is $F(\zeta)$.

b) The action of $\sigma \in G(F(\zeta)/F)$ is completely determined by $\sigma(\zeta)$ which must be one of the conjugates ζ^s of ζ over F. Let $\sigma, \tau \in G(F(\zeta)/F)$, and suppose $\sigma(\zeta) = \zeta^s$ and $\tau(\zeta) = \zeta^t$. Then $(\sigma\tau)(\zeta) = \sigma(\tau(\zeta)) = \sigma(\zeta^t) = [\sigma(\zeta)]^t = (\zeta^s)^t = \zeta^{st} = (\zeta^t)^s = [\tau(\zeta)]^s = \tau(\zeta^s) = \tau(\sigma(\zeta)) = (\tau\sigma)(\zeta)$ so $\sigma\tau = \tau\sigma$ and $G(F(\zeta)/F)$ is abelian.

23. a) Since K is cyclic over F we know that $G(K/F)$ is a cyclic group. Now $G(K/E)$ is a subgroup of $G(K/F)$, and is thus cyclic as a subgroup of a cyclic group. Therefore K is cyclic over E. Since E is a normal extension of F, we know that $G(E/F) \simeq G(K/F)/G(K/E)$ so $G(E/F)$ is isomorphic to a factor group of a cyclic group, and is thus cyclic, being generated by a coset containing a generator of $G(K/F)$. Therefore $G(E/F)$ is cyclic.

b) By Galois theory, we know there is a one-to-one correspondence between subgroups H of $G(K/F)$ and fields $E = K_H$ such that $F \le E \le K$. Since $G(K/F)$ is cyclic, it contains precisely one subgroup of each order d dividing $|G(K/F)| = [K: F]$. Such a subgroup corresponds to a field E where $F \le E \le K$ and $[K: E] = d$, so that $[E: F] = m = n/d$. Now as d runs through all divisors of n, the quotients $m = n/d$ also run through all divisors of n, so we are done.

24. a) For $\sigma \in G(K/F)$, we have a natural extension of σ to an automorphism $\bar{\sigma}$ of $K[x]$ where $\bar{\sigma}(\alpha_0 + \alpha_1 x + \cdots + \alpha_n x^n) = \sigma(\alpha_0) + \sigma(\alpha_1)x + \cdots + \sigma(\alpha_n)x^n$. Clearly the polynomials in $K[x]$ left fixed by τ for *all* $\tau \in G(K/F)$ are precisely those in $F[x]$. For $f(x) = \prod_{\sigma \in G(K/F)} (x - \sigma(\alpha))$ we have

$$\bar{\tau}(f(x)) = \prod_{\sigma \in G(K/F)} (x - (\tau\sigma)(\alpha)).$$

Now as σ runs through all elements of $G(K/F)$, we see that that $\tau\sigma$ also runs through all elements since $G(K/F)$ is a group. Thus $\bar{\tau}(f(x)) = f(x)$ so $f(x) \in F[x]$.

b) Since $\sigma(\alpha)$ is a conjugate of α over F for all $\sigma \in G(K/F)$, we see that $f(x)$ has precisely the conjugates of α as zeros. Since $f(\alpha) = 0$, we know by Theorem 8.3 that $p(x) = \text{irr}(\alpha, F)$ divides $f(x)$. Write $f(x) = p(x)q_1(x)$. If If $q_1(x) \neq 0$, then it has as zero some conjugate of α whose irreducible polynomial over F is again $p(x)$, so $p(x)$ divides $q_1(x)$ and we have $f(x) = p(x)^2q_2(x)$. We continue this process until we finally obtain $f(x) = p(x)^rc$ for some $c \in F$. Since $p(x)$ and $f(x)$ are both monic, we must have $f(x) = p(x)^r$. Now $f(x) = p(x)$ if and only if $\deg(\alpha, F) = |G(K/F)| = [K: F]$. Since $\deg(\alpha, F) = [F(\alpha): F]$, we see that this occurs if and only if $[F(\alpha): F] = [K: F]$ so $[K: F(\alpha)] = 1$ and $K = F(\alpha)$.

25. In the one-to-one correspondence between subgroups of $G(K/F)$ and fields E where $F \leq E \leq K$, the lattice of subgroups of $G(K/F)$ is the inverted lattice of such subfields E of K. Now $E \vee L$ is the smallest field containing both E and L, and thus must correspond to the largest group contained in both $G(K/E)$ and $G(K/L)$. Thus $G(K/(E \vee L))$ is $G(K/E) \cap G(K/L)$.

26. Continuing to work with the one-to-one correspondence and lattices mentioned in the solution to Exercise 25, we note that $E \cap L$ is the largest subfield of K contained in both E and L. Thus its group must be the smallest subgroup of $G(K/F)$ containing both $G(K/E)$ and $G(K/L)$. Therefore $G(K/(E \cap L)) = [G(K/E) \vee G(K/L)]$.

SECTION 9.7 - Illustrations of Galois Theory

1. Recall that if $x^4 + 1$ has a factorization into polynomials of lower degree in $\mathbf{Q}[x]$, then it has such a factorization in $\mathbf{Q}[x]$ (Theorem 5.20). The polynomial does not have a linear factor, for neither 1 nor -1 are zeros of the polynomial. Suppose that

$$x^4 + 1 = (x^2 + ax + b)(x^2 + cx + d)$$

for a, b, c, $d \in \mathbb{Z}$. Equating coefficients of x^3, x^2, x, and 1 in that order, we find that $a + c = 0$, $ac + b + d = 0$, $ad + bc = 0$, and $bd = 1$. If $b = d = 1$, then $ac + 2 = 0$ so $ac = -2$ and $a^2 = 2$, which is impossible for an integer a. The other possibility $b = d = -1$ leads to $a^2 = -2$ which is also impossible. Thus $x^4 + 1$ is irreducible.

2. The fields corresponding to the subgroups $G(K/\mathbf{Q})$, H_2, H_4, H_7 and $\{\rho_0\}$ are either derived in the text or are obvious. We turn to the other subgroups, H_1, H_3, H_5, H_6, and H_8.

Both H_1 and H_3 have order 4 and must have fixed fields of degree 2 over $\mathbf{Q}$. Recalling that the fixed field of H_2 is $\mathbf{Q}(i)$, we note the other two obvious extensions of degree 2, namely $\mathbf{Q}(\sqrt{2})$ and $\mathbf{Q}(i\sqrt{2})$ must must be fixed fields if H_1 and H_3. We find that $\delta_1(\sqrt{2}) = \delta_1((\sqrt[4]{2})^2) = (i\sqrt[4]{2})^2 = -\sqrt{2} \neq \sqrt{2}$. Thus H_3, which contains δ_1, must have $\mathbf{Q}(i\sqrt{2})$ as its fixed field, and H_1 must have $\mathbf{Q}(\sqrt{2})$ as its fixed field.

For the fixed field of $H_5 = \{\rho_0, \mu_2\}$ we need to find some elements left fixed by μ_2. Since $\mu_2(\alpha) = -\alpha$ and $\mu_2(i) = -i$, the product $i\alpha$ is an obvious choice. Now $i(\sqrt[4]{2})$ is a zero of $x^4 - 2$ which is irreducible, so $\mathbf{Q}(i(\sqrt[4]{2}))$ is of degree 4 over $\mathbf{Q}$ and left fixed by H_5.

Since ρ_2 leaves i fixed and maps α into $-\alpha$, it leaves i and $\alpha^2 = \sqrt{2}$ fixed. Thus the fixed field of $H_6 = \{\rho_0, \rho_2\}$ is $\mathbf{Q}(i, \sqrt{2})$.

To find an element left fixed by $H_8 = \{\rho_0, \delta_2\}$, we form $\beta = \rho_0(\alpha) + \delta_2(\alpha) = \alpha - i\alpha = \sqrt[4]{2}(1 - i)$. Now $\beta^4 = 2(-4) = -8$ so β is a zero of $x^4 + 8$. This polynomial does not have ± 1, ± 2, ± 4, or ± 8 as a zero, so it has no linear factors. If $x^4 + 8 = (x^2 + ax + b)(x^2 + cx + d)$, then $a + c = 0$, $ac + b + d = 0$, $ad + bc = 0$, and $bd = 8$. From $a + c = 0$ and $ad + bc = 0$, we find that $ad - ba = 0$ so $a(d - b) = 0$ and either $a = 0$ or $b = d$. Since $bd = 8$, we do not have $b = d$, so $a = 0$. But then $b + d = 0$ so $b = -d$, which again cannot satisfy $bd = 8$. Thus $x^4 + 8$ is irreducible, and $\mathbf{Q}(\sqrt[4]{2}(1 - i))$ has degree 4 over $\mathbf{Q}$ and is left fixed by H_8, so it is the fixed field of H_8.

3. The choices for primitive elements given in the text answers, the corresponding polynomials and the fact that the polynomials are irreducible are obvious or proved in the text or the preceding solution, except for the first and fourth answers given.

For the case $\mathbf{Q}(\sqrt[4]{2}, i)$, let $\beta = \sqrt[4]{2}$ and $\gamma = i$. The proof of Theorem 9.14 shows that $\beta + a\gamma$ will be a primitive element if $(\beta_i - \beta)/(\gamma - \gamma_j) \neq a$, $a \in \mathbf{Q}$, where β_i can be any conjugate of β and γ_j is a conjugate other than γ of γ. Now $\gamma - \gamma_j$ is always $i - (-i) = 2i$, and since $\beta = \alpha = \sqrt[4]{2}$ in Table 9.2, it is clear from the table that $[\alpha - (\text{conjugate of } \alpha)]/(2i)$ is never a nonzero element of $\mathbf{C}$. Thus we can take $a = 1$, and we find that $\sqrt[4]{2} + i$ is a primitive element. Let $\delta = \sqrt[4]{2} + i$. Then $\delta - i = \sqrt[4]{2}$ so $(\delta - i)^4 = \delta^4 - 4\delta^3 i - 6\delta^2 + 4\delta i + 1 = 2$ so $\delta^4 - 6\delta^2 - 1 = (4\delta^3 - 4\delta)i$. Squaring both sides, we obtain $\delta^8 - 12\delta^6 + 34\delta^4 + 12\delta^2 + 1 = -16\delta^6 + 32\delta^4 - 16\delta^2$ so $\delta^8 + 4\delta^6 + 2\delta^4 + 28\delta^2 + 1 = 0$. Thus δ is a zero of $x^8 + 4x^6 + 2x^4 + 28x^2 + 1 = 0$. Since we know that $\mathbf{Q}(\delta)$ is of degree 8 over $\mathbf{Q}$, this must be $\text{irr}(\mathbf{Q}, \delta)$.

For $\mathbf{Q}(\sqrt{2}, i)$, we have $[\sqrt{2} - (\text{conjugate of } \sqrt{2})]/2i$ is never a nonzero element of $\mathbf{Q}$, so $\sqrt{2} + i$ is a primitive element. If $\delta = \sqrt{2} + i$, then $\delta - i = \sqrt{2}$ and $\delta^2 - 2\delta i - 1 = 2$. Then $\delta^2 - 3 = 2\delta i$ so $\delta^4 - 6\delta^2 + 9 = -4\delta^2$ and $\delta^4 - 2\delta^2 + 9 = 0$. Thus δ is a zero of $x^4 - 2x^2 + 9$, and since $\mathbf{Q}(\delta)$ is of of degree 4 over $\mathbf{Q}$, we see that this polynomial is irreducible.

4. a) If ζ is a primitive 5th root of unity, then $1, \zeta, \zeta^2, \zeta^3, \zeta^4$ are five distinct elements of $\mathbf{Q}(\zeta)$ and $(\zeta^k)^5 = (\zeta^5)^k =$

$1^k = 1$ shows that these five elements are five zeros of $x^5 - 1$. Thus $x^5 - 1$ splits in $\mathbf{Q}(\zeta)$.

b) We know that $x^5 - 1 = (x - 1)\Phi_5(x)$ where $\Phi_5(x) = x^4 + x^3 + x^2 + x + 1$ is the irreducible (corollary to Theorem 5.21) cyclotomic polynomial having ζ as a root. Every automorphism of $K = \mathbf{Q}(\zeta)$ over $\mathbf{Q}$ must map ζ into one of the four roots ζ, ζ^2, ζ^3, ζ^4 of this polynomial.

c) Let $\sigma_j \in G(K/\mathbf{Q})$ be the automophism such that $\sigma(\zeta) = \zeta^j$ for $j = 1, 2, 3, 4$. Then $(\sigma_j\sigma_k)(\zeta) = \sigma_j(\zeta^k) = (\zeta^j)^k = \zeta^{jk} = \sigma_m(\zeta)$ where m is the product of j and k in $\mathbf{Z}_5$. Thus $G(K/\mathbf{Q})$ is isomorphic to the group $\{1, 2, 3, 4\}$ of nozero elements of $\mathbf{Z}_5$ under multiplication. It is cyclic of order 4, generated by σ_2.

d) $G(K/\mathbf{Q}) = \{\sigma_1, \sigma_2, \sigma_3, \alpha_4\}$
|
$\{\sigma_1, \sigma_4\}$
|
$\{\sigma_1\}$

Group lattice diagram

$K = K_{\{\sigma_1\}}$
|
$\mathbf{Q}(\sqrt{5}) = \mathbf{Q}(\cos 72°) = K_{\{\sigma_1, \sigma_4\}}$
|
$\mathbf{Q} = K_{\{\sigma_1, \sigma_2, \sigma_3, \sigma_4\}}$

Field lattice diagram

To find $K_{\{\sigma_1, \sigma_4\}}$, note that $\zeta = \cos 72° + i \sin 72°$ and that $\zeta^4 = \cos(-72°) + i \sin(-72°)$. Thus $\alpha = \sigma_1(\zeta) + \sigma_4(\zeta) = \zeta + \zeta^4 = 2 \cos 72°$ is left fixed by σ_1 and σ_4. Alternatively, doing a bit of computation, we find that

$$\alpha^2 = (\zeta + \zeta^4)^2 = \zeta^2 + 2 + \zeta^3,$$
$$\alpha = \zeta + \zeta^4.$$

Now α is a zero of $\Phi_5(x) = x^4 + x^3 + x^2 + x + 1$, so we see that $\alpha^2 + \alpha - 1 = 0$, so α is a zero of $x^2 + x - 1$ which has zeros $(-1 \pm \sqrt{5})/2$. Thus we can also describe $\mathbf{Q}(\alpha)$ as $\mathbf{Q}(\sqrt{5})$.

5. The splitting field of $x^5 - 2$ over $\mathbf{Q}(\zeta)$ is $\mathbf{Q}(\zeta, \sqrt[5]{2})$, since $\sqrt[5]{2}$, $\zeta(\sqrt[5]{2})$, $\zeta^2(\sqrt[5]{2})$, $\zeta^3(\sqrt[5]{2})$, and $\zeta^4(\sqrt[5]{2})$ are the five zeros of $x^5 - 2$. The Galois group $\{\sigma_0, \sigma_1, \sigma_2, \sigma_3, \sigma_4\}$ is described by that table

	σ_0	σ_1	σ_2	σ_3	σ_4
$\sqrt[5]{2} \rightarrow$	$\sqrt[5]{2}$	$\zeta(\sqrt[5]{2})$	$\zeta^2(\sqrt[5]{2})$	$\zeta^3(\sqrt[5]{2})$	$\zeta^4(\sqrt[5]{2})$

We have $(\sigma_j\sigma_k)(\sqrt[5]{2}) = \sigma_j(\zeta^k(\sqrt[5]{2})) = \zeta^k\zeta^j(\sqrt[5]{2}) = \zeta^{j+k}(\sqrt[5]{2})$ from which we see that the Galois group is isomorphic to $\langle \mathbb{Z}_5\ +\rangle$, so it is cyclic of order 5.

6. a)-b) The arguments are identical with those given in the solution of Exercise 4. Just increase all numbers by 2, all lists of powers of ζ by 2, and use $\Phi_7(x)$.

c) As in part (c) of Exercise 4, we let σ_j be the automorphism such that $\sigma_j(\zeta) = \zeta^j$ for $j = 1, 2, 3, 4, 5, 6$. The same argument as in Exercise 4 shows that $\sigma_j\sigma_k = \sigma_m$ where m is the remainder of jk modulo 7. Thus the Galois group is isomorphic to $\{1, 2, 3, 4, 5, 6\}$ under multiplication modulo 7, that is, to $\langle \mathbb{Z}_7^*, \cdot\rangle$. This group is abelian, and is cyclic of order 6.

d)

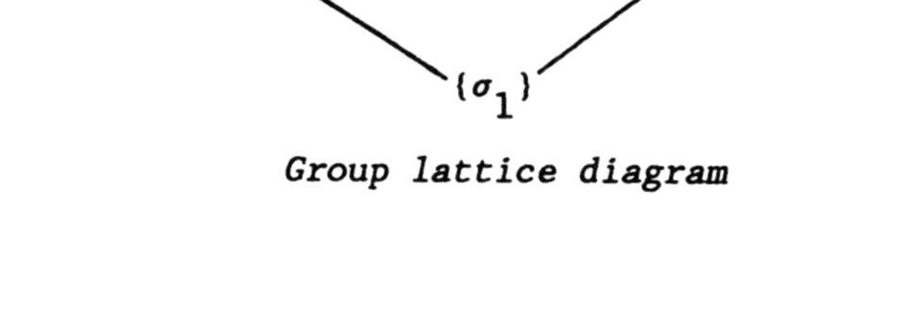

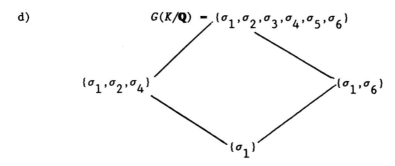

Group lattice diagram

Section 9.7

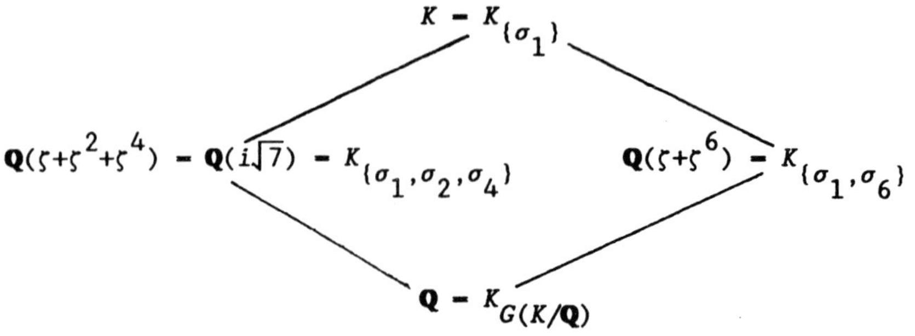

Field lattice diagram

Clearly $\alpha = \zeta + \zeta^2 + \zeta^4$ is left fixed by $\{\sigma_1, \sigma_2, \sigma_4\}$. Doing a bit of computation, we find that

$$\begin{aligned} \alpha^2 &= \zeta^2 + \zeta^4 + \zeta + 2\zeta^3 + 2\zeta^6 + 2\zeta^5, \\ \alpha &= \zeta + \zeta^2 + \zeta^4. \end{aligned}$$

Thus we find that $\alpha^2 + \alpha = 2(\zeta^6 + \zeta^5 + \zeta^4 + \zeta^3 + \zeta^2 + \zeta)$. Since ζ is a zero of

$$\Phi_7(x) = x^6 + x^5 + x^4 + x^3 + x^2 + x + x + 1,$$

we see at once that $\alpha^2 + \alpha + 2 = 0$. The zeros of $x^2 + x + 2$ are $(-1 \pm i\sqrt{7})/2$, and we see that $\mathbf{Q}(\alpha) = \mathbf{Q}(i\sqrt{7})$. Working in an analogous way for the subgroup $\{\sigma_1, \sigma_6\}$, we form the element $\beta = \zeta + \zeta^6$ which is left fixed by this subgroup. Computing, we find that

$$\begin{aligned} \beta^3 &= (\zeta + \zeta^6)^3 = \zeta^3 + 3\zeta + 3\zeta^6 + \zeta^4, \\ \beta^2 &= (\zeta + \zeta^6)^2 = \zeta^2 + 2 + \zeta^5, \\ \beta &= \zeta + \zeta^6. \end{aligned}$$

Recalling that $\Phi_7(\zeta) = 0$ as above, we find that $\beta^3 + \beta^2 - 2\beta - 1 = 0$. Thus β is a zero of $x^3 + x^2 - 2x - 1$ which is irreducible since it has no zero in $\mathbf{Z}$.

7. $x^8 - 1 = (x^4 + 1)(x^2 + 1)(x - 1)(x + 1)$. Example 2 shows that the splitting field of $x^4 + 1$ contains i, which is a zero of $x^2 + 1$. Thus the splitting field of $x^8 - 1$ is the same as the splitting field of $x^4 + 1$, whose group was completely described in Example 2. This is the "easiest way to describe this group."

8. Using the quadratic formula to find α such that $\alpha^4 - 4\alpha^2 - 1 = 0$, we find that $\alpha^2 = (4 \pm \sqrt{20})/2 = 2 \pm \sqrt{5}$ so $\alpha = \pm\sqrt{2 \pm \sqrt{5}} = \pm\sqrt{2 + \sqrt{5}}, \pm i\sqrt{\sqrt{5} - 2}$. We see that the splitting field of $x^4 - 4x^2 - 1 = 0$ can be generated by adjoining in succession $\sqrt{5}$, $\sqrt{\sqrt{5} + 2}$, $\sqrt{2 - \sqrt{5}}$. Thus it has degree $2^3 = 8$ over $\mathbf{Q}$. It can obviously be generated bu adjoining $\alpha_1 = \sqrt{\sqrt{5} + 2}$ and $\alpha_2 = i\sqrt{\sqrt{5} - 2}$. Let $K = \mathbf{Q}(\alpha_1, \alpha_2)$. The eight elements of $G(K/\mathbf{Q})$ are given by the following table.

	ρ_0	ρ_1	ρ_2	ρ_3	μ_1	μ_2	δ_1	δ_2
$\alpha_1 \rightarrow$	α_1	α_2	$-\alpha_1$	$-\alpha_2$	α_2	$-\alpha_2$	$-\alpha_1$	α_1
$\alpha_2 \rightarrow$	α_2	$-\alpha_1$	$-\alpha_2$	α_1	α_1	$-\alpha_1$	α_2	$-\alpha_2$

The group is isomorphic to D_4 and the notation here is taken to coincide with that in Example 5 of Section 2.1. The group lattice diagram is identical with that in Fig. 9.13. Let H_i be as given in Fig. 9.13. The field lattice diagram is as follows:

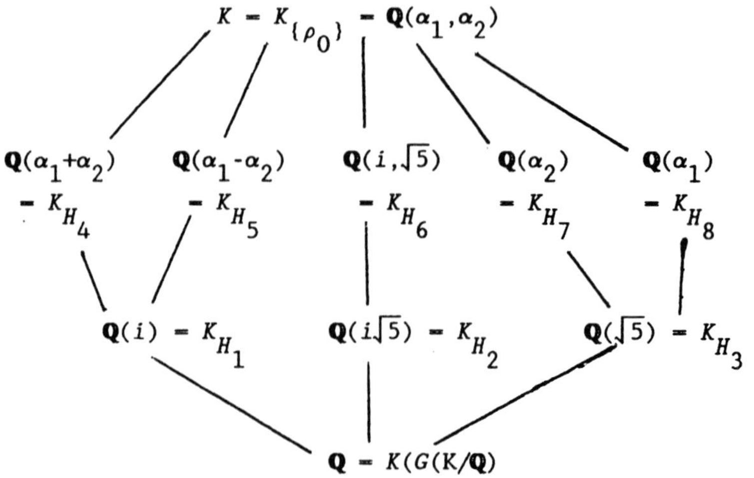

Field lattice diagram

We now check most of this diagram. Note that ${\alpha_1}^2$ is left fixed by $H_3 = \{\rho_0, \rho_2, \delta_1, \delta_2\}$ and that ${\alpha_1}^2 = \sqrt{5} + 2$, so the fixed field of H_3 is $\mathbf{Q}(\sqrt{5})$. Note also that $\alpha_1\alpha_2 = i$ is left fixed by $H_1 = \{\rho_0, \rho_2, \mu_1, \mu_2\}$ so the fixed field of H_1 is $\mathbf{Q}(i)$. Then $H_6 = H_1 \cap H_3$ leaves both i and $\sqrt{5}$ fixed. Also K_{H_2} must be the only remaining extension of $\mathbf{Q}$ of degree 2, and $\mathbf{Q}(i\sqrt{5})$ fits the bill.

The remaining fields are trivial to check since they are described in terms of α_1 and α_2. For example, to see that $\mathbf{Q}(\alpha_1)$ is the fixed field of H_8, we need only note that that $\{\rho_0, \delta_2\} = H_8$ is the set of elements leaving α_1 fixed in the table above.

9. We develop some formulas to use in this exercise and the next one. For simple notation, we denote symmetric expressions in the indeterminates y_1, y_2, y_3 by the notation S(formula) where the formula indicates the nature of *one* summand of the expression. Thus we write

$$s_1 = y_1 + y_2 + y_3 = S(y_i),$$
$$s_2 = y_1y_2 + y_1y_3 + y_2y_3 = S(y_iy_j),$$
$$s_3 = y_1y_2y_3 = S(y_iy_jy_k).$$

Note that the subscripts i, j, k do not run independently through values from 1 to 3; we always have $i \neq j \neq k$. The formula simply indicates the nature of one, typical term in the symmetric expression.

We now express some other symmetric expressions in terms of s_1, s_2, and s_3. Theorem 9.20 asserts that this is possible.

a) $$y_1^2 + y_2^2 + y_3^2 = S(y_i^2) = [S(y_i)]^2 - 2S(y_iy_j)$$
$$= s_1^2 - 2s_2. \tag{1}$$

$$S(y_i^2y_j) = S(y_iy_j)S(y_i) - S(y_iy_jy_k) = s_1s_2 - 3s_3 \tag{2}$$

b) $$\frac{y_1}{y_2} + \frac{y_2}{y_1} + \frac{y_1}{y_3} + \frac{y_3}{y_1} + \frac{y_2}{y_3} + \frac{y_3}{y_2} = S(\frac{y_i}{y_j}) =$$
$$\frac{y_1^2y_3 + y_2^2y_3 + y_1^2y_2 + y_3^2y_2 + y_2^2y_1 + y_3^2y_1}{y_1y_2y_3} = \frac{S(y_i^2y_j)}{y_2y_2y_3}$$
$$= \frac{s_1s_2 - 3s_3}{s_3},$$ using formula (2).

$$S(y_i^2y_j^2) = [S(y_iy_j)]^2 - S(y_i^2y_jy_k) = s_2^2 - 2s_2s_3. \tag{3}$$

$$S(y_i^3) = S(y_i^2)S(y_1) = S(y_i^2y_j) = (s_1^2 - 2s_2)s_1 - (s_1s_2 - 3s_3)$$
$$= s_1^3 - 3s_1s_2 + 3s_3. \tag{4}$$

$$S(y_i^2y_jy_k) = s_3s_1. \tag{5}$$

c) Multiplying things out, we find that

$$(y_1 - y_2)^2(y_1 - y_3)^2(y_2 - y_3)^2$$
$$= S(y_i^4y_j^2) - 2S(y_i^4y_jy_k) + 2S(y_i^3y_j^2y_k) - 2S(y_i^3y_j^3)$$
$$- 6S(y_i^2y_2^2y_3^2). \tag{6}$$

Using some of the formulas (1) through (5) above, we have:

$$S(y_i^4y_j^2) = S(y_i^2)[S(y_iy_j)]^2 - 3S(y_i^2y_j^2y_k^2) - 2S(y_i^4y_jy_k) - 2S(y_i^3y_j^2y_k)$$

$$= (s_1^2-2s_2)s_2^2 - 3s_3^2 - 2s_3(s_1^3-3s_1s_2+3s_3) - 2s_3(s_1s_2-3s_3)$$

$$= s_1^2s_2^2 - 2s_2^3 - 3s_3^2 - 2s_1^3s_3 + 6s_1s_2s_3 - 6s_3^2 - 2s_1s_2s_3 + 6s_3^2 \qquad (7)$$

$$S(y_i^4y_jy_k) = s_3S(y_i^3) = s_3(s_1^3-3s_1s_2+3s_3)$$

$$= s_1^3s_3 - 3s_1s_2s_3 + 3s_3^2 \qquad (8)$$

$$S(y_i^3y_j^2y_k) = s_3S(y_i^2y_j) = s_3(s_1s_2-3s_3) = s_1s_2s_3-3s_3^2 \qquad (9)$$

$$S(y_i^3y_j^3) = S(y_i^2y_j^2)S(y_iy_j) - S(y_i^3y_j^2y_k)$$

$$= (s_2^2 - 2s_1s_3)s_2 - s_2(s_1s_2 - 3s_2)$$

$$= s_2^3 - 3s_1s_2s_3 + 3s_3^2 \qquad (10)$$

$$S(y_1^2y_2^2y_3^2) = s_3^2. \qquad (11)$$

Substituting formulas 7 - 11 in formula 6, we find that the answer to part (c) is

$$s_1^2s_2^2 - 4s_2^3 - 27s_3^2 - 4s_1^3s_3 + 18s_1s_2s_3.$$

10. We have $x^3 - 4x^2 + 6x - 2 = (x - \alpha_1)(x - \alpha_2)(x - \alpha_3)$. Therefore the elementary symmetric expressions in α_1, α_2, and α_3 are given by

$$s_1 = \alpha_1 + \alpha_2 + \alpha_3 = 4,$$
$$s_2 = \alpha_1\alpha_2 + \alpha_1\alpha_3 + \alpha_2\alpha_3 = 6,$$
$$s_3 = \alpha_1\alpha_2\alpha_3 = 2.$$

We feel free to make use of some of the formulas 1-5 of the solution to the preceding exercise.

a) $x - 4$

b) Let $x^3 + b_2x^2 + b_1x + b_0 = (x - \alpha_1^2)(x - \alpha_2^2)(x - \alpha_3^2)$.

Now $b_2 = -\alpha_1^2 - \alpha_2^2 - \alpha_3^2 = -(s_1^2 - 2s_2)$ by formula 1. Evaluating with the values for s_1, s_2, and s_3, we find that $b_2 = -(16 - 12) = -4$. Also $b_1 =$ $\alpha_1^2\alpha_2^2 + \alpha_1^2\alpha_3^2 + \alpha_2^2\alpha_3^2 = s_2^2 - 2s_1s_3$ by formula 3. Evaluating, we find that $b_1 = 36 - 16 = 20$. Finally, $b_0 =$ $-\alpha_1^2\alpha_2^2\alpha_3^2 = -s_3^2 = -4$. Thus the answer is

$$x^3 - 4x^2 + 20x - 4.$$

c) Let $x^3 + c_2x^2 + c_1x + c_0 =$ $[x - (\alpha_1-\alpha_2)^2][x - (\alpha_1-\alpha_3)^2][x - (\alpha_2-\alpha_3)^2]$. Now

$$c_2 = -[(\alpha_1-\alpha_2)^2 + (\alpha_1-\alpha_3)^2 + (\alpha_2-\alpha_3)^2]$$
$$= -[2(\alpha_1^2 + \alpha_2^2 + \alpha_3^2) - 2(\alpha_1\alpha_2 + \alpha_1\alpha_3 + \alpha_2\alpha_3)]$$
$$= -[2(s_1^2 - 2s_2) - 2s_2] = -2s_1^2 + 6s_2 = -32 + 36 = 4.$$
$$c_1 = (\alpha_1-\alpha_2)^2(\alpha_1-\alpha_3)^2 + (\alpha_1-\alpha_2)^2(\alpha_2-\alpha_3)^2 + (\alpha_1-\alpha_3)^2(\alpha_2-\alpha_3)^2.$$

Using the notation of the preceding exercise and formulas 1-5 there, we derive that

$$S(y_i^4) = [S(y_i^2)]^2 - 2S(y_i^2y_j^2)$$
$$= (s_1^2 - 2s_2)^2 - 2(s_2^2 - 2s_1s_3)$$
$$= s_1^4 - 4s_1^2s_2 + 2s_2^2 + 4s_1s_3 \text{ and}$$
$$S(y_i^3y_j) = S(y_i^2)S(y_iy_j) - S(y_i^2y_jy_k) =$$
$$= (s_1^2 - 2s_2)s_2 - s_3s_1 = s_1^2s_2 - 2s_2^2 - s_3s_1.$$

$$c_1 = (\alpha_1-\alpha_2)^2(\alpha_1-\alpha_3)^2 + (\alpha_1-\alpha_2)^2(\alpha_2-\alpha_3)^2 + (\alpha_1-\alpha_3)^2(\alpha_2-\alpha_3)^2.$$

Multiplying things out and using the notation of the preceding exercise and formulas 1-5 of Exercise 9, we find

$$c_1 = S(y_i^4) + 3S(y_i^2 y_j^2) - 2S(y_i^3 y_j) + 4S(y_i^2 y_j y_k) - 4S(y_i^2 y_j y_k)$$

$$= S(y_i^4) + 3S(y_i^2 y_j^2) - 2S(y_i^3 y_j)$$

$$= s_1^4 - 4s_1^2 s_2 + 2s_2^2 + 4s_1 s_3 + 3(s_2^2 - 2s_1 s_3) - 2(s_1^2 s_2 - 2s_2^2 - s_3 s_1)$$

$$= 256 - 4\cdot 96 + 2\cdot 36 + 4\cdot 8 + 3(36-16) - 2(96-72-8) = 4.$$

Finally, $c_0 = (\alpha_1 - \alpha_2)^2(\alpha_1 - \alpha_3)^2(\alpha_2 - \alpha_3)^2$ and Exercise 9c shows that this is equal to

$$-s_1^2 s_2^2 + 4s_2^3 + 27s_3^2 + 4s_1^3 s_3 - 18s_1 s_2 s_3.$$

Evaluating, we find that

$$c_1 = -16\cdot 36 + 4\cdot 216 + 27\cdot 4 + 4\cdot 64\cdot 2 - 18\cdot 48 = 44.$$

Thus the answer is

$$x^3 + 4x^2 + 4x + 44.$$

11. a) If $\Delta(f) = 0$, then $\alpha_i = \alpha_j$ for some $i \neq j$. Thus $\mathrm{irr}(\alpha_i, F) = \mathrm{irr}(\alpha_j, F)$. Since the irreducible factors of $f(x)$ are all separable and do not have zeros of multiplicity greater than 1, we see that $f(x)$ must have $\mathrm{irr}(\alpha_i, F)^2$ as a factor.

b) Clearly $[\Delta(f)]^2$ is a symmetric expression in the α_i, and hence left fixed by any permutation of the α_i, and thus is invariant under $G(K/F)$. Therefore $[\Delta(f)]^2$ is in F.

c) Consider the effect of a transposition (α_i, α_j) on $\Delta(f)$; it is no loss of generality to suppose $i < j$. The factor $\alpha_i - \alpha_j$ is carried into $\alpha_j - \alpha_i$, so it changes sign. $\alpha_k - \alpha_j$ and $\alpha_k - \alpha_i$ for $k > j$ are carried into each other, so they do not contribute a sign change. The same is true of $\alpha_i - \alpha_k$ and $\alpha_j - \alpha_k$ for $k < i$. For $i < k < j$, the terms $\alpha_k - \alpha_i$ and $\alpha_j - \alpha_k$ are carried into $\alpha_k - \alpha_j =$

$-(\alpha_j - \alpha_k)$ and into $\alpha_i - \alpha_k = -(\alpha_k - \alpha_i)$, so they contribute two sign changes. Thus the transposition contributes $1 + 2(j - i - 1)$ sign changes, which is an odd number, and carries $\Delta(f)$ into $-\Delta(f)$. Thus a permutation leaves $\Delta(f)$ fixed if and only if it can be expressed as a product of an even number of transpositions, that is, if and only it is in A_n. Hence $G(K/F) \subseteq A_n$ if and only if it leaves $\Delta(f)$ fixed, that is, if and only if $\Delta(f) \in F$.

12. Let α and β be algebraic integers and let K be the splitting field of $\operatorname{irr}(\alpha,F)\cdot\operatorname{irr}(\beta,F)$. Now

$$g(x) = \prod_{\sigma\in G(K/\mathbf{Q})} (x - \sigma(\alpha))$$

is a power of $\operatorname{irr}(\alpha, F)$, and thus has integer coefficients and leading coefficient 1 since α is an algebraic integer. The same is true of

$$h(x) = \prod_{\sigma\in G(K/\mathbf{Q})} (x - \sigma(\beta)).$$

Now

$$k(x) = \prod_{\sigma,\mu \in G(K/\mathbf{Q})} [x - (\sigma(\alpha) + \mu(\beta))]$$

$$= \prod_{\sigma\in G(K/F)} \left[\prod_{\mu\in G(K/F)} [(x - \sigma(\alpha)) - \mu(\beta)]\right]$$

$$= \prod_{\sigma\in G(K/\mathbf{Q})} h(x - \sigma(\alpha)).$$

Since $h(x)$ has integer coefficients, $h(x - \sigma(\alpha))$ is a polynomial in $x - \sigma(\alpha)$ with integer coefficients. We can view $k(x)$ as a symmetric expression in α and its conjugates over the field $\mathbf{Q}(x)$ involving only integers in $\mathbf{Q}$. By Theorem 9.20, the symmetric expressions in α and its conjugates can be expressed as polynomials in the elementary symmetric functions of α and its conjugates, that is, in terms of the coefficients of $g(x)$ or their negatives. Thus $k(x)$ has integer coefficients. Now one zero of $k(x)$ is $\alpha + \beta$, corresponding to the factor when σ and μ are both the identity permutation. Thus $\operatorname{irr}(\alpha + \beta,\mathbf{Q})$ is a factor of the monic polynomial $k(x)$. Since a factorization in $\mathbf{Q}[x]$ can always be implemented in $\mathbf{Z}[x]$ by Theorem 5.20, we see that $\operatorname{irr}(\alpha + \beta)$ is monic with integer coefficients, and hence $\alpha + \beta$ is an algebraic integer. If α is a zero of $f(x)$, then $-\alpha$ is a zero of $f(-x)$ which again has integer coefficients and is monic, so $-\alpha$ is again an algebraic integer.

One can argue that $\alpha\beta$ is an algebraic integer by the same technique that we used for $\alpha + \beta$, considering

$$\prod_{\sigma,\mu\in G(K/\mathbf{Q})} [x - \sigma(\alpha)\mu(\beta)] = \prod_{\sigma\in G(K/\mathbf{Q})} \left[\prod_{\mu\in G(K/\mathbf{Q})} [x - \sigma(\alpha)\mu(\beta)] \right].$$

Thus the algebraic integers are closed under addition and multiplication, and include additive inverses, and of course 0 which is a zero of x. Hence they form a subring of $\mathbf{C}$.

13. By Cayley's Theorem, every finite group G is isomorphic to a a subgroup of S_n where n is the order of G. Now Theorem 9.20 shows that for each positive integer n, there exists a normal extension K of a field E such that $G(K/E) \simeq S_n$. If H is a subgroup of $G(K/E)$ isomorphic to G, then H is the Galois group of K over K_H, where $F = K_H$ is the fixed field of H. Thus H is isomorphic to G and is the Galois group $G(K/F)$ of K over F.

SECTION 9.8 - Cyclotomic Extensions

1. $\Phi(x) = (x - \zeta)(x - \zeta^7)(x - \zeta^3)(x - \zeta^5)$
$= [x^2 - (\zeta + \zeta^7)x - 1][x^2 - (\zeta^3 + \zeta^5) + 1] \qquad (\zeta^4 = -1,\ \zeta^8 = 1)$
$= (x^2 - \sqrt{2}x - 1)(x^2 + \sqrt{2}x + 1) \qquad (\zeta + \zeta^7 = \sqrt{2},\ \zeta^3 + \zeta^5 = -\sqrt{2})$
$= x^4 + 1.$

2. The group is $G = \{1, 3, 7, 9, 11, 13, 17, 19\}$ under multiplication modulo 20. We find that $3^2 = 9$, $3^3 = 7$, and $3^4 = 1$, so 3 and 7 have order 4 and 9 has order 2. Then $11^2 = 1$, $13^2 = 9$, $17^2 = 9$, $19^2 = 1$. Thus G is isomorphic to $\langle \mathbf{Z}_4 \times \mathbf{Z}_2, +\rangle$.

3. $60 = 2^2 \cdot 3 \cdot 5$; $\phi(60) = 2 \cdot 2 \cdot 4 = 16$.
$1000 = 2^3 \cdot 5^3$; $\phi(1000) = 2^2 \cdot 5^2 \cdot 4 = 400$.
$8100 = 2^2 \cdot 3^4 \cdot 5^2 = 2 \cdot 3^3 \cdot 5 \cdot 2 \cdot 4 = 2160$.

4.

3	12	30	60	102
4	15	32	64	120
5	16	34	68	128
6	17	40	80	136
8	20	48	85	160
10	24	51	96	170

Just use Theorem 9.22 with the prime 2 and the Fermat primes 3, 5, 17.

5. 360 and 180 are divisible by 3^2, so the 360-gon and the 180-gon are not constructible. $360/3 = 120 = 8 \cdot 3 \cdot 5$ so the regular 120-gon is constructible, and an angle of 3° is constructible.

6. a) $\phi(12) = 4$ since the integers ≤ 12 and relatively prime to 12 are 1, 5, 7, and 11.

b) The group is isomorphic to $\{1, 5, 7, 11\}$ under multiplication modulo 12, and 5^2, 7^2, and 11^2 are all congruent to 1 modulo 12. The group is isomorphic to $\mathbb{Z}_2 \times \mathbb{Z}_2$.

7. $\Phi^3(x) = \dfrac{x^3 - 1}{x - 1} = x^2 + x + 1$ for every field of characteristic $\neq 3$ since ζ and ζ^2 are both primitive cube roots of unity.

In $\mathbb{Z}_3[x]$, $x^8 - 1 = (x^4 + 1)(x^2 + 1)(x - 1)(x + 1)$ so the primitive 8th roots of unity must be zeros of $x^4 + 1$, and $\Phi_8(x) = x^4 + 1 = (x^2 + x + 2)(x^2 + 2x + 2)$.

8. $x^6 - 1 = (x^2 - 1)^3 = (x - 1)^3(x + 1)^3$, so the polynomial splits in $\mathbb{Z}_3$, and the splitting field has 3 elements.

9. T T F T T F T T F T

10. The nth roots of unity form a cyclic group of order n under multiplication, for $\zeta^j\zeta^k = \zeta^{j+k} = \zeta^{j+k \bmod n}$ where ζ is a primitive nth root of unity. Each element ζ^j of this group is a primitive dth root of unity for some d dividing n, and conversely, Exercise 40 of Section 2.3 shows that for each d dividing n, there is a primitive dth root of unity among the various ζ^i. Since the dth roots of unity are the zeros of

$\Phi_d(x)$ in a field of characteristic not dividing d, we see that $\Phi_n(x) = \prod_{d|n} \Phi_d(x)$.

11. $\Phi_1(x) = x - 1$, $\Phi_2(x) = x + 1$, $\Phi_3(x) = x^2 + x + 1$, $x^4 - 1 = (x^2 + 1)(x - 1)(x + 1)$ so $\Phi_4(x) = x^2 + 1$, and $\Phi_5(x) = x^4 + x^3 + x^2 + 1$ (see the corollary to Theorem 5.21). A primitive 6th root of unity is a primitive cube root of -1, and hence a solution of $x^3 + 1 = (x + 1)(x^2 - x + 1)$. Since there are two primitive 6th roots of unity, we see that $\Phi_6(x) = x^2 - x + 1$.

12. By Exercise 10, $x^{12} - 1 = \Phi_1(x)\Phi_2(x)\Phi_3(x)\Phi_4(x)\Phi_6(x)\Phi_{12}(x)$
$= (x-1)(x+1)(x^2+x+1)(x^2+1)(x^2-x+1)\Phi_{12}(x)$
$= (x^4 - 1)(x^4 + x^2 + 1)\Phi_{12}(x) = (x^8 + x^6 - x^2 - 1)\Phi_{12}(x)$.
Polynomial long division yields $\Phi_{12}(x) = \dfrac{x^{12} - 1}{x^8 + x^6 - x^2 - 1} =$ $x^4 - x^2 + 1$.

13. Let ζ be a primitive nth root of unity where n is odd. Then $(-\zeta)^n = -1$, and no lower power of $-\zeta$ gives -1. Consequently $(-\zeta)^{2n} = 1$, so the multiplicative order r of $-\zeta$ is either $2n$ or $< n$. If $r < n$, then $(-\zeta)^{2r} = 1$ and ζ^2 has order dividing r and dividing n. Since n is odd, r must be an odd divisor of n, and therefore $2r < n$ and $\zeta^{2r} = 1$, which contradicts the fact that ζ is a primitive nth root of unity. Therefore $r = 2n$, and $-\zeta$ is a primitive $2n$th root of unity.

13. Following the hint, we use induction on k where $n = 2k + 1$. For $k = 1$, $\Phi_3(x) = x^2 + x + 1$ and $\Phi_6(x) = x^2 - x + 1$ by Exercise 11. Thus $\Phi_6(x) = \Phi_3(-x)$ and the formula is true for $k = 1$. Suppose that it is true for $k < m$. Note that the divisors of $4m + 2$ are all of the form d or $2d$ where d is a divisor of $2m + 1$. Note, however, that for $n = 1$, we have $\Phi_{2n}(x) = \Phi_2(x) = x + 1 = -(-x - 1) = -\Phi_1(-x) = -\Phi_n(-x)$. Thus

for $n = 1$, we have a minus sign that is not present for $n = 2k + 1$ for $1 \leq k < m$. By Exercise 10, the observation on divisors of double odd numbers, our induction hypothesis, and this observation on the introduction of a single minus sign, we see that

$$x^{4m+2} - 1 = \underbrace{\left[\prod_{d|(2m+1)} \Phi_d(x)\right]}_{f(x) \,=\, x^{2m+1}-1} \underbrace{\left[-\prod_{\substack{d|(2m+1)\\ d\neq 2m+1}} \Phi_d(-x)\right]\Phi_{4m+2}(x)}_{g(x) \,=\, x^{2m+1}+1}. \qquad (1)$$

Now the polynomial $f(x)$ in the first bracket in Eq.(1) is equal to $x^{2m+1} - 1$ by Exercise 10. By the factorization given in the problem, the rest $g(x)$ of the product in Eq.(1) must be equal to $x^{2m+1} + 1 = -[(-x)^{2m+1} - 1]$. But by Exercise 10

$$x^{2m+1} - 1 = \prod_{d|(2m+1)} \Phi_d(x) \quad \text{so} \quad (-x)^{2m+1} - 1 = \prod_{d|(2m+1)} \Phi_d(-x).$$

Comparing this last equation with $g(x) = -[(-x)^{2m+2} - 1]$, we see that $\Phi_{4m+2}(x) = \Phi_{2m+1}(-x)$ and we are done.

14. Let ζ be a primitive nmth root of unity. Then ζ^m is an nth root of unity and ζ^n is a primitive mth root of unity. Now $\zeta^m\zeta^n = \zeta^{m+n}$ has as order the least positive integer r such that mn divides $r(m + n)$. Now no prime dividing m divides $m + n$ since m and n are relatively prime. Similarly, no prime dividing n divides $m + n$. Consequently, mn must divide r so $\zeta^m\zeta^n$ is a primitive mnth root of unity. Since the splitting field of $(x^m - 1)(x^n - 1)$ contains $\zeta^m\zeta^n$, we see that it contains a primitive mnth root of unity and thus contains the splitting field of $x^{mn} - 1$. We saw above that the splitting field of $x^{mn} - 1$ contains the primitive nth root of unity ζ^m and the primitive *mth root of unity* ζ^n, so the reverse containment also holds. Thus the splitting fields of $x^{mn} - 1$ and of $(x^n - 1)(x^m - 1)$ are the same.

Section 9.8

15. Let $K = \mathbf{Q}(\zeta)$ be the splitting field of $x^{nm} - 1$, so $\zeta^{nm} = 1$. Now form $F = \mathbf{Q}(\zeta^m)$ and $E = \mathbf{Q}(\zeta^n)$ as shown in the diagram. We saw in the solution to Exercise 14 that ζ^m is a primitive nth root of unity and ζ^n is a primitive mth root of unity. Thus $[F: \mathbf{Q}] = \phi(m)$ and $[E: \mathbf{Q}] = \phi(n)$ as labeled on the lower part of the diagram. Formula (1) for $\phi(n)$ in the text shows that since m and n are relatively prime, $\phi(mn) = \phi(m)\phi(n)$. Since $[K: \mathbf{Q}] = \phi(mn)$, we see that $[K: F] = \phi(m)$ and $[K: E] = \phi(n)$ as labeled on the upper part of the diagram. Thus $G(K/F)$ is a subgroup of $G(K/F)$ of order $\phi(m)$ and $G(K/E)$ is a subgroup of order $\phi(n)$.

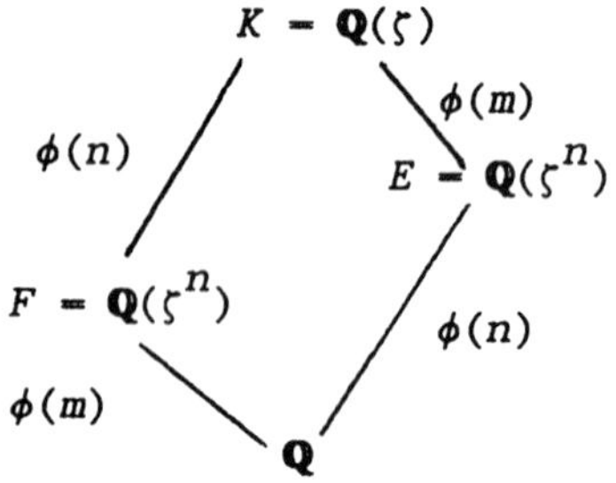

We check the conditions of Exercise 46 of Section 2.4 to show that $G(K/\mathbf{Q}) \simeq G(K/F) \times G(K/E)$. Since $G(K/\mathbf{Q})$ is abelian (see Theorem 9.21), condition (ii) holds. For condition (iii), suppose that $\sigma \in G(K/\mathbf{Q})$ is in both $G(K/F)$ and $G(K/E)$. Then $\sigma(\zeta^m) = \zeta^m$ and $\sigma(\zeta^n) = \zeta^n$. Suppose that $\sigma(\zeta) = \zeta^r$. Then $\sigma(\zeta^m) = \zeta^{rm} = \zeta^m$ so $r \equiv 1 \pmod{n}$. Also $\sigma(\zeta^n) = \zeta^{rn} = \zeta^n$ so $r \equiv 1 \pmod{m}$. Since n and m are relatively prime, we see that $r \equiv 1 \pmod{mn}$, so $r = 1$ and σ is the identity automorphism. Thus $G(K/F) \cap G(K/E)$ consists of just the identity automorphism. To demonstrate condition (i) that $G(K/F) \vee G(K/E) = G(K/\mathbf{Q})$, form the $\phi(m)\phi(n)$ elements $\sigma\mu$ where $\sigma \in G(K/E)$ and $\mu \in G(K/F)$. We claim that these products are all distinct, so that they must comprise all of $G(K/\mathbf{Q})$. Suppose that $\sigma\mu = \sigma_1\mu_1$ for $\sigma, \sigma_1 \in G(K/E)$ and u, $u_1 \in G(K/F)$. Then $\sigma\sigma_1^{-1} = \mu_1\mu^{-1}$ is in both $G(K/E)$ and $G(K/F)$, and thus must be the identity automorphism. Therefore $\sigma = \sigma^{-1}$ and $\mu = \mu^{-1}$.

Exercise 47 of Section 2.4 now shows us that $G(K/\mathbf{Q}) \simeq G(K/F) \times G(K/E)$.

SECTION 9.9 - Insolvability of the Quintic

1. No, the splitting field K cannot be obtained by adjoining a square root to $\mathbb{Z}_2$ of an element in $\mathbb{Z}_2$ since all elements in $\mathbb{Z}_2$ are already squares. However, K is an extension by radicals, for K is the splitting field of $x^2 + x + 1$, so $K = \mathbb{Z}_2(\zeta)$ where ζ is a primitive cube root of unity, and $\zeta^3 = 1$ is in $\mathbb{Z}_2$.

2. Yes. If α is a zero of $f(x) = ax^8 + bx^6 + cx^4 + dx^2 + e$ then α^2 is a zero of $g(x) = ax^4 + bx^3 + cx^2 + dx + e$. Since $g(x)$ is a quartic, $F(\alpha^2)$ is an extension of F by radicals, and thus $F(\alpha)$ is an extension of F by radicals.

3. T T T F T F T F F T

4. $f(x) = ax^2 + bx + c = a(x^2 + \frac{b}{a}x) + c$

$$= a(x + \frac{b}{2\cdot a})^2 + c - \frac{b^2}{4\cdot a} \text{ if } 2\cdot a \neq 0.$$

Thus if $\alpha \in \overline{F}$ satisfies $a(\alpha + \frac{b}{2\cdot a})^2 = \frac{b^2 - 4\cdot ac}{4\cdot a}$, so that

$\alpha + \frac{b}{2\cdot a} = \pm\sqrt{(b^2 - 4\cdot ac)/(4\cdot a^2)}$ and $\alpha = \frac{-b \pm \sqrt{b^2 - 4\cdot ac}}{2\cdot a}$, then α is a zero of $ax^2 + bx + c$.

5. Let $\alpha = \alpha_1$ be a zero of $ax^4 + bx^2 + c$. Then α^2 is a zero of $ax^2 + bx + c$ which is solvable by radicals by Exercise 4. If $\alpha_1, \alpha_2, \alpha_3, \alpha_4$ are the zeros of $ax^4 + bx^2 + c$, then the tower of fields starting with F and adjoining in sequence $\alpha_1^2, \alpha_2^2, \alpha_3^2, \alpha_4^2, \alpha_1, \alpha_2, \alpha_3, \alpha_4$ is an extension by by radicals, where each successive field of the tower is either equal to the preceding field or is obtained from it by adjoining a square root of an element of the preceding field. Thus the splitting field is an extension by radicals, so the quartic is solvable by radicals.

Section 9.9

6. We can achieve any refinement of a subnormal series by inserting, one at a time, a finite number of groups. Let $H_i < H_{i+1}$ be two adjacent terms of the series, so that H_{i+1}/H_i is abelian, and suppose that an additional subgroup K is inserted so that $H_i < K < H_{i+1}$. Then K/H_i is abelian since it can be regarded as a subgroup of H_{i+1}/H_i. By By Theorem 4.3, H_{i+1}/K is isomorphic to $(H_{i+1}/H_i)/(K/H_i)$, which is the factor group of an abelian group, and hence is abelian. For an alternate argument, note that since H_{i+1}/H_i is abelian, H_i must contain the commutator subgroup of H_{i+1}, so K also contains this commutator subgroup and H_{i+1}/K is abelian.

7. Let $H_i < H_{i+1}$ be two adjacent groups in a subnormal series with solvable quotient groups, so that H_{i+1}/H_i is a solvable group. By definition, there exists a subnormal series $H_i/H_i < K_1/H_i < K_2/H_i < \cdots < K_r/H_i < H_{i+1}/H_i$ with abelian quotient groups. We claim the refinement of the original series at this ith level, to

$$K_0 = H_i < K_1 < K_2 < \cdots < K_r < H_{i+1} = K_{r+1}$$

has abelian quotient groups at this level, for by Theorem 4.3 we have $K_j/K_{j-1} \simeq (K_j/H_i)/(K_{j-1}/H_i)$, which is abelian by our construction. Making such a refinement at each level of the given subnormal series, we obtain a subnormal series with abelian quotient groups.

8. a) The generalization of this to an n-cycle and a transposition in S_n is proved in Exercise 51 of Section 2.4.

   b) Since $f(x)$ is irreducible of degree 5 in $\mathbf{Q}[x]$, its splitting field K has degree over $\mathbf{Q}$ that is divisible by 5, for a zero α of $f(x)$ generates $\mathbf{Q}(\alpha) < K$ of degree 5 over $\mathbf{Q}$, and degrees of towers are multiplicative. By Sylow theory, a group of order divisible by 5 contains an element of order 5.

   The automorphism of $\mathbf{C}$ where $\sigma(a + bi) = a - bi$ induces an automorphism of K, which must carry one complex root $a + bi$ of $f(x)$ into the other one $a - bi$ and leave the real roots of $f(x)$ fixed. Thus this automorphism of K is of order 2.

c) We find that $f'(x) = 10x^4 - 20x^3 = x^3(10x - 20)$, so $f'(x) > 0$ where $x > 2$ or $x < 0$, and $f'(x) < 0$ for $0 < x < 2$. Since $f(-1) = -2$, $f(0) = 5$, and $f(2) = -11$, we see that $f(x)$ has one real zero between -1 and 0, one between 0 and 2, and one greater than 2. These are all the real zeros since $f(x)$ increases for $x > 2$ and for $x < -1$. Thus $f(x)$ has exactly three real zeros and exactly two complex zeros so the group of the polynomial is isomorphic to S_5 and the polynomial is not solvable by radicals.